Bimodal Oxidations of Heterogeneous and Homogeneous Reactions

Bimodal Oxidation: Coupling of Heterogeneous and Homogeneous Reactions

Robert Bakhtchadjian

CRC Press
Taylor & Francis Group
Boca Raton London New York

CRC Press is an imprint of the
Taylor & Francis Group, an **informa** business

CRC Press
Taylor & Francis Group
6000 Broken Sound Parkway NW, Suite 300
Boca Raton, FL 33487-2742

First issued in paperback 2021

© 2020 by Taylor & Francis Group, LLC
CRC Press is an imprint of Taylor & Francis Group, an Informa business

No claim to original U.S. Government works

ISBN 13: 978-1-03-223913-2 (pbk)
ISBN 13: 978-0-367-27259-3 (hbk)

Library of Congress Cataloging-in-Publication Data

Names: Bakhtchadjian, Robert, author.
Title: Bimodal oxidation : coupling of heterogeneous and homogeneous reactions /
Robert Bakhtchadjian.
Description: Boca Raton : CRC Press, Taylor & Francis Group, 2019. | Includes
bibliographical references.
Identifiers: LCCN 2019016625 | ISBN 9780367272593 (hardback : alk. paper)
Subjects: LCSH: Oxidation. | Heterogeneous catalysis. | Homogeneous catalysis.
Classification: LCC QD63.O9 B35 2019 | DDC 541/.393–dc23
LC record available at https://lccn.loc.gov/2019016625

Visit the Taylor & Francis Web site at
http://www.taylorandfrancis.com

and the CRC Press Web site at
http://www.crcpress.com

In Memory of My Parents,
Harutyun Bakhtchadjian and Sima Keshishian

Contents

Introduction

The phenomenon of the heterogeneous–homogeneous and homogeneous–heterogeneous reactions of oxidation, often named bimodal oxidation sequences or bimodal oxidation, is the main subject of this work. Previously, bimodal reaction sequences were revealed in the chain-radical and, in part, in heterogeneous catalytic reactions. For a long time, it was considered that this phenomenon was not widespread in chemistry. Only in recent decades did it become evident that these reactions are much more abundant not only in human-made but also in natural chemical systems.

Today, it is hardly necessary to explain the importance of knowledge of oxidation reactions with dioxygen. The use of controllable oxidation processes was always one of the vital necessities of human activity. Apparently, the achievement of the controlled fire, a fast oxidation reaction of different natural materials with air, was one of the first revolutionary findings of prehistoric human beings. The invention of fire could have happened about 1.5 million years ago.[1]

Presently, the investigation of the oxidation phenomena of organic and inorganic matter with dioxygen is one of the largest domains of modern science. Despite the existence of the enormous scientific publications in this domain, presently, many aspects of the oxidation processes related to the bimodal reaction sequences do not have necessary interpretation in the literature. In my knowledge, at least, in the recent decades, there is not any comprehensive work on bimodal reaction sequences that is addressed to a wide audience. The existing articles or reviews, mainly, are either very specialized covering narrow areas or were written from the point of view of the subdisciplines of physics, chemistry, biology, or applied sciences. Obviously, this causes certain problems also in the immediate accessibility of the scientific information for readers who are not specialized in these relatively narrow areas. Unfortunately, often, the authors of scientific publications are not too concerned about the accessibility of the works for a wide range of readers[2, 3] One of the aims of this work is to acquaint the reader with the achievements in this area in an accessible form, however, resting in the frames of the scientific accuracy and terminology, as much as possible.

Many chemical reactions of industrial importance relate to the problems of bimodal reaction sequences (oxidative coupling reactions of hydrocarbons; production of syngas; and selective oxidation of different organic compounds, including ethylene to acetaldehyde, methanol to formaldehyde, ethylene to ethylene oxide, butane to acetic acid, cyclohexane to cyclohexanone, iso-butane to *tert*-butylhydroperoxide, ammoxidation of propylene to acrylonitrile, and catalytic fast pyrolysis of biomass to liquid phase aromatics).

Present developments in the area of heterogeneous–homogeneous reactions are related to the problems that have arisen as a result not only of the historical evolution of investigations of oxidation reactions in fluid phases and partly in heterogeneous catalysis, but also the recent findings in relatively new branches of chemistry, such as photocatalysis, atmospheric chemistry, nanocatalysis, catalytic combustion, and some biochemical processes. Wide application of photochemical reactions on semiconductor materials in different domains of applied chemistry requires a return to the problems of heterogeneous–homogeneous reactions. Atmospheric chemistry also needs more detailed investigations of homogeneous–heterogeneous or heterogeneous–homogeneous interactions of reactive atmospheric species with atmospheric solid or liquid particles. In any case, the opinion that these reactions had a very limited prevalence in chemistry may no longer be considered valid.

The controllable combination of the heterogeneous and homogeneous constituents of the complex bimodal reaction sequences has great importance for heterogeneous catalysis, oxidation reactions in the gas or liquid phases, decomposition, and polymerization processes. The presence of the homogeneous constituent in the heterogeneous reaction and the heterogeneous constituent in the homogeneous reaction remarkably changes the nature of the overall processes. Often, it permits the achievement of synergistic effects, which are favorable for selective formation of the desired products. Convincing examples may be found in a number of oxidative coupling and oxidative dehydrogenation reactions on the different catalysts, as well as in several catalytic combustion processes.

The pivotal problem of the bimodal reaction sequences remains that of revealing the detailed reaction mechanism. Obviously, it is key to controllable chemical processes. The majority of the mechanism suggestions in combination or decomposition reactions are based on the formation, interactions, and destruction of intermediate species, such as atoms, radicals, excited species, and ions. Therefore, the problems associated with heterogeneous–homogeneous reactions are strongly related to studies of their reactivity in the fluid phases, as well as on the surfaces of solid substances. Although the chemical literature of recent years is rich in data on the reactivity of the reaction intermediates, particularly that of the free radicals, both in the fluid phases and on the solid surfaces (for instance, serial publications[4, 5] and collections of works[6, 7]), existing works generally do not pay the requisite attention to their interactions in bimodal reaction sequences.[8–12]

In the present work, different aspects of the problems of heterogeneous–homogeneous and homogeneous–heterogeneous reactions were involved in Chapters 2–5. The bimodal reaction sequences are exemplified by the oxidation reactions of hydrogen and organic compounds, such as hydrocarbons, alcohols, and aldehydes with dioxygen, as well as by the decomposition reactions of compounds, such as hydrogen peroxide, organic peroxides, and light hydrocarbons (pyrolysis). Chapter 1 includes some fundamentals of the bimodal reaction sequences. This chapter, as well as some sections of Chapter 2, are destined to readers who are less familiar with the problems in this area and may be bypassed by specialists. The last chapter, Chapter 6, is devoted to a brief discussion of some general problems of heterogeneous–homogeneous reactions. There an attempt is made to reveal the problem of the origin of the driving force in heterogeneous–homogeneous reactions based on the thermodynamics of nonequilibrium or irreversible processes. The analogies between the spatial separations of stages or steps of the stepwise processes in the heterogeneous catalytic and heterogeneous–homogeneous oxidation are briefly discussed. Also examined is the problem of why and how the overall heterogeneous–homogeneous process may be irreversible and "shifted to the right," overcoming the thermodynamic restrictions. Some examples of the emergence of self-organization phenomena in the complex heterogeneous–homogeneous reaction of oxidation are also presented in this chapter.

In addition, I would like to mention some considerations, which have willingly or unwillingly arisen during the preparation of this work. The investigation of the heterogeneous–homogeneous reactions is an interdisciplinary area. Presently, it is linked to advances in a number of other branches of chemistry and physics, and some discussions involve different aspects of the problems that may emerge from the frameworks of the main subject. They are presented very briefly, only in a subtextual form. Another consideration in writing this book is connected with the scientific information in this area, information that is less known or even unknown to English-language readers. In this regard, it has been attempted to complete state-of-the-art data for a number of problems, particularly, with the available data from Russian language literature.

A final consideration is that during the historical development of theoretical notions about reaction mechanisms, a number of concepts, approaches, models, and even conclusions were considered "excepted" or "unverified ideas." Over time, however, as a result of the publication of new experimental data, many of these ideas have become explicable and acceptable. In this regard, some sections of the present work include scientific approaches that have not yet gained complete acceptance by the wider scientific community. However, these approaches are on the way to becoming established (for example, the known problem of the feasibility of the so-called heterogeneous radical-chain mechanisms in catalysis, which are briefly presented in Sections 2.2 and 3.5).

In the context of all the above-mentioned problems, the aim of this work is to present the bimodal (heterogeneous–homogeneous or homogeneous–heterogeneous) reaction sequences in various chemical systems against the background of historical developments in this area, destined for a wide range of readers. However, this work does not presume to be a comprehensive review of knowledge in this area; rather, it is an introduction to the bimodal reaction sequences in chemical systems.

"Life is a slow combustion" or oxidation with air; this famous expression was made about two centuries ago by Lavoisier and Laplace.[13] The biological oxidation, being an enzymatic catalytic reaction, occurs in a heterogeneous–homogeneous environment as a living cell. However, it is different from the oxidation reactions having a bimodal reaction sequence as discussed here. That is why it is the subject for a separate discussion and why the biological aspects of bimodal oxidation sequences were not included in this work.[4]

Presently, the coupling of the heterogeneous and homogeneous reactions in a unique process opens new, attractive perspectives for the creation of controllable oxidation processes involving the advantages of two types of reactions. This is a subject for new investigations in many branches of the chemical sciences, involving catalysis, electrochemistry, photochemistry, biochemistry, and others, leading to the cheaper and "greener" production of chemicals. From the cognitive point of view, they have special importance in the deep knowledge of processes, beginning from simple fire and progressing to astrochemical processes.

REFERENCES

1. Cartmill M., Smith F. H., Brown K. B. The Human Lineage, Chapter 5. 2009, Wiley-Blackwell, Hoboken, NJ, 624 pages.
2. Brush S. G. Making 20th Century Science: How Theories Became Knowledge. 2005, Oxford University Press, New York, 552 pages.
3. Stichweh R. Differentiation of scientific disciplines: Causes and consequences. In: Encyclopedia of Life Support Systems (EOLSS), Unity of Knowledge in Transdisciplinary Research Sustainability, v. 1. EOLSS Publishers, 2003, UNESCO, Paris, 7 pages. http://greenplanet.eolss.net.
4. Series: Specialist periodical reports on Electron Paramagnetic Resonance, Royal Society of Chemistry, Series editors: V. Chechik, D. M. Murphy, 16 volumes.
5. Encyclopaedia of Radicals in Chemistry, Biology and Materials, editors: C. Chatgilialoglu and A. Studer, John Wiley and Sons, Hoboken, NJ, volumes 1999–2014.
6. Lund A., Rhodes C. J. (eds). Radicals on Surfaces. 1995, Springer Science + Business Media, B. V., 318 pages.
7. Lund A., Shiotani M. EPR of Free Radicals in Solids: Trends in Methods and Applications, 2003, Springer Science + Business Media, B. V., 644 pages.
8. Stephenson C. R. J., Studer A., Curran D. P. The renaissance of organic radical chemistry – deja vu all over again. Beilstein J. Org. Chem., 2013, v. 9, p. 2778–2780. DOI: 10.3762/bjoc.9.312
9. Epstein I. R., Pojman J. A. An Introduction to Nonlinear Chemical Dynamics, Oscillations, Waves, Patterns, and Chaos (Topics in Physical Chemistry), 1998, Oxford University Press, New York, 408 pages, p. 12.
10. Lazar M., Rychly J., Klimo V., Pelican P., Ladislave V. Free Radicals in Chemistry and Biology, Chapter 1, Introduction. 1989, CRC Press, Boca Raton, FL, 307 pages, p. 1–3.
11. Koelsch C. F. Syntheses with triarylvinylmagnesium bromides. α,γ-bisdiphe-nylene-β-phenylallyl, a. Stable Free Radical. J. Am. Chem. Soc., 1957, v. 79, 16, p. 4439–4441. DOI: 10.1021/ja01573a053
12. Müllegger S., Rashidi M., Fattinger M., Koch R. Surface-supported hydrocarbon π-radicals show Kondo behavior. J. Phys. Chem. C, Nanometer Interfaces. 2013, 117, 11, p. 5718–5721. DOI: 10.1021/jp310316b
13. Magner L. N. A History of the Life Sciences, Chapter 6, 2005, Third Edition, revised and expanded, Marcel Dekker, New York, Basel, 503 pages, p. 228.

1 Bimodal Reaction Sequences Occurring through the Active Intermediates

The vast majority of the oxidation reactions of organic and inorganic compounds, both catalytic or noncatalytic, are multistage and complex processes. They occur through the formation and destruction of the chemical entities named the active intermediate species. From the point of view of modern notions about the mechanisms of complex reactions, both homogeneous and heterogeneous reactions, mainly, occur via the active intermediate species, such as radicals, radical-like species, ions, electronically excited atoms and molecules, and other species. Obviously, the complexity of heterogeneous–homogeneous or homogeneous–heterogeneous processes is conditioned by the interacting reactions in different phases and the physical phenomena accompanying them. Thus, knowledge of the mechanism, first of all, is related to the detection, characterization, and investigation of the chemical properties of the active intermediates.

1.1 ABOUT THE CLASSIFICATION OF CHEMICAL REACTIONS

In general, the classification of chemical reactions may be based on the nature of the chemical transformation or different physical and chemical properties of the reaction system. Therefore, there is no general principle for the classification of all chemical reactions in a unified form.

The accepted general classification of the enormous number of reactions is based on the differences in the phase state of their reaction components. From this viewpoint, chemical reactions may be classified as homogeneous or heterogeneous, as well as heterogeneous–homogeneous and homogeneous–heterogeneous, in the case of reactions that consist of two constituents.

In a homogeneous reaction all reactants, products, and other components of the chemical system (intermediates, catalyst, and solvent) are in the same phase state. Homogeneous reactions occur in a single phase, mainly in the gas or liquid phase and, in rare cases, in the solid phase. In turn, a homogeneous reaction may be a catalytic or noncatalytic process. In a heterogeneous reaction, one or more components of the chemical system react on the surfaces of the solid phases.

There exist a great number of complex reactions, which occur in/on two or more phases, involving both homogeneous and heterogeneous reactions of components, including intermediates. These reactions are known as heterogeneous–homogeneous or homogeneous–heterogeneous reactions. Usually, they are multistep and multiphase chemical processes. In the majority of the chemistry textbooks or handbooks, even in present-day editions, the heterogeneous–homogeneous or homogeneous–heterogeneous reactions are not mentioned among the main classes of chemical reactions, while several aspects of them are the subject of separate investigations. As will be shown later in this chapter, an overall heterogeneous–homogeneous or homogeneous–heterogeneous reaction is not a mechanical combination of single homogeneous and heterogeneous reactions.

Heterogeneous–homogeneous or homogeneous–heterogeneous reactions have great importance, particularly in a great number of oxidation reactions with dioxygen.[1] In the present work, these reactions will be discussed as an individual class of chemical reactions. Sometimes, in classification of the complex, multistep and multiphase reactions, some authors use the term *homogeneous–heterogeneous reaction*, giving the same definition that is usually given for heterogeneous–homogeneous reactions.[2] Historically, the term *heterogeneous–homogeneous reactions* predates the term *homogeneous–heterogeneous*. The multiphase and multistage reactions occurring in the gas or liquid phases by participation of solid substances were named heterogeneous–homogeneous, when the reaction was initiated on the surface of the solid substance, and then, it was transferred into the fluid phase. Indeed, further interaction between the processes in two phases was not excluded. It was considered that the reaction was homogeneous–heterogeneous if it was initiated in the fluid phase, after which one or more stages of the overall reaction occurred on the surfaces of the solid phases. It is to be noted that "the logic of studies forces us to take into account the formation of homogeneous constituents in heterogeneous catalytic reactions and heterogeneous constituents in homogeneous processes."[3]

Elucidating the term *heterogeneous–homogeneous*, Kiperman considered that "it denotes different variants of a process in which a solid surface participates collectively with a surrounding volume phase, regardless of the consequences of its action."[4] Although this definition does not reflect all the complexity of heterogeneous–homogeneous or homogeneous–heterogeneous reactions, it reveals their essential differences from other types of reactions.

Presently, the heterogeneous–homogeneous or homogeneous–heterogeneous reactions in the chemical literature are often named bimodal reactions or bimodal reaction sequences.[5] For example, Iglesia and Reyes describe the "bimodal (heterogeneous–homogeneous) oxidative coupling of methane" as follows: "Oxidative coupling of methane to form ethylene and ethane products via bimodal reaction pathways that include radical generation on metal oxide surfaces and coupling and oxidation reaction of radicals and stable species in the surrounding gas phase."[6]

These terms are not widely used, although they characterize both heterogeneous–homogeneous and homogeneous–heterogeneous reactions. It is also convenient to denote both types of reactions by a unique and shortest form of the term.

Thus, the bimodal (heterogeneous–homogeneous or homogeneous–heterogeneous) reaction, is a complex, multiphase, and multistage chemical process involving the interacting stages of the heterogeneous and homogeneous processes, via the transfer of different intermediates from one phase to the other, as well as energy exchanges between them.

If the components of the reaction mixture are in the same phase but the reaction occurs on the interface, the process is said to be heterogeneous and heterophase—for instance, the oxidation of the $H_2 + O_2$ gaseous mixture on the solid Pt-catalyst. In the homogeneous heterophase reaction, as in the case of the oxidation of organic compound in the liquid phase with gaseous O_2, the reactants are in different phases, but the reaction occurs in the same phase. Finally, in heterogeneous–heterophase reactions, the reactants, as well as products, are in different phases, and the reaction occurs at the interface of components, for example, the reaction system $CaCO_3 + HCl$.

Another classification of chemical reactions may be based on changes in the oxidation numbers of atoms in chemical compounds participating in the reactions, which are named redox (reduction–oxidation) reactions. In general, they are electron transfer reactions from one atom to the other in the same or different phases.

The classification of chemical reactions in the chemical thermodynamics takes into consideration either energetic effect (exogenous, endogenous) or the thermodynamic state of the system and the reversibility of the chemical process (reversible, irreversible, equilibrium, nonequilibrium).

In investigations of kinetic parameters, reactions may be divided, on the one hand, into fast and slow reactions, as combustion and slow oxidation, and, on the other hand, into mono, bi-, and trimolecular reactions, if one takes into consideration the mechanism of the elementary interactions.

From the point of view of the type of chemical transformation, in the courses of chemistry all reactions are usually divided into combination, decomposition, replacement, double placement, and exchange reactions.

Although the frontiers between organic and inorganic chemistry have nearly been destroyed in modern chemistry, the traditional separation into reactions of inorganic compounds and organic compounds remains in the chemical literature (as "organic reactions" or "organic synthesis").

Unlike multistep (stepwise) reactions, in concerted reactions, the breaking of old chemical bonds and the formation of new bonds occur simultaneously, involving only a transition state complex but not other active intermediates, which can exist as individual chemical entities. In other words, the concerted reaction is a single-step chemical transformation.

In addition, the appearance of nanomaterials gave birth to a new class of catalytic reactions, known as nanocatalysis. From the point of view of the phase state classification of chemical reactions, the catalysis by nanomaterials is neither a homogeneous nor a heterogeneous process. Nanocatalysts can be classified as materials exhibiting quasi-homogeneous and quasi-heterogeneous catalytic properties.[7] In other words, nanocatalysis occupies an intermediate position between heterogeneous and homogeneous catalysis. Obviously, nanocatalysis of this nature is different from heterogeneous–homogeneous catalysis.

In this work, we will frequently be guided by the most generalized classification of all chemical reactions into heterogeneous, homogeneous, and, from the point of view of the multiphase processes, heterogeneous–homogeneous or homogeneous–heterogeneous reactions, very often, using also the term *bimodal reaction sequence* (bimodal reaction) as a unified term for these two types of reactions.

1.2   ACTIVE INTERMEDIATE SPECIES

What is an active intermediate? Complex oxidation reactions are multistep processes. The majority of homogeneous and heterogeneous complex reactions occur through the formation and further destruction of intermediates. Gold Book, the compendium of chemical terms published by the International Union of Pure and Applied Chemistry (IUPAC), defines an intermediate as "a molecular entity with a lifetime appreciably longer than a molecular vibration (corresponding to a local potential energy minimum of depth greater than RT) that is formed (directly or indirectly) from the reactants and reacts further to give, either directly or indirectly, the products of a chemical reaction; also the corresponding chemical species."[8]

The existence of active intermediates was first hypothesized by F. Lindemann, in 1922, and became one of the major subjects of scientific discussion for many famous scientists of that time (including Nobel Prize winners S. Arrhenius, I. Langmuir, J. Perrin, and C. Hinshelwood).[9] However, the term *active intermediate* was not immediately accepted in the scientific literature. Lindemann's mechanism explained the appearance of the kinetic second order in unimolecular reactions, via the generation of an "activated" intermediate. The adjective "active" characterizes the highly energetic and reactive state of the intermediate species.

As a rule, oxidation with dioxygen of organic and inorganic matter in the gas or liquid phases, as well as in the interphases occurs by the formation of active intermediates. They may be atoms, molecules, ions, radicals, excited species, complex compounds, conformers, active sites on the solid substances, etc. Their lifetimes vary depending on the nature of the compound and conditions under which the chemical system exists. The lifetime is the time interval in which the concentration of an intermediate decreases to $1/e$. If no reaction intermediates, naturally, the reaction is an elementary chemical act and occurs as a concerted reaction. Understanding the reaction mechanism is based mainly on the detection and investigation of the key intermediates. The more active reaction intermediates, the lower is their concentration in the system, as their lifetime is short. The modern experimental methods permit detection of intermediates, whose lifetimes are about 10^{-15} s (femtosecond; Zewail received the Nobel Prize in Chemistry in 1999 "for his studies of the transition states of chemical reactions using femtosecond spectroscopy"[10]). However, application of more sophisticated spectroscopic techniques permits survey of the reactions at timescale, even 10^{-18} s (attosecond)![11]

Intermediate is a kinetic term, and it loses its sense out of the chemical reaction. In the simplest case, intermediates (X) are chemical species that form and then disappear during the chemical reaction.

$$A \rightarrow X \rightarrow C$$

At this point, a simple question arises: Is the transition state-activated complex an intermediate in the transient state theory of chemical reactions?[12] According to the above definition, an intermediate has a definite lifetime, which is longer than the time of the vibrational motion in an activated complex in the theory. Moreover, being only a transition state configuration, the activated complex, according to theory, has a local energy maximum, while in contrast, a reaction intermediate has an energy minimum. Evidently, it is impossible to isolate the activated complex as an intermediate.

Short-living intermediates often are named transient species. This relative term has a meaning in the timescale of the definite reaction kinetics, relative to the reactants and products. Transient species, as intermediates of the reaction, are also thermodynamically metastable or unstable species. According to Gold Book of the IUPAC, "the term unstable specie should not be used in place of reactive or transient, although more reactive or transient species are frequently also more unstable." Note that metastable species are stable kinetically but not thermodynamically.[8] Thus, each of the above-mentioned terms has its own "nuance."

In the fluid phases, the properties of intermediates, radicals, excited species, ions, or others were long a subject of investigation in different subdisciplines of chemistry. However, nature, electronic and geometric structures, physical characteristics, and chemical reactivity of active intermediates in heterogeneous reactions are relatively less investigated than those in the fluid phases, and their nature is not clear for many important oxidation reactions in the chemical literature.

Among the above-mentioned different types of reaction intermediates, the free radicals play an important role in homogeneous, as well as in a great number of heterogeneous reactions of oxidation with dioxygen. The chemistry of radicals began with the discovery of the thriphenylmethyl $(C_6H_5)_3C$ radical by M. Gomberg (1866–1947) in 1900.[13] At that time, the concept of molecules containing trivalent carbon atoms with one "dangling" bond was a daring, original idea. This major discovery in the history of chemistry later revolutionized the development of other sciences, especially biology, physics, and astronomy. Unfortunately, Gomberg did not receive the Nobel Prize for his discovery, although he was repeatedly nominated for this award.[14] In 1971, the Nobel Prize in Chemistry was awarded to Gerhard Herzberg (1904–1999) for his contribution to "knowledge of the electronic structure and geometry of molecules, particularly, of free radicals."[9] The chemical literature on free radicals of different classes, including the methods of their synthesis, detection, physical properties, and reactivity in chemical systems, is very rich.[14–16]

In the present work, we will briefly cover only some of the reactions of organic or inorganic radicals on the solid surfaces. The main attention will be paid to the reactivity of H, O, OH, HO_2, RO, RO_2, etc., as well as ion–radicals, as O^-, O_2^-, CO_2^-, O_3^- on the solid surfaces of different substances in oxidation with dioxygen.

1.3 HISTORICAL BACKGROUND OF THE EARLY NOTIONS ABOUT HETEROGENEOUS–HOMOGENEOUS REACTIONS

In general, the presence of the solid phase(s) in the gaseous or liquid reaction media either may affect (accelerating or reducing) or may not affect the overall rate of chemical reactions. As will be shown below, an absence of the experimentally observable changes of the reaction rate and composition of the products in the presence of solid substances is not evidence that the solid substances in overall chemical transformations are inactive. Experimentalist-chemists well know that in certain conditions, many "ordinary" chemical reactions can be sensitive to the presence of solid-phase substances (impurities on the walls of reactors, grease on the valves of an experimental setup, etc.) in the reaction media. In certain cases, the experimental results may even become unreproducible. Evidently, the acceleration or inhibition of the chemical reaction on the solid phase(s) is a result of the interactions of the components of the homogeneous reaction mixture, including initial reactants, reaction intermediates, and products with the solid surface.

In the oxidation reactions of organic and inorganic compounds with dioxygen, in the presence of solid substances in the fluid phases, the surface effects are often well manifested. Here, the response to the question, "Why and how do the solid-phase substances affect the overall process of oxidation?" is related to understanding the detailed mechanism of these reactions.

The initial revelation of the existence of heterogeneous–homogeneous reactions was related with the observations of the influence of the chemical nature and surface area of the reactor walls to the rate of reaction.[17] Historically, although some notes on the interactions of the occurring reaction with the walls of the reaction vessel may be found in the works of nineteenth-century chemists, the first extended study in this area was carried out by J. H. Van't Hoff, the first winner of the Nobel Prize in Chemistry (1901) for his works in chemical dynamics. In his work *Études de dynamique chimique*, published in 1884,[18] Van't Hoff described the polymerization reaction of anhydrous cyanic acid (HCNO) in two glass vessels with the same volumes but with different internal surface areas. One of the vessels was a round bulb, and the other a spiral tube. Van't Hoff remarked that the reaction rate was greater in the spiral tube with the greater surface area than in the bulb. Moreover, by coating the glass vessel with cyamelid, a product of polymerization, he observed triple acceleration of the reaction rate in comparison with the uncoated vessel.

J. W. Mellor, the author of *Chemical Statics and Dynamics*[19] (1904), referring to works written at the end of the nineteenth and beginning of the twentieth centuries, noted many interesting facts about the investigations into the influence of the walls of the vessel to the reaction rate. According to Mellor, in the reaction of decomposition of phosphine, Kooij[20] observed an effect that was later described as "aging" of the walls of the reactor. In this reaction, the "velocity coefficient" (the rate constant of reaction, in modern terminology) in the "old" vessel was greater than that in the little used vessel. Some other examples Miller described concern the reaction of hydrogen with oxygen. In the ordinary glass vessel, the reaction starts at 448°C, but in the silver-covered reaction vessel it starts at 182°C. Similarly, according to Bone and Wheeler's[21] experiments with "electrolytic gas" (a mixture of hydrogen with oxygen), in six glass vessels, among seven similar vessels of different quality, kept at 350°C, no reaction was observed. In one of the vessels that was devitrified, however, traces of water were seen. In 1897, investigating the reaction of hydrogen with oxygen, Berthelot[22, 19(p. 365)] found that this reaction was accelerated when the walls of the vessel were coated with barium hydrate, alkali, or traces of manganese salts. He thought that the various glass materials, containing some substances, might be decomposed during the reaction, giving different amounts of these substances. Therefore, the experimental results might be different in these vessels.

Van't Hoff suggested some other explanations for these observations.[18] He assumed that the materials of walls of the reactor, carefully cleaned, might contain different "irregularities" (which also might be formed in the course of the reaction), causing some "discordant experimental results." In general, at that time nearly all suggested explanations of the effect of walls on the rate were related either to the "irregularities" or to the "secondary occasional effects," such as the presence of moisture on the surface and differences of heat conductivity of the materials of the walls. That is why the chemists of those times attempted to make the glass of the reaction vessels "as "smooth" as possible," often, covering the internal surface of the walls with the metals, as Ag, Pb, Sn.[23] Later, it became evident that the mechanical treatment of glass could not eliminate the effect of the walls. Moreover, coverage of the walls with metals, changed the occurrence of the overall reaction, acting as catalyst.[23] Among the different explanations of that time were suggestions regarding the catalytic role of the walls, especially when the internal surface of the reactor walls was covered with different materials, as metals. This explanation seems relatively closer to modern understanding.

The gas-phase reaction between hydrogen and oxygen was investigated by C. N. Hinshelwood (1897–1967), who provided more detail in his early work (since 1927).[24, 25] It was revealed that the effect of the walls in this reaction was more complex and even more "variegated." Hinshelwood observed that the rate of reaction strongly depended not only on the chemical nature of the walls (the coating of vessels by KCl and B_2O_3), but also on the diameter (or shape) of vessel, as well as on the differences of the cleaning ways of the walls of the reactor. Surprisingly, even, in the same reaction vessel, in repeated experiments, the results were often nonreproducible. That is why the reaction of hydrogen with oxygen was described as "extremely capricious" and "greatly affected by the character of the surface of the vessel."[23–25] Investigating the recombination of hydrogen atoms and reaction H + OH in a gas mixture passed over thermometers, the bulbs of which were covered by different substances, Taylor found that potassium chloride and especially alumina were more efficient in these reactions than the clean glass.[26] The source of these radicals was water in the discharge tube.

Further attempts to explain the wall effects were related to the creation of the theory of branched-chain reactions in the 1930s.[27, 28] In 1956, the creators of this theory, N. N. Semenov and C. N. Hinshelwood, were awarded the Nobel Prize in Chemistry for their discovery. On the other hand, the development of the theory of heterogeneous catalysis by I. Langmuir made it possible to distinguish some similarities and differences between the heterogeneous catalytic reactions and the actions of solid surfaces as initiators or inhibitors in gas or liquid phase reactions.[29]

I. Langmuir (1881–1957) was an American chemist and physicist who received the Nobel Prize in Chemistry in 1932 "for his discoveries and investigations in surfaces chemistry".[64]

In connection with these advances in the theory of chemistry, the first thorough explanations of the surface effects in the gas-phase reactions, based on experimental kinetic data, were given in the Semenov–Hinshelwood theory.[27, 28, 30, 31] In this regard, to understand the historical development of notions about heterogeneous–homogeneous reactions and surface effects, the gas-phase hydrogen–oxygen reaction serves as a classical example of the branched-chain reaction. According to the Semenov–Hinshelwood theory, the principal stages of this reaction are as follows[25, 27, 30]

$H_2 + O_2 \rightarrow HO_2 + H$	chain initiation stage
$H + O_2 \rightarrow OH + O$	chain branching stage
$OH + H_2 \rightarrow H_2O + H$	chain propagation stage
$O + H_2 \rightarrow HO + H$	chain branching stage
$H + wall \rightarrow H_{ads}$	heterogeneous chain termination stage
$O + wall \rightarrow O_{ads}$	heterogeneous chain termination stage
$H + O_2 + M \rightarrow HO_2 + M$	homogeneous chain termination stage
$HO_2 + wall \rightarrow (HO_2)_{ads}$	heterogeneous chain termination stage

It had already been well established that, in this reaction, the position of the ignition limits on the pressure–temperature diagram (Figure 1.1) was strongly dependent on heterogeneous factors. For example, according to the experimental data,[32, 33] the maximal rate of combustion of hydrogen with oxygen (500°C, $P_{total} = 500$ mm Hg) was 100 times greater in the reaction vessel, the walls of which were coated with boric acid, than that in a KCl-coated vessel. The substances on the walls of the vessels, influencing to the maximal rates of chain reaction and, therefore, to the position of the ignition limits, conditionally, were divided into two main groups: (i) the group including KCl, $BaCl_2$, Na_2WO_4, $K_2B_2O_4$, LiCl, and RbCl[32, 34, 35] and (ii) the group including quartz, Pyrex glass, boric acid, $MnCl_2$, and inorganic acids.[32, 36, 37] The early versions of the theory maintained that in a chain-radical branched reaction, the surface effects were generally explained by the chain

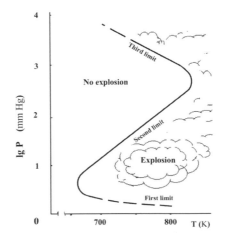

FIGURE 1.1 Ignition limits of the hydrogen + oxygen mixture in the Pyrex reaction vessel, coated by KCl. (Adapted from Ref.[32]).

termination reactions of the active centers on the wall of the reaction vessel. In the combustion reaction of hydrogen–oxygen gaseous mixtures, in a reaction vessel precoated with the substances of the first group, both the lowest and the upper ignition limits were related to the heterogeneous terminations of leading active centers, such as the radicals H, O, OH, and HO_2. It had also been known that the values of pressure of the upper ignition limit for an identical reaction, carried out in the reaction vessels precoated with the substances of the second group (the so-called reflecting surfaces), were higher than those in the case of the first group.[24, 34, 37] These observations were explained on the basis of the fact that the probability of heterogeneous (linear) termination of the chains, in this case, is small and that homogeneous (nonlinear) terminations of active centers make a significant contribution to the rate of overall reaction. All these explanations, in their quantitative expressions, satisfactorily described the conditions of the existence of the ignition limits in hydrogen–oxygen reaction.[27] In many other chain-radical reactions (e.g., the oxidation of hydrocarbons), however, the surface effects were not satisfactorily explained within the framework of the theory. After analyzing the kinetic data on a large number of surface effects, it was concluded that, in certain cases, the initiation stage of the radical chains also could be a heterogeneous reaction.[27(p. 197), 34, 38] Convincing examples were known in nonbranched radical reactions as the $H_2 + Cl_2$ reaction. Here, the initiation of chains takes place by heterogeneous reaction on the walls of the reactor[39]

$$Cl_2 + wall \rightarrow Cl + Cl\text{-wall}$$

(For discussion of the heterogeneous initiation of chains, see Section 1.4.)

Since the 1970s, new experimental data have been obtained in investigations of the combustion of hydrogen–oxygen mixtures, showing that the surface of the solid substance undergoes reversible changes during the reaction.[40–43] Azatyan named such behaviors of a solid surface during the chain-branched reaction the "reversible instability of a solid surface."[41, 42] It therefore follows that the probability of the linear termination of the chains cannot be constant in this reaction. It had also been known that the efficiency of the heterogeneous chain termination was in nonlinearly dependence on the concentration of atomic hydrogen. More detailed investigations[41–45] revealed that, in certain cases, some species (e.g., H-atoms[45]), previously adsorbed on the surface, can expand the ignition limits on the P-T diagrams (Figure 1.1), while other adsorbed species may inhibit chain reaction. In other words, the heterogeneous reactions of the adsorbed species and, first of all, the reactions of the adsorbed H-atoms play a determining role in advancing the overall process. Apparently, the reactions of the adsorbed hydrogen atoms with dioxygen, as well as with the radicals and other species formed on the surface or in the gas phase, may be considered important stages of the heterogeneous process. As shown in a number of works,[34, 41–46] all mentioned facts and many other general kinetic regularities of the combustion of $H_2 + O_2$ mixtures might be quantitatively described, assuming that the surface may participate in the additional propagation and branching of chains. In this regard, some of the following elementary reactions[47] may complete the above scheme of the principal stages of this combustion reaction.

$$H \rightarrow H_{ads} \tag{1}$$

$$H_{ads} + O_2 \rightarrow (HO_2)_{ads} \tag{2}$$

$$H_{ads} + O_2 \rightarrow (OH)_{ads} + O \tag{3}$$

$$(HO_2)_{ads} + H \rightarrow (OH)_{ads} + OH \tag{4}$$

$$(HO_2)_{ads} + H \rightarrow 2OH \tag{5}$$

$$HO_2 + H \rightarrow 2OH \tag{6}$$

$$HO_2 + O \rightarrow OH + O_2 \tag{7}$$

$$HO_2 + H \rightarrow H_2O + O \tag{8}$$

$$HO_2 + H \rightarrow H_2 + O \tag{9}$$

$$OH + H_2 \rightarrow H_2O + H \tag{10}$$

According to this concept,[41–46] the acts (1–3) are completely heterogeneous reactions. Here, in fact, reaction (3) is a stage of the heterogeneous branching of chains. Apart from the heterogeneous acts (2–5), the reactions (6–8 and 10) may be partially heterogeneous or completely homogeneous stages of the propagation of chains. Reaction (3) "can be responsible for only possible heterogeneous additional branching."[46] Among the reactions of HO_2 species (4–9), chain termination by reaction (9) may also be either heterogeneous or homogeneous.

The mentioned conception was corroborated by numerous experimental data and modeling studies of the combustion processes.[34, 41–47] According to some authors,[34, 41, 42 46, 47] this conception was very useful, equally, for the explanation and description of some other important observations in combustion processes, such as isothermal processes, including multiple self-ignition in a confined volume; heterogeneous propagation of flame; induction of other chain reactions via the reactions of adsorbed active centers; heterogeneous chain branching in the isothermal flame near the surface; vibrational modes of combustion in reactors pretreated with flame; and transfer of lattice atoms of the crystal to the gaseous phase stimulated by the heterogeneous chains. Some experimental evidence and additional elucidations of this concept are also given in Section 3.5.

Recent research interest in bimodal oxidation sequences has been focused not only to industrially important reactions, but also to numerous environmental problems whose solutions have been the subject of investigations in photocatalysis, atmospheric chemistry, catalytic combustion, nanocatalysis, electrochemistry, and many other domains. This recent stage of development on the bimodal reaction sequences, in both heterogeneous-catalytic and homogeneous oxidation reactions, is briefly presented in Chapter 3.

1.4   THE REVELATION OF THE HETEROGENEOUS STAGES IN HOMOGENEOUS REACTION

The influence of heterogeneous factors on the reaction rate may be observed by the simple method of changing the surface/volume (S/V) ratio (the ratio of the geometrical surface area of the walls to the volume of the reaction vessel) in given reaction conditions.[17] For a spherical vessel, S/V = 3/r, where r is the radius of the vessel. The S/V ratio may be changed not only by variations in the diameter or shape of the reaction vessel, but also by packing the reaction vessel with pieces of different geometric forms and sizes, made from the same material. It is expected that the rate of overall reaction may be changed with the variation of the S/V ratio, if the reaction involves hetero-geneous stages. Note that the overall geometric surface area of the reaction vessel is proportional to the real ("working") surface area. In early kinetic investigations, the influence of the S/V ratio was examined for numerous chemical reactions, in both the gas and liquid phases,[17, 48] some examples of which will be presented below.

In the 1930s, it was already known[23, p. 28] that the increase of S/V in gas-phase chain-radical reactions caused different and, even, contradictory influences on the overall rate of oxidation of organic compounds. For example, in the chain reaction of benzaldehyde, as well as formaldehyde, with dioxygen, in the gas phase, the reac-tion rate decreases with increase in S/V,[23, 48] while in the oxidation of acetaldehyde it increases.[23, 49] Rice,[23] investigating the influence of S/V on the rate of homoge-neous chain reaction of acetylene with oxygen in empty and packed glass vessels at 250–315°C (the main product of which was glyoxal), found that the increase of S/V "inhibited" the oxidation. However, "when the homogeneous reaction (formation of glyoxal) was suppressed, a new heterogeneous oxidation of acetylene on the glass surface directly to carbon monoxide and water appeared." In other words, with the increase of S/V, changes were observed not only in the reaction rate, but also in the composition of products.

The rates of homogeneous and heterogeneous reactions are different and are not comparable. In addi-tion, their effective rate constants have different units of measurement. In kinetic calculations, for-mally, they may be redimensioned.[50, 51]

$$k = k_{het} (S/V) = k_{0(het)} \exp(-E_{het}/RT) (S/V)$$

where k is the effective constant of the homogeneous reaction, recalculated from the data on heteroge-neous reactions, k_{het} and $k_{0(het)}$ are effective heterogeneous constants and its pre-exponential factor, S surface of the solid substance (catalyst or inhibitor), and V free volume, respectively. E_{het} is the activa-tion barrier of the heterogeneous reaction.[2] In this case, the rate of the overall reaction, W (it may be called the rate of pseudo-homogeneous reaction), in respect to any reaction product, formed or con-sumed by both the heterogeneous and homogeneous pathways, may be expressed as

$$W = W_b + W_s$$

where W_b is the rate of reaction in the bulk and W_s is the rate of heterogeneous reaction, redimensioned into homogeneous one, correspondingly.

The dependencies of the reaction rate on S/V, as well as the dependencies of the effective rate constant of reactions on S/V, are not linear functions. Heterogeneous–homogeneous oxidation may involve more than one heterogeneous stages, each of which has its own dependence on the ratio of S/V, and, therefore, also makes differ-ent contributions to the rate of overall reaction. For instance, the rate of overall reac-tion of the oxidation of propionic aldehyde, in the gas phase, in the packed reactors, pretreated by KCl, increases about three times in about eight-times increase of S/V, while it increases only five times, in a 20-time increase of S/V.[52] The dependence of the rate of overall reaction (in arbitrary units) on the ratio of S/V in this reaction is shown in Figure 1.2.

Another example of the dependence of the reaction rate on S/V is the hetero-geneous catalytic reaction of the oxidative coupling of methane on the surfaces of different catalysts described by Pyatnitsky et al.[53] The suggested mechanism for this reaction includes the following homogeneous and heterogeneous stages.

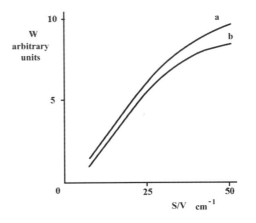

FIGURE 1.2 Dependence of the overall rate W (in arbitrary units) of reaction of the oxidation of propionic aldehyde on the ratio of S/V (surface/volume, cm^{-1}) in the KCl-treated vessel, a-437 K, b-403 K. (Adapted from Ref.[52]).

$$CH_4 + ZO \rightleftarrows CH_3 + ZOH$$
$$2CH_3 \rightarrow C_2H_6$$
$$CH_3 + O_2 \rightarrow CH_3O_2$$
$$2CH_3O_2 \rightarrow CH_3OH + CH_2O + O_2 \overset{ZO}{\rightarrow} CO_2, CO$$
$$2CH_3O_2 \rightarrow 2CH_3O + O_2 \overset{ZO}{\rightarrow} CO_2, CO$$
$$CH_3O_2 + CH_3 \rightarrow 2CH_3O \overset{ZO}{\rightarrow} CO_2, CO$$
$$2ZOH \rightleftarrows H_2O + Z + ZO$$
$$O_2 + Z \rightleftarrows 2ZO$$

where Z is the surface active site. As the above scheme shows, the two types of heterogeneous reactions, the heterogeneous initiation and heterogeneous termination of chains on the surface, were taken into consideration in oxidative coupling of methane. Through kinetic analysis of the above scheme, using quasi-equilibrium approximations, it was found that

$$k_1 C_{CH4} [ZO] \, S = k_{-1} [ZOH] [CH_3] S + 2k_{hom} [CH_3]^2 V$$

or

$$2k_{hom} [CH_3]^2 = S/V \, (k_1 C_{CH4} [ZO] - k_{-1} [ZOH] [CH_3])$$

The effective rate constant of the homogeneous reaction (k_{hom}) has the following expression:

$$k_{hom} = k_2 + K_3 [C_{O2} (k_4 + k_5) \, K_3 C_{O2} + k_6]$$

Note that the heterogeneous rate constants were calculated per unit of the surface area, whereas the homogeneous rate constants were calculated per unit of the volume. By using these equations, specific and total rates of the consumption of methane were calculated in the presence of catalysts with different surface areas. The calculations demonstrated the dependence of the reaction rate on the S or S/V of the reaction vessel and then allowed comparison of the obtained results with the

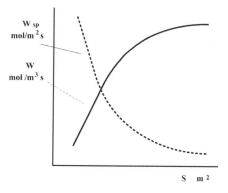

FIGURE 1.3 Simplified presentation of the dependence of the reaction rate (W) on the surface area (S) in oxidative coupling of methane over different catalysts. The unified curves were obtained by combining the experimental and calculation results. The dotted curve corresponds to the specific rate W_{sp} (mol/m²s). (Altered and adapted from Ref.[53]).

experimental data. Use was made of experimental data on the oxidative coupling of methane on the catalysts, such as Ni_3B, Ni_2B, Co_3B, TiB_2, $TiSi_2$, at 1023 K, and on a mixture of methane with oxygen (CH_4 − 30.4 kPa + O_2 − 14 kPa).[53] Both the experimental and calculation data of the W (or W_{sp}) − S(S/V) dependencies were monotonously decreasing or increasing curves, which in general, was characteristic for a number of heterogeneous–homogeneous reactions. It is noteworthy that the calculation data on the basis of the above kinetic scheme and experimental data, obtained in this reaction, on the catalysts with the different surface area, were in good agreement, according to the authors. Both the experimental and calculation data were included in one unique curve, represented in Figure 1.3.[53] This dependence shows that the rate for catalysts with the higher surface area tends to be limited. Therefore, higher surface areas of the catalysts are not effective in this reaction, and there exists an optimal interval of the surface areas for the reaction of oxidative coupling of methane.

The decrease in the reaction rate together with the increase in S/V was observed in oxidative dehydrogenation of n-butane in the quartz vessel.[54] The variation of S/V was achieved by packing the reactor with crushed quartz. In the empty reaction vessel, the conversion of butane was 62%, while in the packed vessel, it was only about 2.5%. In the latter case, the rate of reaction decreased by at least one order of magnitude, when the residence time of mixture in the reactor decreased only two times, at 580°C. It is therefore obvious that the chain termination reaction has a predominant role on the walls.[54]

The kinetic investigations show that in the thermal nonbranched chain-radical reaction of $H_2 + Cl_2$ (at T = 286°C, P = 115 mm Hg), the initiation and termination of chains occur on the walls of the reaction vessel.[35, 39] Surprisingly, at first glance, the changes in S/V have no essential effect on the reaction rate. In this regard, Semenov[27] cited the results of Chaikin's experiments showing that in about an eight times increase in S/V, the rate was changed only about 10%. Thus, the increase in S/V is not a reliable basis for disclosing the heterogeneous stages in homogeneous reactions. Studies on the dependence of the reaction rate on the variation in S/V may be only the first step in revealing the heterogeneous stages in any complex reaction. The validation of their existence requires additional evidence, obtained by the application of other experimental methods.

To determine the heterogeneous–homogeneous nature of complex reactions, in 1946, Kowalski and Bogoyavlenskaya[56] applied the so-called method of differential calorimetry whereby two thermocouples, isolated from the medium by glass capillary tubes, were placed in the vessel, one in the center of the reaction vessel and the other on the wall of the reactor. Then, by measuring the differences in temperatures, the overall rate of reaction, and the heat conductivity of the mixture, the method permitted determination of the contribution of the heterogeneous reaction in heat release in the overall reaction. The released heat in the heterogeneous reaction was shown to be proportional to the rate of the reaction on the surface.

According to this method,[17, p.46] the easily measurable value of the difference of temperatures (ΔT), obtained by thermocouples, in a chemical reaction, is

$$\Delta T = T_c - T_w$$

where T_c is the temperature measured by the thermocouple, placed in the center of the reaction vessel, and T_w is that on the wall. In a cylindrical reactor for the homogeneous reaction, the initial rise of temperature (ΔT_g) is

$$\Delta T_g = q_g / 4\pi k$$

where q_g is heat released in the reactor per second and per cm^3 and k is the coefficient of the thermal conduction of gas. In a heterogeneous reaction, the difference of temperatures (ΔT_w) depends on the amount of heat released in the reaction (q_w) and the radii of the reactor (R), as well as capillary tubes (r) of the thermocouple

$$\Delta T_w = q_w r (\ln R - \ln r) / 2\pi k (R + r)$$

For the heterogeneous–homogeneous reaction

$$\Delta T = \Delta T_g + \Delta T_w$$

If the rate of homogeneous reaction is W_g and the rate of heterogeneous reaction is W_w, the rate of the overall reaction, W, will be

$$W = W_g + W_w$$

Denoting by α the part of the contribution of the heterogeneous reaction, and by Q, Q_g, and Q_w the molar heat of the overall, homogeneous and heterogeneous reactions, correspondingly, it may be considered that

$$\Delta q_g = (1 - \alpha) W Q_g \quad and \quad q_w = \alpha W Q_w$$

Finally, the following equation for ΔT may be obtained:

$$\Delta T = (WQ/4\pi k) \left[1 - \alpha + \alpha (2r/R + r) \ln(R/r) \right]$$

where α may be determined measuring ΔT, the overall rate of reaction W, the molar heat of reaction Q, and the coefficient of the thermal conductivity k.

Application of this method had great historical importance in investigations of the chain-radical reaction. This method showed that the walls of reactors participated not only in the termination stage, but also in initiating the chain-radical reaction, transferring the radicals, formed on the solid substances of walls, to the gaseous mixture. Creation of the active centers in the gas phase, in the chain reaction, by the heterogeneous reactions was evidenced in Kowalski's experiments in a number of oxidation reactions.[57, 58] In particular, in applying this method, Kowalski and Bogoyavlenskaya showed that the oxidation of SO_2, at Pt (500°C), was a "purely" heterogeneous reaction, but the reduction of SO_2 by hydrogen on Al_2O_3 was a homogeneous reaction.[17]

Application of this method is more complicated in cases where the "hot" molecules or atoms, energetically excited species or radicals, are transferred to the gas phase from the surface, continuing the energy exchange with other species by collisions in the homogeneous medium. In these cases, according to Krilov, the determination of α is not correct.[51] Thus, the variation of S/V, diameter, or share of the reaction vessel, as well as the chemical nature of the wall or solid material in the reaction medium, may serve for the initial exploration of the existence of the heterogeneous stages in the heterogeneous–homogeneous reaction of oxidation. Investigation of the heterogeneous constituent of complex reactions requires the "arsenal" of the experimental methods of the surface science and heterogeneous catalysis.

For a long time, the effects of walls on the reaction rate were recognized as a difficulty in kinetic measurements, and overcoming this difficulty required special efforts to separate the "purely" homogeneous reaction. In order to eliminate or, at least, to minimize the heterogeneous constituent in the rate coefficient measurements in homogeneous reactions, in the 1970s, specially designed experimental set-ups, known as wall-less like reactors, were proposed.[59] In these reactors, the reaction mixture was heated not through the reactor walls, but through the separately supplied hot inert gases, the streams of which contain initial reactants. The walls of the reactor were cooled, and the inert gases were heated in other sections of the reactor. Subsequently, this method was improved. A new wall-less like reactor was named the "homogeneous front" reactor,[60] where the mixed gases flow through a large tube forming a "front," in the midstream of which the occurring reaction was homogeneous. Investigations in this kind of reactor showed that the kinetic results obtained for a number of pyrolysis and oxidation reactions of the organic compounds were remarkably different from those in ordinary reactors. For example, by application of this method, the well-manifested surface effects were proven in pyrolysis of ethane and isobutene in the presence of oxygen.[60]

The difficulties related to the influence of the walls on the rate and distribution of products are known in a number of industrial processes. In the interest of minimizing the effects of the walls, in the chemical industry large-size reactors (larger than 200 liters) have been used.[61]

1.5   FRONTIERS OF THE HETEROGENEOUS, HETEROGENEOUS–HOMOGENEOUS AND HOMOGENEOUS REACTIONS

The reactivity of solid-phase substances in a reaction medium may be very different regarding the components of the fluid phases, including reaction intermediates. The overall effect may be either acceleration or retardation of the rate of reaction, depending on reaction conditions. In the first case, when the presence of the solid-phase substance(s) in the reacting medium accelerates the chemical reactions in the gas- or liquid-phase mixture, it is said that the solid substance plays the role of catalyst. In the second case, when the presence of the solid phase(s) reduces the rate of chemical reaction in the fluid phases, often, it is named the "negative catalysis," despite the recommendation of IUPAC's Terminology Commission, which states that the term *negative catalysis* must be replaced with the term *inhibition*.[8]

The heterogeneous catalysis phenomenon was discovered long before the effect of the walls on the rate of reaction in the fluid phases was recognized.

In 1823, J. W. Dobereiner (1780–1849) observed spontaneous combustion of hydrogen premixed with air at room temperature over the sponged platinum.[62] He mentioned that "the platinum became red-hot, then white-hot, and the jet ignited spontaneously." This sensational discovery was soon applied for construction of Dobereiner Feuerzeug *(lighter).[63]*

Although the number of heterogeneous catalytic reactions continuously increased beginning in the nineteenth century, the role of solid-phase substances as catalysts was not clear. For a longtime, investigations of heterogeneous catalytic reactions were developed separately from those of homogeneous reactions. One of the first descriptions of bimodal reaction sequences, related to the transfer of active intermediates from one phase to another, came from an early work by I. Langmuir (1912).[64] He observed that the tungsten lamp (tungsten wire, heated electrically in vacuum), improves the vacuum conditions in the system. This observation was named "clean-up" of gas. He found that at between 1300 and 2500 K of wire temperature, the small amounts of hydrogen (0.001–0.01mm Hg) "disappear" in the system. Langmuir offered the following explanation: Molecular hydrogen dissociates to hydrogen atoms on the tungsten wire, and part of the atoms escape from the tungsten surface and diffuse to the gas phase. Then, a part of the hydrogen atoms may recombine in the volume, forming molecular hydrogen. As the pressure is low, only a small part of the hydrogen atoms may recombine in the volume. Other part of the hydrogen atoms, reaching to the internal surface of the glass bulb, adsorb on the glass, where it may either react with the components of glass material or recombine on its surface. Later investigations confirmed the hypothesis of the dissociation of hydrogen molecules into atoms on the surface of tungsten and their escape from the solid surface.

As mentioned above, the bridges between these areas in chemistry were found with the creation of the theory of chain-radical reactions in the gas phase in the 1930s. The majority of the chain-radical reactions in the gas phase, especially, at low temperatures, almost indispensably include heterogeneous stages on the walls of the reaction vessels (initiation or termination of chains).[65–67] However, only in the 60-70s of the XX century, it was established the heterogeneous–homogeneous nature of a great number of reactions, which earlier were suggested as "purely" homogeneous or "purely" heterogeneous–catalytic reactions.[68]

In 1927, Polyakov[69] was the first scientist to theoretically predict the existence of complex heterogeneous–homogeneous reactions as a class of the chain reactions, initiated by the surface and developed in the gas or liquid phases.[9, 10] Indeed, from today's point of view, this is a particular case of heterogeneous–homogeneous reactions, revealing when the gas-phase chain reaction may be initiated by the radicals formed on the surface. Presently, there are other known types of heterogeneous–homogeneous reactions, and the role of the surface is not limited solely by the generation of radicals or other intermediates, transferred into the gas or liquid phases. As mentioned above, further investigations in this area showed that in many reactions, there are more complex interactions between surface reactions and processes in the homogeneous phases (heterogeneous branching and heterogeneous propagation of the homogenous chains) and heat transfers between the solid and fluid phases (see Chapter 6).

The difficulties in simultaneously investigating the heterogeneous and homogeneous stages were related to the problems of separating the individual heterogeneous stages in chain-radical reactions. Note again, the complexity of this kind of problem, taking into account the feasibility of the interactions of the components, including reaction intermediates in the fluid phases and on the surface of the solid substance, as well as heat and mass transfer processes. Moreover, the experimental methods applied in the investigations of the homogeneous reactions, mainly, are not applicable for the heterogeneous reactions. On the other hand, the theoretical or computational simulations also are complicated, as the kinetic data for radicals or other intermediates in heterogeneous reactions are known in rare cases.

The theoretical suggestions about the chain-radical mechanisms in heterogeneous catalysis (Semenov,[31] Volkenstein[65]), in analogy with the homogeneous chain-radical reactions, done beginning from the 1950s, were not directly evidenced by the experimental investigations. In this occasion, in the 1980s Boreskov[68] mentioned that, although there were many theoretical suggestions of chain-radical mechanisms in heterogeneous catalysis, "there is not even one proven reaction, occurring by 'purely' radical mechanism, by the accumulation of non-steady concentrations of intermediates." This statement, as will be shown in Chapter 6 (discusses the analogies between heterogeneous catalysis and chain-radical reactions), from the modern point of view, has lost its exactitude and actuality. The main analogies between chain-radical reactions and reactions in heterogeneous catalysis are expressed not in the paramagnetic nature of intermediates or, often, not only in their free-radical character, but also in more profound phenomena governing both types of reactions.

A bridging area between chain-radical reactions and heterogeneous catalysis became the class of reactions known as catalytic combustion.[71–74] These reactions are the best examples of heterogeneous–homogeneous processes.[71–73] Catalytic combustion is mainly flameless oxidation of fuel by dioxygen or air in the presence of the heterogeneous catalyst. In the 1960s, it was appeared as an alternate way of combustion, permitting to replace the conventional (homogeneous) combustion of fuel, usually, accompanied by the release of heat and light in the form of flames, and posed many environmental problems. Catalytic combustion occurs at remarkably lower temperatures than conventional combustion, without producing significant quantities of NO_x, CO, and soot. Conditionally, it may be divided into low- (<300°C), medium- (300–900°C), and high-temperature (>900°C) processes. Presently, catalytic combustion has a very wide application in heat production, as well as in control of pollution. Note only one of the wide applications of catalytic combustion processes, the so-called three-way catalysts are used to reduce pollutants (NO_x, CO, hydrocarbons) in the exhaust gases of internal combustion engines. The mechanism of catalytic combustion has been investigated in many systems,[68, 71–74] some of which will be covered in Chapter 3.

Many catalytic oxidation reactions at low temperatures, and especially at low pressures, are "purely" heterogeneous processes. However, at a certain range of temperatures, they become typically heterogeneous–homogeneous reactions. For example, silver is known as a catalyst of the oxidation of alcohols to aldehydes.[75–76] In general, this is a heterogeneous catalytic reaction transformed into a heterogeneous–homogeneous reaction at relatively elevated temperatures.[75] For instance, methanol and ethanol oxidation on silver with dioxygen to formaldehyde and acetaldehyde, at temperatures of 600–700 K, are heterogeneous–homogeneous processes. One of the evidences of this is the detection of radicals in gaseous reaction mixtures by the EPR (electronic paramagnetic resonance) method, whereby the products of the reaction are frozen at cryogenic temperatures.[75]

The frontiers between heterogeneous, heterogeneous–homogeneous, and homogeneous reactions were discussed in Boreskov's monograph[68] about a half century ago. Generalizing the dependencies of the reaction rate on the temperature (at a wide range of temperatures, from cryogenic to 1300–1500 K) for a number of reactions, he proposed a hypothetical dependence of a portion of the homogeneous reaction ("bulk reaction") on the temperature. This dependence, involving the transition of the heterogeneous catalytic reaction to a heterogeneous–homogeneous one, and, then, into a noncatalytic homogeneous reaction, with the increase of temperature, is presented in Figure 1.4. The curve was obtained based on analysis of data taken mainly from the works performed by use of differential calorimetry. The heterogeneous–homogeneous reaction is intermediary between the homogeneous and heterogeneous reactions. The region (**a**) in the figure corresponds to the heterogeneous reaction, which is gradually transformed to the heterogeneous–homogeneous reaction in the region (**b**), and, predominantly, to the homogeneous reaction at higher temperatures (**c**). This is a much-generalized case and may not always correspond to the heat or temperature effects of the mentioned transfers for reactions in real systems. It does not take place for a chain-radical reaction, when the heterogeneous stages play the role of inhibitor, accelerating the decay of radicals from the gas phase and reducing the rate of overall reaction. In certain cases, initially, the rate of the overall heterogeneous–homogeneous reaction increases with the increase in temperature but, then, it decreases or rests practically unchanged, related to termination of the homogeneous chains on the surface of the solid substance (Figure 1.5).[77]

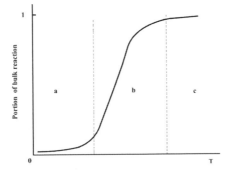

FIGURE 1.4 Hypothetical curve of the dependence of the portion of the bulk reaction on the temperature proposed by Boreskov. (Adapted from Ref.[68]).

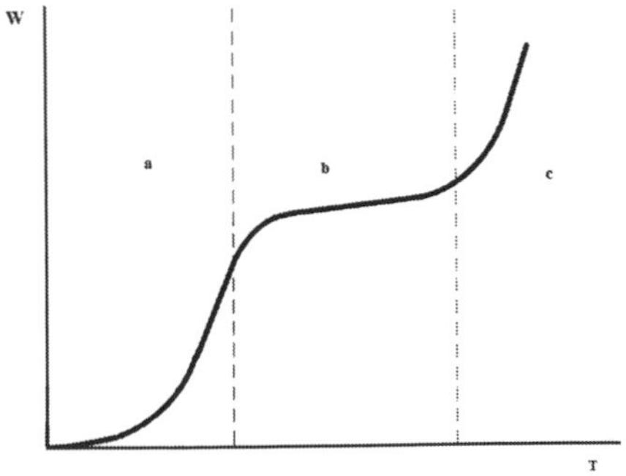

FIGURE 1.5 Curve of the dependence of the rate of catalytic combustion (W) on the temperature (T). (Adapted from Ref.[77]).

In this case, the overall process is dependent on the lengths of the homogeneous chains.[78, 79] Three regions on the curve in Figure 1.5 correspond to the ignition (a), mass transfer control (b), and homogeneous reaction (c) in catalytic combustion. More or less enough agreement between this curve and experimental results may be observed in a number of catalytic combustion reactions, for example, in the catalytic combustion of methane[73] and hydrogen[74] over Pt.

Since the 1990s, theoretical notions about heterogeneous–homogeneous reactions have advanced in two ways. One of them was revelation of more profound theoretical interdependencies between gas-phase reactions and heterogeneous catalytic reactions developed by a number of authors.[66, 80–83] Application of the thermodynamic considerations of nonequilibrium processes in heterogeneous and homogeneous oxidation reactions, occurring by the accumulation of nonsteady-state concentrations of intermediates, changed the approaches in interpreting the existing experimental data.[83] Establishing the important analogies between these two classes of reactions permitted the development of more general notions about heterogeneous–homogeneous reactions.[80] Second way was the accumulation of experimental data expanding knowledge of the kinetic peculiarities of the heterogeneous constituent of reactions[25, 53, 82, 86, 87] in heterogeneous–homogeneous reactions. In this regard, some authors newly developed the perceptions about the possibility of surface chain-reactions on the catalysts. Investigating a number of nonlinear phenomena (the so-called "hysteresis, domain instability at Lewis numbers below unity, and plateau of zero velocity of autowave transitions"),[53, 84] they assumed the feasibility of the stages of propagation and branching of the chain reactions on the active centers of the surface by their multiplication, under certain conditions. Two sections, Chapter 2 (2.2) and Chapter 6 (6.2) of this work, are mainly devoted to the discussions of these developments.

One aim of investigations in the area of heterogeneous–homogeneous reactions is to combine the advantages of the heterogeneous reactions with those of the homogeneous catalytic or autocatalytic, often, even noncatalytic reactions.[88, 89] In this regard, the appropriate combination of these two types of reactions in a unified process is one way to create chemical systems having desired advantages. Another, principally different, way to solve this kind of problem is to use the nanosized substances as catalysts. The main differences of the properties between the nanocatalysts and "ordinary" heterogeneous catalysts are related to the sizes of particles in their action. In a bulk material, the discrete energy levels of individual atoms or molecules nearly disappear, and therefore, the effects of commune interactions, the properties characteristic for the bulk, are predominant. At the nanometer scale, the clusters or particles have intermediate sizes between individual atoms or molecules and bulk materials. Nanocatalysts, being demi-heterogeneous and demi-homogeneous in microscopic scale, open new frontiers between two types of processes.[7, 88, 89] The nanoparticles usually exhibit enhanced catalytic activity, in comparison with the "ordinary" heterogeneous catalysts. For instance, gold is not active in oxidation of CO with oxygen to CO_2, but the gold nanoparticles exhibit catalytic activity in the same reaction.[88] Another example, the Pt-nanocatalyst, may activate the π-bond after treatment by $PhICl_2$, which is usually available only through homogeneous catalysts, using Pt-complexes.[89] These and many other examples[7, 88] indicate the possibility of transferring the heterogeneous or homogeneous catalytic reaction to another type of catalysis, by application of nanosize catalysts. Knowledge of the detailed mechanism of heterogeneous–homogeneous reactions also has great importance for the creation of new catalytic processes at the nanosize scale.

1.6 THERMODYNAMIC AND KINETIC ASPECTS OF THE BIMODAL REACTION SEQUENCES

Thermodynamics is a universal science. Thermodynamic laws govern processes in both macro and micro worlds. Evidently, the thermodynamic considerations applied for bimodal reaction sequences are not specific or not very different from those in other chemical systems. Thermodynamic systems, including the systems involving chemical reactions, can be defined as a region of the universe, taken into investigation in different sciences. Thus, in thermodynamics, the system and its environment are hypothetically separated. The system becomes isolated if it does not exchange energy and substance with the environment. In contrast, an open system interacts with the surroundings. Thermodynamic analysis begins by determining the state of the system.[9] The system is in an equilibrium state when its parameters do not change in time. In a nonequilibrium state, processes, including chemical reactions, continue.

The term *stationary state of the system* has a sense other than equilibrium. In the stationary state, in a chemical system, the concentrations of reacting components are unchanged over time. In an equilibrium state, the concentrations of reactants and products also are unchanged. However, the steady state is different from the equilibrium state, being a more general concept than equilibrium. The equilibrium state and the reversibility of the thermodynamic system should not be confused with the equilibrium state and the reversibility of the chemical reaction. In the equilibrium state, forward and reverse chemical reactions occur at the same rate, in given conditions. In a stationary state, in general, the reversibility of processes is not a necessary condition for the existence of constant concentrations of components. In this case, the process may also contain irreversible reactions. In the steady state, the concentrations can be balanced through the flux of reactants and products from the system to the environment or, contrariwise, from environment into the system. The above mentions also refer to other variable parameters of the system (enthalpy, entropy, etc.).

In Chapter 6, it will be shown that the formation of intermediates in superequilibrium concentrations both in the homogeneous and heterogeneous complex reactions occurring in conditions far from equilibrium is a common phenomenon, conditioning some analogies between them. Nearly all complex chemical systems with oxidation processes (homogeneous thermal, heterogeneous catalytic and enzymatic), in certain conditions, may be considered as self-organizing systems of a different level.

All scientific experience of humanity confirms the existence of the thermodynamic laws in the universe. Two general tendencies obviously derive from thermodynamic considerations: the creation of order and disorder. Philosophically, it is the exhibition of the unity of contraries or opposites, as even the ancient Greek philosophers, such as Heraclitus (ca. 535–475 BC), observed.[91] Perhaps it is one of the most universal laws of nature that leads to mysterious and contradictory conclusions. The self-organization is characteristic not only of living matter, but also of nonliving matter. Both biological and many chemical complex processes occur in open systems, also containing irreversible reactions by the accumulation of intermediates in nonstationary concentrations, accompanied by the importation of entropy: local fluctuations of concentrations; and formation of dissipative structures and chemical waves.

1.6.1 CONJUGATE REACTIONS: THERMODYNAMIC AND KINETIC CONJUGATION

The conjugation of reactions, historically known as chemical induction (in English literature, often referred to as coupled reactions), is one of the most interesting and widespread phenomena in chemistry. Moreover, in some aspects, it may be considered to be a more general expression of the phenomenon of the catalysis and chain reactions proceeding by the formation of chemically active intermediate species. As mentioned earlier, the chemical reactions in multicomponent and open systems are usually complex and multistep processes. The conjugate reactions are a result of the interactions of the different reactions in the same system and under the same conditions.

The phenomenon was first investigated by the Russian chemist N. A. Shilov[92] on oxidation reactions with dioxygen in 1905. He studied the behaviors of some couples of reactions of oxidation with dioxygen. The general observation was the following. Let us consider a chemical system, where the oxidation of a substance (C) with dioxygen does not occur. In modern terminology, it is an unfavorable reaction for thermodynamic or kinetic reasons, under the given conditions. Nevertheless, this oxidation reaction may occur in addition to the system of a new substance B that can enter into oxidation reaction with dioxygen. In other words, the new reaction $B + O_2$ induces the reaction $C + O_2$. In Shilov's terminology, the phenomenon was named chemical induction and the reacting substances were named actor (A or O_2), inductor (B), and acceptor (C), correspondingly,

$$B + O_2 \rightarrow \text{primary reaction}$$

$$C + O_2 \rightarrow \text{secondary reaction}$$

Let us consider that B is an organic compound containing the H-atom and that in certain conditions it is able to react with oxygen to give radical $B^\bullet$. In this case, the following mechanism may be suggested:

$$B + O_2 \rightarrow HO_2 + B^\bullet$$
$$B^\bullet + O_2 \rightarrow BOO^\bullet$$
$$BOO^\bullet + B \rightarrow BOOH + B^\bullet$$
$$BOOH + C \rightarrow CO + BOH$$
$$BOOH + B \rightarrow 2BOH$$

One simple example of the conjugate reaction system, corresponding to the above scheme, is oxidation of indigo with dioxygen. In ordinary conditions, the reaction between two substances is practically absent in solution. The addition of small amounts of benzaldehyde induces an oxidation reaction of indigo that is observable by the loss of the blue color of solution.[55]

It is obvious that, here, an intermediate BOOH, reactive toward C and B substances, becomes an oxidant agent. In this case, the direct reaction C + A is excluded. The induced reaction in Shilov's simple version is a system of consecutive and parallel reactions, where the common intermediates are able to react with the primary reagent that does not happen in direct oxidation of any substance by a single reaction. Shilov clearly showed that the chemical induction occurred only when at least one common intermediate for a conjugate couple of reactions formed and the overall process was energetically favorable. It is understandable that, at that time, Shilov could not entirely explain the conjugate reactions. However, his observations were of great historical importance in chemistry.

A system of the conjugate reactions may be characterized by the so-called induction factor (Φ).[39, 55, 93] This is a ratio equal to the changes in concentrations of acceptor to inductor

$$\Phi = d[C]/d[B]$$

Consider the following reactions in the simplest case of the conjugate system:

$$B + A \rightarrow X \qquad k_1$$

$$X + B \rightarrow P_1 \qquad k_2$$

$$X + C \rightarrow P_2 \qquad k_3$$

where X is intermediate, P_1 and P_2 are products of reactions, and k_1, k_2, and k_3 are kinetic constants of reactions. Then, considering the quasi-stationarity of the concentration of X, it may be shown that

$$\Phi = k_3[C]/2k_2[B] + k_3[C] \qquad (I)$$

In a more general case, considering that the stoichiometric coefficients of inductor (B), and acceptor (C), consumed in the reaction with intermediate X are m and n, the conjugate reaction may be represented as

$$B + A \rightarrow X$$

$$X + mB \rightarrow (m+1)\,P_1$$

$$X + nC \rightarrow P_1 + P_2$$

Summarizing these reactions, it may be obtained

$$A + (m+1)B \rightarrow (m+1)P_1 \qquad (v)_1$$

$$A + B + nC \rightarrow P_1 + P_2 \qquad (v)_2$$

where $(v)_1$ and $(v)_2$ are the rates of reactions. Φ may be expressed as

$$\Phi = n(v)_2/(m+1)\,(v)_1 + (v)_2 \qquad (II)$$

It is obvious that two expressions, (I) and (II), are analogous.

The quantitative determination of Φ permits us to characterize the complex chemical reactions of different classes.

In equations (II), $\Phi \to n$, if $(v)_2 \gg (m+1) (v)_1$, and $\Phi \to 0$, if $(v)_2 \ll (m+1) (v)_1$. Thus, in general, Φ is a variable in the range $0 < \Phi < n$. The values of Φ increase with the increase in the concentration of acceptor, or, more exactly, with the increase in the rate of the reaction $(v)_2$. Therefore, at high concentrations of acceptor, relative to inductor, the $\Phi \gg n$. This takes place, for instance, in the case of the initiated chain-radical reactions, where, theoretically, each molecule of the inductor is able to transform n moles of the acceptor.

In another case, $\Phi \to \infty$, when the concentration of inductor rests constant during the reaction. It means that the inductor is regenerated in the reaction system. In other words, the inductor plays the role of catalyst. From this point of view, the phenomenon of catalysis is similar to the induced reactions as one of its particular cases.

Finally, in some chemical reactions, the concentration of the inductor increases during the reaction, and Φ decreases ($\Phi \to a < 0$). It is obvious that the chemical system creates its own inductor and that the reaction becomes self-induced over time. Self-induced reactions in chemical kinetics are known as autocatalytic processes.

Understanding the mechanism of conjugate reactions, first of all, relates to the analysis of the thermodynamic aspects of the reaction system.[39, 80] In particular, in the simplest case, the induced reaction we presented as

$$A + B \to X$$
$$X + B \to P_1$$
$$X + C \to P_2$$

Let us consider that the Gipps energy changes in the primary and secondary reactions are ΔG_1 and ΔG_2, respectively.

$$A + B \to \qquad \Delta G_1 < 0$$
$$C + A \to \qquad \Delta G_2 > 0$$

According to the above description, the secondary reaction is endergonic and $\Delta G_2 > 0$; therefore, it is a nonspontaneous process. When a thermodynamically unfavorable reaction, with $\Delta G_2 > O$, is conjugated with an exergonic reaction $\Delta G_1 < 0$, the thermodynamic "restriction" may be overcome if $|\Delta G_1| > |\Delta G_2|$ (or $\Delta G_1 + \Delta G_2 < 0$). In other words,

$$\Delta G_\Sigma < 0$$

This type of conjugation of reactions is known as **thermodynamic conjugation**. Thus, one way to realize the thermodynamically unfavorable reaction is its conjugation with another thermodynamically favorable reaction, compliant to the condition $\Delta G_\Sigma < 0$, in the same system and under the same conditions. Kinetic analysis of this system gives the following relationship between Φ and ΔG_Σ:[39]

$$\Phi \Delta G_1 = |\Delta G_\Sigma|$$

One of the more important properties of conjugate or, particularly, induced reaction systems is the feasibility of the creation of **superequilibrium concentrations** in the reacting system. How is it possible? Let us assume that the reaction in chemical equilibrium state conditions (or near to equilibrium state conditions) produces certain products (or intermediates) in equilibrium concentrations. If this reaction is conjugated or induced by other reaction(s), the formation of the additional amounts of products (or intermediates) becomes possible using the energy of the inducing reactions.[94] Why these reactions become favorable, under certain conditions will be briefly discussed also in Chapter 6. Superequilibrium concentrations of active intermediates may be formed both in chain-radical and heterogeneous-catalytic reactions. For example, in the reaction of hydrogen with oxygen, the superequilibrium concentrations of H-atoms in the reaction zone is about 10^6 order more than their equilibrium concentrations.[80] As an example of the formation of intermediates in superequilibrium concentrations in heterogeneous catalysis, the reaction of dehydrogenation of methylcyclohexane to toluene on Pt is often mentioned. According to the experimental data, the benzene intermediate forms in greater quantities than equilibrium concentrations.[80]

The thermodynamic conjugation is not the only way to obtain favorable reactions from the thermodynamically unfavorable one.[34, 80, 94, 95] Moreover, in certain conditions, the thermodynamic conjugation is not realizable. For example, in the case of the stepwise process, when the chemical reaction is multistage or multistep and consecutive in the stationary state, the thermodynamic conjugation is not feasible. (The reason for this "restriction" is discussed in the next section.) Krilov[80] illustrated this statement by the noncatalytic reaction of oxidative dehydration of butane in stationary state by the following mechanism:

$$C_4H_8 \rightarrow C_4H_6 + H_2 \qquad A_1v_1 < 0$$
$$H_2 + 1/2\,O_2 \rightarrow H_2O \qquad A_2v_2 > 0$$

where A_1 and A_2 are affinities of the steps (the definition of A is given in Section 1.7.2); v_1 and v_2 are the rates of these steps. The summary reaction is

$$C_4H_8 + 1/2O_2 \rightarrow C_4H_6 + H_2O \qquad A_1v_1 + A_2v_2 > 0$$

where $A > 0$, signifying that the reaction is thermodynamically favorable; however, it is not realizable, as the consecutive reactions contain a step with negative affinity $A_1 < 0$.

One way "to circumvent the restrictions" imposed by thermodynamics on the chemical reaction is **kinetic conjugation** or **coupling** in the chemical system.[96] It may be revealed by the microkinetic analysis[5] of the reaction schemes, involving elementary reactions of catalytic cycles. The kinetic coupling arises in the system, when common intermediates may form and participate in the different reactions in and between cycles,[97] which obviously, occur at different rates. As a result, the equilibrium concentrations may be different (lower or higher) from the steady-state concentrations created by elementary reactions. Thus, in the systems with chemical reactions, the kinetic conjugation occurs when either the initial reagents are in concentrations more than the equilibrium concentrations or the products are in concentrations less than equilibrium concentrations.[94] In the case of kinetic coupling, an active intermediate "can be accumulated as a reactant or be depleted as a product in steady-state concentrations."[94]

In the 1960s, Temkin,[94, 95] based on the Shilov–Ostwald theory of the conjugate chemical reaction,[55] developed the kinetic theory of multiroute chemical processes and named it the "kinetic conjugation of steps and routes related to the existence of common intermediates."[94] He presented a kinetic analysis of the general schemes of the conjugated systems for single (as $A \rightleftarrows X \rightarrow P$) and multiroute noncatalytic, as well as catalytic reactions[92] (as $M + A \rightleftarrows X_1 \,(\rightleftarrows X_2) \rightarrow P + M$, where M is a catalyst). Temkin[94] also showed the differences and similarities between chemical induction and kinetic conjugation phenomena. In the simplest case, the important reactions are

1. $A + B \rightarrow P_1$ primary reaction $\Delta G_1^0 < 0$

2. $A + C \rightarrow P_2$ secondary reaction $\Delta G_2^0 > 0$

where A is actor, B is inductor (Ind), C is acceptor (Ac), and P_1 and P_2 are products of reactions, according to the "classical" terminology. The combination of these two reactions, which make possible the formation of P_2, is a conjugate reaction. Using the examples of two kinetic schemes involving the formation and destruction of a common intermediate for a couple of chemical reaction, Temkin concluded that kinetic conjugation was a more general case than the chemical induction.

Following Temkin,[94] let us consider a simple example: a couple of reactions 1–2 has a common intermediate X (let us use also the "classical" terminology of the chemical induction). The simplest scheme of reactions may be the following:

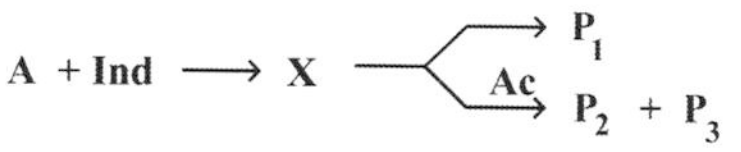

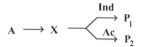

$$A + \text{Ind} \longrightarrow X \underset{\text{Ac}}{\overset{}{\diagdown}} \begin{matrix} \rightarrow P_1 \\ \rightarrow P_2 + P_3 \end{matrix}$$

Scheme 1

The overall reactions are

3. $A + \text{Ind} \rightarrow P_1$

4. $A + \text{Ind} + \text{Ac} \rightarrow P_2 + P_3$

Comparing reactions 1 and 2 with reactions 3 and 4, we find that in order to obtain reaction 4, reaction 5 must be added to reaction 2

5. $\text{Ind} \rightarrow P_3$

Thus, the stoichiometry of the secondary reaction is modified by 5. It is obvious that there is no product P_1 in reaction 4. Therefore, the conjugation takes place before the formation of P_1 via a primary reaction $X \rightarrow P_1$, which in Temkin's expression "does not perform useful work for the secondary reaction."[94]

Now let us consider another scheme containing intermediate X and corresponding to reactions 1 and 2

$$A \longrightarrow X \underset{\text{Ac}}{\overset{\text{Ind}}{\diagdown}} \begin{matrix} \rightarrow P_1 \\ \rightarrow P_2 \end{matrix}$$

Scheme 2

Scheme 2 also contains consecutive and parallel reactions that do not change the stoichiometry of the secondary reaction. On the other hand, in Scheme 2 there is nothing common with the chemical induction. However, according to this scheme, the use of same initial reactants gives products P_1 and P_2 via the kinetic conjugation of the routes of two reactions. Thus, comparison of Schemes 1 and 2 shows that the kinetic conjugation is the more general case.

The essential difference between the chemical induction and kinetic conjugation arises as a result of the modification of the secondary reaction stoichiometry by the addition of new reagents "that are foreign for the induced secondary reaction" and are absent in the case of kinetic conjugation. In other words, chemical induction creates "the new route with the corresponding new overall equation" in the kinetic analysis of systems.[94, 95]

Generally, microkinetic analysis permits us to obtain the ratios of the rates of elementary steps, which allows us "to remove" the thermodynamic restrictions due to kinetic coupling in the system. The next section describes some quantitative relationships in the thermodynamics of irreversible processes. Note again that knowledge of the mechanism of bimodal reactions is essentially related to the investigation of their thermodynamic aspects.[80]

A great number of examples of the kinetic conjugation were discussed in the works of Boudart,[97] Temkin,[94] Krilov[80] and others in irreversible chemical processes, far from the equilibrium conditions. Here, it is useful to briefly note some fundamental definitions, presenting their mathematical forms and signification.

1.6.2 THERMODYNAMICS OF THE IRREVERSIBLE CHEMICAL PROCESSES APPLIED IN BIMODAL REACTIONS

The thermodynamics of irreversible chemical processes was developed by, among others, Prigogine,[98] Onsager,[99] De Donder,[100] Boudart,[101] Krilov,[80] and Parmon.[102] The Nobel Prize in Chemistry in 1977 was awarded to I. Prigogine for his contributions to nonequilibrium thermodynamics, particularly the theory of dissipative structures.

In the closed chemical system, a spontaneous chemical reaction takes place, when $\Delta G < 0$. The accumulation of superequilibrium concentrations is possible only in the nonequilibrium state. There are no absolute reversible processes in nature. The irreversibility of the processes may be introduced in the nonequilibrium state of the system. The reversible process is an idealization. To predict the feasible direction of the process, the thermodynamics of irreversible processes uses inequalities. The thermodynamics of irreversible processes states that in open systems

$$dS = d_iS + d_eS$$

where dS is change of entropy in the system, d_iS is internal production of entropy, and d_eS is incoming or outcoming entropy.

In the adiabatic system (where there is no heat and matter exchange with the surrounding environment), $d_eS = 0$; therefore,

$$dS = d_iS$$

In the equilibrium state, $d_iS = 0$, but $d_iS > 0$, when a spontaneous process (physical or chemical) occurs in the system.

In an open system, if d_eS is not zero, it can be either $d_eS > 0$ or $d_eS < 0$. Accordingly, the different combinations of d_iS and d_eS may therefore take place in an open system, and so the changes of entropy may be $dS = 0$, $dS > 0$, and $dS < 0$.

If $d_iS \geqslant 0$, therefore, $d_iS/dt \geqslant 0$

$$d_iS/dt = \int \sigma \, dV \geq 0$$

where σ is the dissipation function, t is time, V is volume, and d_iS/dt is accordingly the rate of the production of entropy. The processes occur when there exists a difference of forces creating flux, for example, differences of temperatures, and concentrations can become the cause of the emergence of the driving forces.

In the 1920s, T. De Donder developed the thermodynamics of irreversible processes that was also applicable for systems with chemical reactions.[100] He created the concept of affinity in irreversible processes. For example, if a chemical reaction takes place in the system, σ is determined as

$$\sigma = v\, A/T$$

where v is the rate of the chemical reaction and A is called the affinity of the chemical reaction.

Here, the affinity is assigned to the elementary chemical act (below denoted by index i), and it is different from the affinity in classical thermodynamics. For any chemical reaction, one of the more important parameters is the variation of entropy. According to De Donder,[100] in the case of the chemical reaction in a system, the affinity A_i is determined as

$$A_i = -dG/d\xi \quad \xi = n_i/v_i$$

where G is Gibbs energy, ξ is the extent of reaction, n_i number of moles in the system, and v_i stoichiometric coefficients.

The rate of entropy production in the system is equal to σ

$$\sigma = dS/dt = \left(A_i/T\right)v_i$$

For an equilibrium state $\sigma = 0$ and for a nonequilibrium process (even in the steady state), $\sigma > 0$. Thus, for a system where the occurring chemical reaction is only an irreversible process,

$$A_i v_i > 0$$

This condition takes place when

$$A_i > 0 \text{ and } v_i > 0 \quad \text{or} \quad A_i < 0 \text{ and } v_i < 0$$

For multistep successive reactions

$$\sum_i A_i v_i > 0$$

$$A = \sigma_i A_i$$

In conjugated successive reactions, the affinity may not be negative ($A < 0$), accordingly, A_i of every state must be positive. If the successive chemical process contains a reaction with $A_i < 0$, the thermodynamic conjugation becomes impossible. Obviously, in an equilibrium state, the chemical affinity is not related to the reaction pathway. In contrast, in irreversible processes, it is a function of the reaction pathway, even in the steady or quasi-steady state. De Donder's equation connects the affinity with the rate of reaction. Thus, if a chemical reaction occurs in the system, the production of entropy takes place. The reaction occurs by the decrease of free energy (equally, the chemical potentials or other analogous thermodynamic parameters will be changed). The rate of chemical reaction "plays the role of flux." The measure of the dissipation function σ is [mol/time • energy/mol • volume • temperature] = [energy / time • volume • temperature].[103] It is the driving force of the chemical reaction multiplied by flux.

Returning to the conjugate processes, let us consider that there exist two irreversible reactions that occur far from the equilibrium state in an open system. In this case, the following equation may be written:[103]

$$\sigma = J_1 X_1 + J_2 X_2 \geq 0$$

where J is flux and X is force. Let us also include coefficients of proportionality L_{11}, L_{12}, L_{22}, L_{21}. The flux J may be expressed as

$$J_1 = L_{11} X_1 + L_{12} X_2$$

$$J_2 = L_{21} X_1 + L_{22} X_2$$

according to the Onzager theorem, showing the symmetry with respect to indexes 1 and 2, in conditions that are not far from equilibrium[99]

$$L_{12} = L_{21}$$

$$\sigma = L_{11} X_1^2 + 2 L_{12} X_1 X_2 + L_{22} X_2^2 > 0$$

In the conjugated process, σ must be positive $\sigma > 0$). As $L_{11} X_1^2$ and $L_{22} X_2^2$ are positive, σ will be positive if $L_{12}^2 < L_{11} L_{22}$. Here L_{12} may be either positive or negative.

Let us consider that $J_1 X_1 > 0$, but $J_2 X_2 < 0$. In this case, σ will be positive only when

$$J_1 X_1 > |J_2 X_2|$$

The condition $J_2 X_2 < 0$ means that the second irreversible process is not possible separately, but it is possible in the overall process, joint with the first process, when $J_1 X_1 > |J_2 X_2|$.

The conclusions from the results of these mathematical manipulations are very important. In the thermodynamic system, which is far from the equilibrium state, as a result of conjugation of the separate processes, the overall process may occur by the decrease of entropy. This is not possible in the case of the completely reversible processes.

Let us also observe the feasibility of kinetic conjugation in the case of the consecutive steps of a complex reaction, following the elucidations, given in the work of Krilov.[80] For a given step, the reversible reaction is

$$X_1 + M_1 \rightleftarrows X_2 + M_2$$

where X_1 and X_2 are intermediates, and M_1 and M_2 are molecules. Let us consider that this step has an affinity $A < 0$, indicating that it may not be a spontaneous process. This reversible reaction may be "shifted" to the right if the concentrations of products are lower than their "equilibrium" (equally, pseudo equilibrium) concentrations or if the concentrations of the initial reactants are higher than their "equilibrium" concentrations. In a very simple case, when $[M_1] = [M_2]$, the affinity of this reaction will be

$$A = A^0 + RT \ln \left\{ [X_1]_{\text{stationary}} / [X_1]_{\text{equilibrium}} \right\} = RT \ln \left\{ [X_2]_{\text{equilibrium}} / [X_2]_{\text{stationary}} \right\}$$

(where A^0 is the standard affinity of the given step), according to De Donder,[100]

$$\ln (v_i / v_{-i}) = A_i / RT$$

(where v_{-i} is the rate of the reverse reaction). When $[X_1]_{\text{equilibrium}} = [X_1]_{\text{stationary}}$, and $[X_2]_{\text{stationary}} < [X_2]_{\text{equilibrium}}$, $A > 0$. On the other hand, when $[X_2]_{\text{equilibrium}} = [X_2]_{\text{stationary}}$, A will be positive ($A > 0$) if $[X_1]_{\text{equilibrium}} < [X_1]_{\text{stationary}}$. Thus, in these cases, the reaction may be shifted to the right[80, p. 193] even if $A^0 < 0$. It is evident that one of the main conditions of the feasibility of the kinetic conjugation is accumulation of intermediates in superequilibrium concentrations. Correspondingly, other analogous conditions for kinetic conjugation will be the removal of the reaction product, creating concentrations less than equilibrium concentration.

The thermodynamics of irreversible processes had revolutionary importance for understanding many natural phenomena. Even the first attempts of the consideration of irreversible and conjugate processes in chemical systems in the early 1950s, long before the creation of the above-mentioned theory, showed the efficacity of this approach. For example, British computer scientist and mathematician Alan M. Turing (1912–1954), in his only work in chemistry and biology, "The Chemical Basis of Morphogenesis," proposed the theory of morphogenesis, explaining biological phenomena through chemical reactions.[104] Conjugating autocatalytic reaction with diffusion, he theoretically showed that in the thermodynamically open chemical system, a spatiotemporal ordered pattern of chemical reaction might be obtained. The theory was generally based on the kinetic analysis of chemical reactions, where two components, initiator and inhibitor, had very different diffusion coefficients. In fact, Turing's model was based on the autocatalytic reaction, considering diffusion phenomena of species in the system.

1.7 SUGGESTIONS OF THE REACTION MECHANISMS IN HETEROGENEOUS–HOMOGENEOUS REACTIONS

1.7.1 NOTES ABOUT THE CLASSIFICATIONS OF REACTION MECHANISMS

Different oxidation reactions occur at a wide range of temperatures. In rough approximation, the majority of combustion reactions occur at temperatures above 900°C; homogeneous and heterogeneous catalytic reactions, in the gas phase, at 200–500°C; in the liquid phase at 25–150°C; and in biological processes at 10–40°C.[105] Therefore, considering the nature and types of intermediates, evidently, the mechanisms of these reactions are different.

In particular, as mentioned earlier, the oxidation of organic and inorganic compounds with molecular oxygen in the fluid phases has a chain-radical mechanism. Although oxidation reactions were investigated beginning in the nineteenth century, the first mechanism suggestions, in a modern understanding, were appeared only after the discovery of the chain-radical reactions by Bodenstein, who, in 1913, proposed the term *chain* for photochemical reaction $H_2 + Cl_2$.[106] In 1925–1927, Semenov proposed his theory of the branched chain-radical reactions in the gas phase, and in 1928–1930, Hinshelwood suggested the chain-branched mechanism of the $H_2 + O_2$ reaction. (The chronology of the early development of the theory of chain-radical reactions was given by Denisov et al.[107]). In the liquid phase oxidation reactions with dioxygen, the first chain and radical mechanism was suggested by Backstrom in 1927, who investigated the thermal and photochemical reactions of benzaldehyde.[108] The chain mechanism he proposed was based on the observation of very high quantum yields of reaction ($\Phi \gg 1$) and on the formation of unstable intermediates having a free valence. In fact, it was not much different from today's notions of chain-radical reactions:

$$PhC(=O)H \rightarrow PhC^{\bullet}(=O) + H$$

$$PhC^{\bullet}(=O) + O_2 \rightarrow PhC(=O)O\text{-}O^{\bullet}$$

$$PhC(=O)O\text{-}O^{\bullet} + PhC(=O)H \rightarrow PhC(=O)O\text{-}OH + PhC^{\bullet}(=O)$$

In 1946, Bolland and Gee[109] proposed a mechanism of the chain degenerate-branching reaction of the oxidation of hydrocarbons by dioxygen in the liquid phase. It was analogous to gas-phase branched-chain reactions, involving all the main stages (initiation, propagation, degenerate (partial) branching and termination of chains)[107] of branched chain-radical reactions. These and many other investigations revealed the important role of organic peroxy compounds, as main intermediates responsible for the degenerate branching of chains. Nearly a century of extensive investigations on oxidation reactions fully proved the existence of radical and chain mechanisms in the fluid phases. Seemingly, the only comment that may be made in this occasion, relates to the role of singlet oxygen, which often, to some extent, changes the "classical" schemas of oxidation reactions with dioxygen.

A much more simplified version of the mechanism suggested in thermal oxidation of hydrocarbons with oxygen in the fluid phases, at not very high temperatures, may be presented as:[10, 20, 34, 39]

a. Initiation of chains in any way (generation of free radicals)

$$RH + O_2 \rightarrow R + HO_2$$

b. Propagation

$$R + O_2 \rightarrow RO_2$$

$$RO_2 + RH \rightarrow RO_2H + R$$

c. Branching of chains

$$ROOH \rightarrow RO + OH$$

or

$$ROOH + RH \rightarrow RO + R + HO_2$$

$$ROOH + ROOH \rightarrow RO + RO_2 + H_2O$$

d. Termination of chains

$$R + R \rightarrow R\text{-}R$$

$$RO_2 + RO_2 \rightarrow ROOR + O_2$$

$$RO_2 + R \rightarrow ROOR$$

or disproportionation as

$$R\text{-}CH_2\text{-}CH_2 + R\text{-}CH_2\text{-}CH_2 \rightarrow R\text{-}CH = CH_2 + R\text{-}CH_2\text{-}CH_3$$

Note again that this is much generalized and very simplified scheme and each of hydrocarbons have their own peculiarities of oxidation. On the other hand, the influence of the heterogeneous factors, more or less contributing to the overall rate of the oxidation reactions, here, were not taken into consideration.

The theory of the heterogeneous catalytic oxidation was developed separately from the theory of chain-radical reactions. (This historical development will be presented in Chapters 2 and 6.) The classical approach to the mechanism notions in heterogeneous catalysis usually involves three basic and general types of the overall reactions on the surface: the Langmuir–Hinshelwood, Rideal–Eley, and precursor mechanisms.[34, 63, 68, 110, 111] In the Langmuir–Hinshelwood mechanism, the initial reactants from the fluid phases (A and B) adsorb on the surface of the catalyst. Then, the reaction occurs on the surface layer of the catalyst, after which the products of reactions desorb from the surface. In the Rideal–Eley mechanism, at least one of the initial reactants adsorbs on the surface of the catalyst (A or B). The second reactant from the fluid phase reacts with the surface-bonded reactant (without being adsorbed on the surface), and the formed products desorb from the surface. In the precursor mechanism, at least one of the initial reactants is adsorbed on the surface of the catalyst. The second initial reactant, interacting with the surface, forms a mobile precursor state that is able to react with the adsorbed reactant, and then the product(s) of reaction diffuse to the fluid phases.

According to Missono,[110] although these mechanisms, as well as the rate expressions of reaction that follow from them, "are based on super simplified models" and may not quantitatively describe real reactions, they are useful for understanding some dependencies of the rate of catalytic reaction on the concentrations, partial pressures of reactants, and so on.

At first glance, according to the above mechanisms, the catalyst is only an "active support"[110] of the reaction between reactants. In reality, the heterogeneous catalyst undergoes reversible or irreversible changes during interactions with the components of the catalytic system. For example, the surface of catalyst undergoes relaxation, reconstructing and restructuring during the catalytic action. That is why this classification of the mechanism gives a basic, but not profound division of the heterogeneous catalytic reactions. Note also that there is no other unique categorization of all heterogeneous reactions than the three types of reactions mentioned above.

There are different approaches to classification of the mechanisms involving both the heterogeneous and homogeneous (catalytic or noncatalytic) reactions of oxidation.[111] Below, we survey the classification of oxidation reactions given by Sheldon,[111, 112] who proposed that classification of oxidation reactions be based on some analogies between the basic mechanisms in heterogeneous and homogeneous catalytic reactions. According to Sheldon,[111] from the point of view of reaction mechanism suggestions, all homogeneous and heterogeneous catalytic oxidation reactions with dioxygen or other oxidants may be divided into three types: (i) autocatalytic, (ii) direct oxidation of the coordinated substrate, and (iii) oxygen transfer reactions.

1. In *autocatalytic oxidation* of organic compounds in both the gas and liquid phases, which occur mainly through the radical pathway, self-acceleration of the reaction is usually conditioned by the rates of the formation and decomposition (disproportionation) of main intermediates.[34, 39] Decomposition (disproportionation; sometimes, isomerization) of the main intermediates (usually, peroxy compounds) in oxidation with dioxygen is responsible for the self-acceleration of reaction. Principally, in certain conditions, the oxidation of organic compound may occur without a catalyst. In an autocatalytic reaction that is either homogeneous or heterogeneous, acceleration is

often related to one or more stages or steps of the complex process. The best examples of this kind of reaction are heterogeneous–homogeneous reactions of oxidation with dioxygen, where a heterogeneous catalyst may participate in certain stages of the process (initiation, degenerate branching, or even propagation of chains). A more detailed discussion of this subject will be given in Chapter 3.

Let us now briefly discuss a general case of the oxidation of organic compounds in the presence and absence of the catalyst. As has been mentioned, in the absence of the homogeneous or heterogeneous catalyst, an organic compound may be oxidized with dioxygen, mainly by the free radical mechanism. The chain-radical reaction may be initiated by different ways (heat, light, or electromagnetic waves of different length, catalyst) and propagated by the above-mentioned general scheme, until the termination of chains, again, in any way. Obviously, this kind of oxidation is indiscriminate.

In homogeneous catalysis, in which metal ions of variable valance are added to the system, the degenerate branching stage may be accelerated by the following scheme

$$RO_2H + M^{n+} \rightarrow RO_2 + H^+ + M^{(n-1)+}$$

$$RO_2H + M^{(n-1)+} \rightarrow RO + OH^- + M^{n+}$$

A famous example of this is the Fenton reaction, the mechanism of which was first proposed by Willstatter, Haber, and Weis.[113]

In heterogeneous catalysis, the analogous role of catalysts may play metals, metal oxides, or other solid substances, accelerating the degenerate branching of chains. Known examples are from the oxidation of alcohols, aldehydes, aromatic compounds, etc., on metal oxide catalysts.

2. *Metal or metal ion oxidations of coordinated substrate*. Oxidation of the substrate or ligand occurs due to the metal or metal ion. An example of this kind of oxidation is the step (stage) of oxidation by metal ion (Pd^{2+}) coordinated ethylene, producing acetaldehyde in water, in the presence of dioxygen (using tetrachloropalladate(II) and $CuCl_2$ as a catalyst). This is known as the Wacker process.[114]

The simplified mechanism suggestions in this reactions are the following

$$[PdCl_4]^{2-} + C_2H_4 + H_2O \rightarrow CH_3CHO + Pd + 2HCl + 2Cl^-$$

$$Pd + 2\,CuCl_2 + 2\,Cl^- \rightarrow [PdCl_4]^{2-} + 2CuCl$$

$$2CuCl + \tfrac{1}{2}O_2 + 2HCl \rightarrow 2CuCl_2 + H_2O$$

In generalized form for other olefins the important stage may be written as

$$R\text{-}CH = CH_2 + [Pd^{2+}] + H_2O \rightarrow R\text{-}CH\text{-}OCH_3 + Pd^0 + 2H^+$$

Another example is oxidative dehydrogenation of alcohols by metal ions. In dehydrogenation of alcohols, the reaction may be catalyzed by M^{n+} (for instance, Pd^{2+} or Rh^{3+} and others). The important stage of this process is

$$2RR'C\text{-}OH + M^{n+}X \rightarrow 2RR'C = O + M^{(n-2)+} + 2HX$$

3. The third type of the mechanism suggestions, spread in the heterogeneous and homogeneous catalytic oxidation reactions, is known as *oxygen transfer reaction*.

Let us consider that S is the substrate, X-O-Y the oxygen atom donor and that metal M or metal oxide M_xO_y is a catalyst in any coordination environment. The main reaction in this system is

$$S + X\text{-}O\text{-}Y \overset{M}{\to} SO + XY$$

The oxidant may be H_2O_2 or other peroxide compound (ROOH, ROOR, ROOR', etc.). In the presence of metal, it may form a metal-oxo or metal peroxo compound:

$$MX + RO_2H \to MO_2R + HX$$
$$\downarrow S$$
$$SO + MOR$$

$$MX + RO_2H \to M\text{-}OX + ROH$$
$$\downarrow S$$
$$SO + MX$$

where MOR and MO_2R ($O = M\text{-}R$ and $\overset{O-O}{\underset{}{M}}\text{-}R$) may also be metalorganic compounds.

In addition, the catalytic cycle may be performed by reoxidation of the reduced metal by the reaction of any strong oxidant—in the simplest case, by dioxygen

$$MX + O_2 \to (\text{-O-O-})MX\,(\text{peroxo})$$
$$MX + O_2 \to (O = M = O)X\,(\text{di-oxo})$$

The oxo-atom transfer mechanism is suggested for a number of reactions in both heterogeneous and homogeneous catalysis. An example of the liquid-phase heterogeneous catalytic process occurring through the mechanism of oxygen transfer is the epoxidation of ethylene, using ethylbezene hydroperoxide as the oxidizing agent, over the catalyst Ti^{4+}/SiO_2 in butanol solution that is known as the Shell process.[112]

According to Sheldon,[111] the first investigated oxygen atom transfer reaction was the catalytic hydroxylation of ethylene (alkene) giving vicinal diol, performed by Milas,[115] in 1936. Mixing the metal oxides OsO_4, MoO_3, WO_3, V_2O_5, and CrO_3 with H_2O_2 solution in *tert*-butanol solution, he prepared a reagent, which is now known as Milas reagent.

$$RCH = CR' + H_2O_2 \to RC(OH)\text{-}C(OH)R'$$

Although the majority of the mentioned metal oxides are not soluble in *tert*-butanol, they may form peroxidic or peracidic compounds with hydrogen peroxide, which serve as oxygen atom donor agents. For example, in acidic solution, V_2O_5 forms pervanadic acid HVO_4, which is a peroxyacid:

$$V_2O_5 + 2H_2O_2 \to 2HVO_4 + H_2O$$

The suggested mechanism of this reaction involves primarily formation of ethylene oxide that undergoes hydrolysis, giving vicinal diol in acidic solution.[116]

Investigation of oxo-atom transfer reactions in the oxidation of organic compounds is one of the intensively developing areas not only in chemistry, but also in biology, as these reactions are widespread in enzymatic oxidation. Different aspects of the mechanism suggestions of the O-atom transfer reactions are discussed in a review by Arzoumanian and Bakhtchadjian.[117]

1.7.2 HETEROGENEOUS–HOMOGENEOUS PATHWAY IN MECHANISM SUGGESTIONS

Desorption of the active intermediates, particularly those of radicals, from the surfaces of the solid substances, generated in oxidation reactions, is a key stage in determining the bimodal nature of the processes. The presence or absence of this stage divides complex chemical reactions into homogeneous, heterogeneous, and mixed heterogeneous–homogeneous processes.

Why may the effects of the walls, expressed through desorption of radicals, be an energetically allowed pathway in oxidation reaction? Let us take the simplest case of the thermolysis of small organic molecules in the presence of solid substances. In general, the energy required for desorption of radicals from the surface is lower than that for breaking the chemical bond in molecules in the gas phase. Obviously, in the case of the heterogeneous formation of radicals with their further transfer to the fluid phases, the energetic balance includes heat of adsorption (chemisorption), thermal effect of the reactions (breaking of chemical bonds, formation of the intermediates, for instance, radicals, their structural stabilization on the surface, relaxation and reconstruction of the solid surface, etc.), and energy required for their desorption.[118] Let us compare the generation of radicals in thermolysis of hydrogen in the gas phase and on the surface of the heterogeneous catalyst as Pt. The energy of the homolytic dissociation of H_2 in the gas phase is 436,4 kJ/mol. The energy required for breaking the Pt-H bond on the surface of Pt is 249 kJ/mol. Note that for a diatomic molecule the homolytic dissociation energy is equal to the bond energy that does not take place for polyatomic molecules for a number of reasons.[119] It is also known that the bond energy of metal–hydrogen does not depend on the crystalline plan and is nearly the same for monocrystalline or polycrystalline samples.[105] The energetic difference between the gas-phase dissociation of hydrogen and rupture of the chemical bond of hydrogen atom with Pt is approximately double.

The metal surface, adsorbing H_2, notably weakens the chemical bond and contributes to its dissociation. A similar situation takes place in the decomposition of the polyatomic molecules, peroxides, amines, etc., on the solid surface. In any case, at least, one of the new adsorption (chemisorption) bonds on the solid surface, forming in decomposition of the polyatomic molecule, giving radicals, is weaker than the bond in the original molecule in the gas or liquid phase.[119] Thus, the activation barrier of the overall process, including the formation and further desorption of radicals from the surface, in thermal decomposition, depends mainly on the strength of the adsorption bonds. In the case of hydrogen dissociation on Pt, it is obvious that the spontaneous desorption of hydrogen atoms from the surface into the fluid phases, requiring at least an energy of more than about 249 kJ/mol, is hindered at low temperatures, for instance, at room temperature. Therefore, at low temperatures, heterogeneous–homogeneous mechanisms via the escape of H–atoms to the gas phase, from the Pt-catalyst, may not be expected.

The desorption of species may be favored if the newly formed surface species are weakly adsorbed, or if the energy produced on the surface, due to other physical and chemical processes, is sufficient to overcome the energetic barrier of desorption. As an example, demonstrating the essential role of the solid surface in energetic balance of the formation and desorption of radicals, may serve the thermal decomposition of H_2O_2. The vapors of hydrogen peroxide thermally decompose at about 670–720 K, producing HO_2 radicals (Section 2.3). In contact with the solid surface of SiO_2, the vapors of H_2O_2 decompose even at room temperature and above.[120] Generation of adsorbed HO_2 radicals on SiO_2 was proven by in-situ detection of radicals using the EPR method. Desorbed radicals from the solid surface of SiO_2 may be detected above about 473 K by the method of matrix isolation of radicals combined with the EPR spectroscopy.[121] Thus, the presence of the solid surface in the reaction zone may reduce the temperature of the generation radicals in the gaseous mixture in hundreds of degrees, initiating chain reactions in the homogeneous reaction mixture. As a result, the oxidation reaction in the gas phase may occur at dozens, often, hundreds of degrees lower temperatures than in the case of the "purely" homogeneous oxidation reaction.[119] Evidently, the heterogeneous–homogeneous pathway of oxidation of many organic or inorganic compounds is energetically more favorable than a "purely" homogeneous pathway, if it is chain and radical reaction in the fluid phase. On this occasion, in 1923, Hinshelwood wrote, "The number of gas reactions, which proceed without disturbance by the walls of the containing vessel, is so limited that the homogeneous change must be regarded as an exceptional case."[122]

The mechanism suggestions in heterogeneous–homogeneous oxidation processes will be discussed in Chapter 3. They are based on experimental kinetic data, obtained mainly by EPR investigations, verifying the transfer of radicals from the surface to the gas phase.[123] The formation of radicals, by their transfer from the surface to the volume, may occur through different ways: without combination of the primary radical with oxygen species, as CH_3 radicals in the case of the oxidative coupling of methane (see Section 3.5.2.2) or through formation of oxygen containing radicals, as RO_2 radicals in the case of the oxidation of aldehydes in the capillary quartz reactor.[124] Depending on the nature of the solid surfaces and the mobility of surface oxygen species, the existence of both pathways simultaneously through the generation of active intermediates of both types is not excluded.

It is known that the mobility of lattice oxygen is lower for MnO_2 than that for V_2O_5. In oxidation of alcohols, it has been suggested that greater amounts of radicals, subsequently escaping from the surface and diffusing to the gas phase, might be formed rather in the case of V_2O_5 samples than in oxidation over MnO_2.[123] However, there is no strong experimental evidence proving this suggestion.

1.8   DESORPTION OF ACTIVE INTERMEDIATE SPECIES. RELATION OF THE HETEROGENEOUS–HOMOGENEOUS REACTION TO THE PRINCIPLE OF SABATIER

Paul Sabatier (1854–1941) was a French chemist and recipient of the Nobel Prize in Chemistry in 1912 (together with another famous French chemist, Victor Grignard (1871–1935), for his works in catalytic hydrogenation of the organic compounds. Sabatier is also known for a principle in catalysis that bears his name. He formulated a principle qualitatively described the volcano-shaped plot of the dependence of the catalytic activity on the heat of adsorption of the substrate on the catalyst, indicating that the maximum activity of the catalyst may be observed in the middle points of this plot (Figure 1.6).[125, 126] This is conditioned by the limitations of too weak and too strong interactions of substrate with the catalyst. In other words, the principle indicates how to choose the best catalyst for a given reaction. Thus, the adsorption complex formed between the surface and substrate, as a result of the chemical transformation, should have an intermediate stability. According to this principle, among a number of catalysts of the same group, there exist catalysts, having an optimal energy of bonding of an intermediate or reactant on the surface. After the desorption of the intermediate or product, the active site regenerates itself, in modern perceptions, often via self-reorganization of the surface structure, in reaction conditions. Thus, the kinetic competition between the rates of formation of an intermediate complex on the active sites, on the one hand, and the desorption and regeneration of the active site, on the other hand, are determining factors in catalysis.[126] The Sabatier principle connects the energy of the bonding of the compound with the catalytic activity. Indeed, here, the steric and many other factors play an important role. The Sabatier principle is a qualitative and empirical statement.

The Sabatier principle also motivates the required activity of the intermediate in heterogeneous initiation of homogeneous reaction occurring by the

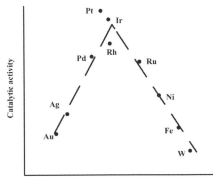

FIGURE 1.6   Volcano curve dependence of the catalytic activity on the binding energy of formic acid in its decomposition reaction on the metals, as an exhibition of the Sabatier principle. (Adapted from a number of handbooks on heterogeneous catalysis[68, 126, 127, 132, 133]).

heterogeneous–homogeneous mechanism. The transfer of the reaction from the surface to the fluid phases depends on the ability of desorption of the surface-generated intermediates, particularly, radicals. The probability of the desorption of radicals is a function of the bonding energy of radicals with the solid surface or active site of the surface. Therefore, a certain relationship is expected between the rate of the surface generation of radicals and parameters of the solid surfaces related to their chemical nature. Although a number of other parameters influence the rate of desorption of radicals, generated on the surface (adsorbent–adsorbate and adsorbate–adsorbate interactions, coverage of the surface), however, among them, the chemical nature of the solid substances is an essential factor in heterogeneous–homogeneous reactions.

Regarding to the heterogeneous–homogeneous reactions and the problem of the probability of the desorption of radicals, note that the experimental dependencies of the desorption of radicals from the surface[84] in many features remind us of the Sabatier principle. Clearly, the strongly adsorbed radicals or radical-like species, as a rule, will not be able to be desorbed from the surface; to the contrary, weakly adsorbed species will be desorbed from the surface more easily. Graphically, in this case, the Sabatier principle may be observed as a dependence of the catalytic activity on the different parameters of the system, mainly heat of adsorption or other forms of energy related to the formation of intermediates. According to Bhatia,[127] "the Sabatier principle is nicely displayed by Boudart for formic acid decomposition over diverse metals" (Figure 1.6). As a classical example, at present it is represented in most handbooks on heterogeneous catalysis. Analogously, in heterogeneous–homogeneous reactions, the above elucidations may be exemplified by the catalytic oxidation of alcohols on metal oxide catalysts. Long-term investigations of the catalytic oxidation of alcohols on noble metals, transition metal oxides, and other heterogeneous catalysts confirm the heterogeneous–homogeneous character of the overall reaction, in certain conditions.[123] A characteristic behavior of this kind of chemical system is its ability to generate free radicals on the surface of the oxide material and to transfer them to the gas phase. The organic peroxy radicals were detected by the EPR method in the gas phase, using the matrix isolation technique. Investigating the oxidation reaction of the n-propanol on TiO_2, V_2O_5, Cr_2O_3, MnO_2, Fe_2O_3, Co_3O_4, NiO, CuO, ZnO and some other surfaces, Ismagilov at al.[123] found that the rate of the generation of RO_2 radicals depends on the nature of the solid surfaces. The latter may be exhibited as a dependence of the rate of generation of radicals on certain chemical parameters of the surface.

Figure 1.7 represents the dependence of lgW_{RO_2} (logarithmic values of the rate of the generation of radicals) on the E_b, energy of binding of oxygen with the surface in the transition metal oxide. E_b is changed at a wide interval of 15–60 kcal/mol (the chemisorbed oxygen species are thought to be involved in the formation of radicals). Willingly or unwillingly, Figure 1.7 resembles the graphical illustrations of Sabatier's principle, the "volcano curves." Let us compare the "volcano-shaped" curve obtained for illustrating Sabatier's principle (Figure 1.6) with the curve represented in Figure 1.7. The former volcano-shaped curve illustrates the dependence of the catalytic activity, for instance, on the heat of reaction (heat of adsorption of reactant or other determining parameter, exhibiting the strength of the chemical bonding of reactant or intermediate with the surface). Analogously,

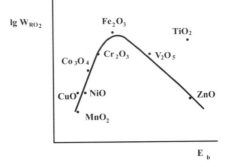

FIGURE 1.7 Simplified presentation of the dependence of the rate (in logarithmic form) of generation of RO_2 radicals (lgW_{RO_2} [particle/m²s]) on the energy of the binding of oxygen in oxides of the metals (E_b [kcal/mol]) of the IV group, propanol oxidation with molecular oxygen (1:1) and at 773 K. (Adapted from the work of Ismagilov et al.[123])

Figure 1.7 suggests some ideas about the feasibility of the transfer of radicals to the gas phase, making the overall process a heterogeneous–homogeneous reaction. In both cases, an optimal interval of the bond energy with the metal exists, corresponding to chemical bonding that is neither too weak nor too strong. Note that in the case of the heterogeneous–homogeneous reaction, the existence of at least two stages must be taken into consideration: (i) the generation of radicals and (ii) their desorption from the surface to the fluid phase.

Vasileva and Buyanov[128] investigated pyrolysis of hydrocarbons on quartz, $BaCl_2$, MgO, and defected MgO and generation of hydrocarbon radicals from these surfaces during the heterogeneous–homogeneous process. They found that "the higher is the bond energy of the surface active center with hydrocarbon radical D_{S-R}, the higher the probability for the reaction to pass from the surface into the gas phase volume." However, this conclusion was based on a comparison of just three surfaces and so may not be generalized for all oxide surfaces or for a group of catalysts. According to Garibyan and Margulis,[84] the catalyst's low activity is favorable for transfer of the reaction to the gas phase. Launsfort[129] considered that on MgO, where the transfer of methyl radicals to the gas phase in methane oxidation was observed, the radicals have a weak attachment to the surface and "small sticking coefficients, even, for relatively reactive oxides."[125] In this regard, among SiO_2, Al_2O_3, MgO (and number of others), MgO is recognized as a better catalyst in this reaction. The increase in temperature usually, favors the desorption of radicals to the gas phase, making the process heterogeneous–homogeneous.

Golodets[130] formulated the following relationship between the catalytic activity and specific surface area of the catalyst for a number of heterogeneous–homogeneous reactions, assuming that within the group of similar catalysts, r_{sp} (specific catalytic activity) is determined mainly by the surface area S,

$$r_{sp} = A/(1 + BS)$$

where, A and B are constants. Here, the specific catalytic activity decreases with the increase of S, while the Boreskov rule of catalytic activity for heterogeneous catalytic reaction shows that specific catalytic activity is relatively constant at different surface sizes and conditions of the preparation of catalysts. However, the Boreskov rule has a number of exceptions, which are named "structure-sensitive reactions" (demonstrating at least a 1–2 order of magnitude shifting in r_s dependence on the surface area or other equivalent parameter (BET area, porosity, etc.).[68, 130]

If the entropy effect in the changes of free energy is neglected, a relationship like Bronsted relations between the heat or energy changes of reaction (ΔH) and the activation barrier (E) may be written as[125, 126, 131]

$$\Delta E = \alpha \, \Delta H$$

α is the coefficient of the proportionality ($1 > \alpha > 0$). For the same group of reactions, replacing $\Delta E = E_a - E_0$ (E_a and E_0 are the internal energetic barriers of reaction)

$$E_a = E_0 + \alpha \, \Delta H$$

(This relationship is known as a combination of several names, such as Bronsted–Polanyi, Bell–Evans–Polanyi, Brønsted–Evans–Polanyi, Evans–Polanyi–Semenov, Polanyi–Horiuti and Polanyi–Semenov).[125, 131–133] Semi-empirically or theoretically, the volcano plot may be obtained as a combination of two linear dependencies, corresponding to this equation. The heterogeneous catalytic reactions can be treated kinetically as consisting of two steps: adsorption and desorption.[133] In the case of the overall heterogeneous–homogeneous reaction, the linear dependencies of the rate of generation and desorption of radicals from the surface versus any characteristic parameter exhibiting properties of the solid surface must be taken into consideration.[133] Unfortunately, to our knowledge, there are no examples of the Sabatier analysis for heterogeneous–homogeneous reactions. However, there are a number of this kind of analyses for other similar reaction systems, *a priori*, evidencing the feasibility of the theoretical prediction of the volcano plots ("strength of bond" versus "reactivity")[132, 133] in bimodal reaction systems.

1.9 THE ROLE OF ACID–BASE AND REDOX PROPERTIES OF SOLID SUBSTANCES IN GENERATION OF ACTIVE INTERMEDIATES

All the above examples of heterogeneous–homogeneous reactions indicate the determining role of the chemical nature of the solid substances for this class of reactions, as well as the importance of the proportion of initial reactants, temperature, and other parameters. The solid materials, used in investigations of these reactions are often either metals or metal oxides of the different elements. In particular, the metal oxides, as well as metal oxide-supported materials, consist of a main class of catalysts in heterogeneous–homogeneous reactions. Reactions of organic compounds with dioxygen, in the presence of these materials, always begin with their adsorption on the surface. At least one of the two main actors of oxidation, organic compound and dioxygen, may be adsorbed and activated on the surface. More often, their co-adsorption takes place. Two important properties of the solid materials, acid–base and redox properties, play a pivotal role in oxidation reactions on the metal oxides. Conditionally, the metal oxide surfaces may be divided into two main groups: metal oxide surfaces containing reducible metal ions (as transition metal ions) and nonreducible metal ions (as earth-alkali metal ions), the surface properties of which are different in the oxidation reactions of organic substances.[126, 134–139]

The typical feature of nonreducible metal oxides in the oxidation of hydrocarbons or a number of other organic compounds is that the products of the heterogeneous reactions usually involve the formation of free radicals, part of which may be desorbed from the surface and diffused to the gas phase, where a radical reaction may be initiated.[126] In general, in the presence of nonreducible metal oxide materials in the reaction zone of oxidation, the heterogeneous–homogeneous occurrence of reaction is one of the expected pathways. A classical example of the generation of radicals on the surface of nonreducible metal oxides is the reaction of the oxidative coupling of methane on MgO, already, more than once mentioned in different sections. The heterogeneous–homogeneous nature of this reaction is well established experimentally. In oxidative coupling of light alkanes, MgO is known as a typical catalyst, possessing basic surface properties and surface sites (O^- and O^{2-}), readily permitting abstract proton (Sections 3.5).

In this division, the group of nonreducible metal oxide catalysts also includes alkali metal oxides, rare earth metal oxides (as oxides of lanthanide metals), zeolites, different compositions of the mixed metal oxides, silica, and alumina. Usually, they are isolator materials with a wide energetic gap between valence and conductive bands of metal oxides. In the case of saturated hydrocarbons, the primary reaction is usually abstraction of the hydrogen from the hydrocarbon-forming alkyl radical. The radicals may either react on the surface or desorb from the surface. The further reactions of the radicals resting on the surface may be recombination, β-elimination of the second hydrogen atom by formation of the unsaturated hydrocarbon, and a combination with one form of the active oxygen species on the surface, finally, forming oxygenates.[134] Note that the heterogeneous reaction may also propagate as a radical process.

From the viewpoint of the detailed mechanism, there are different and, even, controversial suggestions for the heterogeneous reactions on the nonreducible metal oxides. However, two different mechanism suggestions are predominant: either the oxidation via the participation of the lattice oxygen or the progression of the heterogeneous reaction via oxygen active species adsorbed on the surface of the metal oxides, in dependence on the reactivity of the different forms of oxygen active species on the surface.[126, 134]

In the case of the reducible metal oxides (as transition metal oxides), the characteristic feature of oxidation, particularly in hydrocarbon oxidation, relates to the redox properties of the surface, suggesting a Mars–van Krevelen type of mechanism. In this case, the lattice oxygen participates in the primary formation of intermediates.[134] On the reducible metal ion oxides, the breaking of C-H usually takes place by the formation of new C-O or O-H bonds with the surface ions.[136] As a result, the alkoxy species may be formed. In the next stage, the alkoxy group, attached on the surface, may either lose the second hydrogen or desorb from the surface as an oxygenate. Here, the oxygen vacancy and cation in reduced forms may be formed.

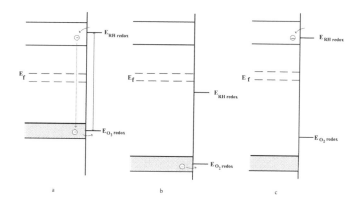

FIGURE 1.8 Electron transfer processes in oxidation of the hydrocarbon with gaseous dioxygen on the surface of the catalyst, in dependence on the energetic position of the redox potentials of the hydrocarbon and oxygen relative to the Fermi level. (a) Redox catalytic cycle is possible, (b) and (c) redox catalytic cycle is impossible. (Adapted from Ref.[136, p 282]).

Oxygen, adsorbed on the surface, incorporates to the lattice of oxide, oxidizing cation to its initial state.[136]

$$RH + 2O^{2-} \rightarrow RO\cdot + HO\cdot + 2e$$

$$1/2O_2 + 2e \rightarrow O^{2-}$$

Usually, the authors of handbooks give the following elucidations in oxidation of the hydrocarbon with the gas-phase oxygen on the surface of the catalyst by the redox mechanism (Figure 1.8). As shown here, successful catalysis depends on the position of redox potentials of hydrocarbon and oxygen, relative to the Fermi level of metal oxide catalysts. The Fermi level may be shifted upon doping the surface of the catalyst by other metal ions of an electro-acceptor or electro-donor nature. This theoretical concept is based on the interactions of the electrophilic and nucleophilic reactions of oxygen, as well as on the notions of surface-sensitive reactions. It is supported in part by experimental data (for instance, by STM data on TiO_2), some examples of which were given in the review of Haber.[136]

Note that both concepts (acid–base and redox reactions) are based on electron transfer effects and, principally, are neither different nor opposing. They supplement one another, and, in activation of reactants, the acid–base properties are usually exhibited as the redox capacity of the surface.

The acid–base properties, electronegativity, and nature of the cation in chemical bonding on the solid phase play a central role in heterogeneous reactions. Electronegativity increases with the increase of the oxidation state of cations. In high-oxidation states (M^{5+}, M^{6+}), the chemical bond mainly has a covalent character, and correspondingly, the acidic properties prevail over the basic properties.[137] In the case of a low-oxidation state of the cation in metal oxides, the chemical bond has a more ionic character, and it exhibits basic rather than acidic properties. In the metal oxides, the cations act as Lewis acids and anions as bases. The Lewis acidic site is an acceptor of the electron pair, and the Lewis base site is an electron pair donor, while the Bronsted acidic site is a proton donor and the Bronsted basic site is a proton acceptor. The strength of the Lewis active site depends mainly on the ionic radius, coordination number, and polarity of bond.

Any oxidation reaction occurs by partial or complete transfer of the electron, electron pair, or proton. The formation of radicals may be expected both in homolytic and heterolytic breaking of the C-H on the metal oxides. The heterolytic breaking of the C-H bond occurs through the formation of carbanion or carbocation. Carbocation may be formed in hydrid abstraction by a strong basic catalyst. Carbanion may be formed on the Lewis acid sites by abstraction of the proton. Correspondingly, coordinatively unsaturated cations on the surface react as the Lewis acid site and hydroxyl groups as the Bronsted acid site.[134, p. 283]

In the context of these elucidations, the electrophilic surface oxygen species may participate mainly in deep oxidation, while lattice oxygen (nucleophilic specie) may participate in partial oxidation.[137b] Evidently, lattice oxygen is less reactive than expected surface oxygen species.

Correspondingly, SiO_2 (having covalent Si-O bonds) is a weak Bronsted acid (i.e., it has no basicity), whereas γ-Al_2O_3 is a strong Lewis acid and weak base; TiO_2 (ionic crystal) is a medium Lewis acid and medium base; ionic crystals MgO, CaO, CuO are strong bases and very weak acids; Mo_2O_5, V_2O_5 are Bronsted and Lewis acids of medium strength.[134] The testing of the base–acid nature of the surface active sites may be carried out by adsorption experiments, for example, NH_3 or pyridine for acidic active sites and CO_2 or SO_2 for base active sites.[139]

The heterogeneous–homogeneous pathway of oxidation on semimetal oxides, as SiO_2, is a frequently observed case. Due to the covalent bond, in the absence of the surface uncoordinated cations, on the hydroxylated surface, the active site usually has Bronsted acidity.[137]

The acid–base properties of the active sites are important in understanding the catalytic properties of transition metal oxides in many oxidation reactions with dioxygen.[140–142] For example, the basic and acidic properties of metal oxides (cations as Lewis acid active sites, and O^{2-} and OH as base active sites) play a decisive role in the formation of intermediates or products in oxidation or oxidative dehydrogenation reactions of alkanes. For example, on the catalyst V-Mg-O, the oxidative dehydrogenation of propane occurs through the participation of lattice oxygen.[138] According to the authors, the highest propene yields correspond to the adjusted redox potential combined with a strong Lewis acidity and a mild basicity. The acid–base properties of the oxides of transition metal oxides were discussed in Kung's monograph.[135, p. 57]

Note that, according to some authors, as a rule, on the surfaces of acidic metal oxides the acidic products or intermediates of oxidation (as carboxylic acids) favorably may be formed, while on the surfaces of basic metal oxides the products or intermediates formed are species of basic character (as olefins in some oxidative dehydrogenation reactions).[139] Almost all oxide surfaces of the transition metal oxides exposed in air, containing moisture, are covered with adsorbed water and hydroxyl groups. According to the mainly accepted mechanism suggestions,[135, 141]

$$M^{n+} + H_2O \rightleftarrows M^{n+}OH_2$$

where M^{n+} is coordinative unsaturated metal ion. H_2O might undergo dissociative adsorption on the transition metal oxide

$$\begin{array}{ccc} O^{2-} & OH & OH \\ | & | & | \\ \end{array}$$
$$H_2O + M^{n+} \text{- - -} M^{n+} \rightleftarrows M^{n+} \text{- - -} M^{n+}$$

A simple way of removing adsorbed water and surface hydroxyls involves heating the samples and then pumping. Heat treatment at 200–350°C is sufficient for desorption of the weakly adsorbed water and, partially, of hydroxyl groups. Usually, this is insufficient, however, for activation of metal oxides as catalysts. The dehydroxylation of different extent occurs by the surface reactions with hydrogen, CO, O_2. The coordinative unsaturated cation formed may play the role of active site in the adsorption of small molecules, possessing a lone pair of electrons. Cationic active centers of this kind are known as Lewis acidic centers[140]

$$M^{n+} + : A \rightleftarrows M^{n+} : A$$

and the charge transfer are in dependence on the oxidation number of cations.

On the other hand, the surface hydroxyl groups usually exhibit Bronsted acidity (conjugate base):

$$S\text{-}OH + B \rightleftarrows S\text{-}O^- + BH^+$$

For instance, in the adsorption of ammoniac on hydroxylated surface, NH_4^+ is formed.

Examples of the exhibition of the acid–base and redox properties of metal oxides, as well as their role in the heterogeneous–homogeneous occurrence of oxidation for main classes of organic compounds (hydrocarbons, alcohols, aldehydes), are discussed in Chapter 3.

REFERENCES

1.1

1. Shaofen L. Reaction Engineering (Translated and updated by Lin L. and Feng X.). Chemical Industry Press, Oxford, UK, 2017, 665 pages, p. 2.
2. Chernyi G. G., Losev S., Macheret S., Potapkin B. Physical and Chemical Processes in Gas Dynamics: Physical and Chemical Kinetics and Thermodynamics, vol. 2 (Progress in Astronautics and Aeronautics, no. 197). 2004, AIAA, Virginia, 300 pages, p. 115.
3. Yablonskii G. S., Bykov V. I., Elokhin V. I., Gorban A. N. Kinetic models of catalytic reactions. In: Comprehensive Chemical Kinetics, vol. 32. 1991, Elsevier Science Publishers, Amsterdam, 392 pages, p. 79.
4. Kiperman S. L. Kinetic peculiarities of the gas phase heterogeneous-homogeneous reactions. *Kinetika i Kataliz* (Kinetics and Catalysis, in Russian), 1994, v. 35, p. 37–53.
5. Poissonnier J., Thybaut J. W., Marin G. B. Understanding and optimization of chemical reactor performance for bimodal reaction sequences." AIChE J., 2017, v. 63, 1, p. 112–119.
6. Iglesia E., Reyes S. C. Structural and reaction models for the design and optimization of catalytic sites, pellets and reactors. In: M. Doyama, J. Kitara, M. Tanaka, R. Yamamoto (eds). Computer-Aided Innovation of New Materials II: Proceedings of the

Second International Conference and Exhibition on Computer Application to Materials and Molecular Science and Engineering, September, 1992, Yokohama, Part II. 1993, Elsevier Science Publishers, Amsterdam, 1813 pages, p. 1057.

7. Polshettiwar V., Asefa T. (eds). Nanocatalysis: Synthesis and Applications. 2013, John Wiley and Sons, New Jersey, 731 pages, p. 11.

1.2

8. IUPAC. Compendium of Chemical Terminology, 2nd ed. (Gold Book). Compiled by A. D. McNaught and A. Wilkinson. 1997, Blackwell Scientific Publications, Oxford. XML online corrected version: http://goldbook.iupac.org (2006), created by M. Nic, J. Jirat, B. Kosata; updates compiled by A. Jenkins.

9. Lindemann F. A., Arrhenius S., Langmuir I., Dhar N. R., Perrin J., Lewis W. C. McC. Discussion on 'the radiation theory of chemical action'. Trans. Faraday Soc., 1922, v. 17, p. 598–606.

10. Zewail A. H. Femtochemistry. Past, present, and future. Pure Appl. Chem., 2000, v. 72, 12, p. 2219–2231.

11. Baykusheva D., Wörner H. J. Theory of attosecond delays in molecular photoionization. J. Chem. Phys., 2017, v. 146, 124306.

12. Levine R. D. Molecular Reaction Dynamics. 2005, Cambridge, Cambridge University Press, 555 pages, p. 202.

13. Gomberg M. An instance of trivalent carbon: triphenylmethyl. J. Am. Chem. Soc., 1900, v. 22, 11, p. 757–771.

14. Hicks R. G. (ed). Stable Radicals: Fundamentals and Applied Aspects of Odd-Electron Compounds. 2010, John Wiley, New Jersey, 588 pages.

15. Forrester A. R., Hay J. M., Thomson R. H. Organic Chemistry of Stable Free Radicals, 1st ed., 1968, Academic Press, New York, 405 pages.

16. Encyclopedia of Radicals in Chemistry, Biology and Materials. 2012, John Wiley and Sons, Ltd., New Jersey, series of books.

1.3

17. Kondratev N. V. Chemical Kinetics of Gas Reactions. 1964, Pergamon Press, Oxford, UK, 400 pages, p. 44.

18. Van't Hoff J. H. *Études de Dynamique chimique* (Studies in Chemical Dynamics). 1884, Amsterdam, Frederic Muller, 215 pages.

19. Mellor J. W. Chemical Statics and Dynamics Including the Theories of Chemical Change, Catalysis, and Explosions. 2013 (original work published in 1904), Forgotten Books, London, 584 pages.

20. Kooij D. M. Inaugural Dissertation, Amsterdam, 1892. Zeit. Phys. Chem., 1893, v. 12. p. 155.

21. Bone W. A., Wheeler R. V. J. Chem. Soc., 1902, v. 81, p. 538.

22. Berthelot M. Compt rend. acad. sci., 1897, v. 125, p. 271.

23. Rice F. O. Kinetics of gases and gas mixtures. In: Annual Survey of American Chemistry, v. 5. C. J. West (ed), Chemical Catalog Company, New York, 1931, p. 21–35.

24. Hinshelwood C. N. The theory of unimolecular reactions. Prog. Roy. Soc. A, 1927, v. 113, p. 230.

25. Hinshelwood C. N., Williamson A. Reaction Between Hydrogen and Oxygen. Oxford University Press, Oxford, UK, 1934.

26. Taylor H. S., Lavin G. I. Surface reactions of atoms and radicals. A new approach to the problem specific surface action. J. Am. Chem. Soc., 1930, v. 52, p. 1910–1918.
27. Semenov N. N. Some Problems of Chemical Kinetics and Reactivity, v.1, 2. 1959, Pergamon Press, London, 686 pages.
28. Hinshelwood C. N. The Kinetics of Chemical Change in Gaseous Systems, 4th ed. 1940, Clarendon Press, Oxford, UK, 274 pages.
29. The Collected Works of Irving Langmuir (with contribution of memoriam including a complete bibliography of these works). Editors: C. Guy Suits and Harold E. Way. Vol.12, Langmuir, The Man and the Scientist. Pergamon Press, 1962, Oxford, 473 pages.
30. Nalbandyan A. B., Voevodsky V. V. Mechanism of Hydrogen Oxidation and Combustion (in Russian). 1949, Moscow-Leningrad: Acad. Nauk SSSR Publishing.
31. Semenov N. N., Voevodski V. V. Heterogeneous Catalysis in Chemical Industry (in Russian). 1955, Moscow, Acad. Sci. USSR, 233 pages.
32. Lewis B., Elbe G. V. Combustion, Flame and Explosion in Gases, 4th ed. 1987, London, Academic Press.
33. Baldwin R. R., Mayor L. 7th Symposium on Combustion (Aug. 29–Sep. 3, 1958). 1959, London, Oxford University Press, p. 8.
34. Denisov E. T., Sarkisov O. M., Likhtenshteïn G. I. Chemical Kinetics: Fundamentals and New Developments. 2003, Elsevier Science, Amsterdam, 566 pages.
35. Willbourn A. H., Hinshelwood C. N. Proceedings of the Royal Society of London. Series Mathematical and Physical Sciences, A185. 1946, London, Royal Society, 353 pages, p. 369–376.
36. Pease R. N. J. Am. Chem. Soc., 1930, v. 52, p. 5106–5110.
37. Nalbandyan A. B., Mantashyan A. A. Elementary Processes in Slow Gas-Phase Reactions (in Russian). 1975, Izd. Aked. Nauk Arm. SSR: Yerevan, 258 pages.
38. Nalbandyan A. B., Vardanyan I. A. Modern State of the Problem of Gas Phase Oxidation of Organic Compounds (in Russian). 1986, Izd. Akad. Nauk Arm. SSR, Yerevan, 227 pages.
39. Emanuel N. M. Knorre D. G. Chemical Kinetics. 1984, *Vyshaya Shkola*: Moscow, 4th ed. (in Russian), 463 pages.
40. Azatyan V. V., Semenov N. N. The mechanism of burning hydrogen at low pressures. *Kinetika i Kataliz* (Kinetics and Catalysis, in Russian), 1972, v. 13, p. 17–25.
41. Azatyan V. V. Chain processes and the non-stationary nature of the state of the surface. Russian Chem. Rev., 1985, v. 54, p. 21.
42. Azatyan V. V. The new regularities of chain branching processes and some new aspects of theory. *Chimicheskaya Phizika* (Chem. Physics), 1982, v. 1, p. 491–508.
43. Denisov E. T., Azatyan V. V. Inhibition of Chain Reactions. 2000, Gordon and Breach, London, 341 pages.
44. Azatyan V. V., Kislyuk M. U., Tret'yakov I. I., Shavard A. A. Role of chemisorption of atomic hydrogen in the chain combustion of H_2. *Kinetika i Kataliz* (Kinetics and Catalysis, in Russian), 1980, v. 21, p. 583.
45. Azatyan V. V., Rubtsov N. M., Tsvetkov G. I., Chernysh V. I. The participation of preliminarily adsorbed hydrogen atoms in reaction of chain propagation in the combustion of deuterium. Russian J. Phys. Chem, 2005, v. 79, p. 320–324.
46. Kozlov S. N., Markevich E. A., Aleksandrov E. N. Catalytic oxidation of hydrogen on the surface of quartz, stainless steel, and MgO near the third ignition limit. Russian J. Phys. Chem. B, 2017, v. 11, p. 255–260 (Russian version published in *Phisicheskaya Khimia*, 2017, v. 36, p. 13–19).
47. Aleksandrov E. N., Kozlov S. N., Kuznetsov N. M. Heterogeneous chain propagation on a quartz surface. Combust. Explos. Shock Waves, 2006, v. 42, p. 281–292.

1.4

48. Compton R. G., Bamford C. H., Tipper C. F. H. The theory of kinetics. In: Comprehensive Chemical Kinetics, v. 2. Bamford C. H., Tipper C. F. H. (eds). 1969, Elsevier, Amsterdam, 468 pages. p. 102.
49. Askey P. J. The oxidation of benzaldehyde and acetaldehyde in the gaseous phase. Am. Chem. Soc., 1930, v. 52, p. 974–980.
50. Pyatnitsky Yu. I., Pavlenko N. V., Il'chenko N. I. Modeling of the kinetics of the hetero-geneous-homogeneous conversion of methane into ethane and ethylene in the absence of oxygen in the gas phase. Theoret. Exper. Chem., 2000, v. 36, p. 204–207.
51. Krilov O. V., Shub B. R. Non-equilibrium Processes in Catalysis. 1990, Khimia, Moscow, 285 pages.
52. Bakhchadjyan R. H., Vardanyan I. A., Nalbandyan A. B. Possibility of the chain mechanism in the oxidation reaction of propionaldehyde on the surface. *Dokladi AN SSSR* (Reports of the Academy of Sciences of USSR, in Russian), 1985, v. 281, p. 611–615.
53. Pyatnitsky Y. I., Pavlenko N. V., Ilchenko N. I. Introduction to Non-linear Kinetics in Heterogeneous Catalysis. 2000, Nova Science Publishers, Huntington, NY, 204 pages, p. 166–168.
54. Lemonidou A. A., Stambouli A. E. Catalytic and non-catalytic oxidative dehydrogena-tion of n-butane. Appl. Catal. A, 1998, v. 171, p. 325–332.
55. Beyleryan N. M. Chemical Kinetics, Part I (in Armenian), 1978, Yerevan State University Publishing Hous, Yerevan, 336 pages, p. 210.
56. Kowalski A. A. Bogoyavlenskaya M. L. *Zhur. Fiz. Khim.* (J. Phys. Chem.), 1946, v. 20, p. 1325.
57. Emanuél N. M., Kritsman V. A., Zaikov G. E. Chemical Kinetics and Chain Reactions: Historical Aspects. 1995, Nova Science Publishers, Huntington, NY, 429 pages, p. 281.
58. Rollefson G. K. Annual Review of Physical Chemistry, v. 19. 1968, Annual Reviews, Inc., Michigan, p. 8.
59. Taylor J. E., Kulich D. M., Hutchings D. A., Frech J. K., Kenneth J. F. Abstracts, 163rd National Meeting of the American Chemical Society, Boston, MA, April 1972, No. Petr 033.
60. Taylor J. E. The wall-less reactor. Technique for the study of gas oxidation and pyrolysis of hydrocarbons. In: Characterization of High Temperature Vapors and Gases, v. 2. 1979, National Bureau of Standards, Special Publication, 948 pages, p. 1351–1358.
61. Skinner G. Introduction to Chemical Kinetics. 1974, Elsevier, New York, 224 pages, p. 174.

1.5

62. Kauffman G. B. Johann Wolfgang Dobereiner's Feuerzeug, Platinum Metals Rev., 1999, v. 43, 3, p. 122–128.
63. Kauffman G. B. The origins of heterogeneous catalysis by platinum: Johann Wolfgang Döbereiner's contributions. Enantiomer. 1999, v. 4, 6, p. 609–619.
64. Langmuir I. A chemically active modification of hydrogen. J. Am. Chem. Soc., 1912, v. 34, 10, p. 1310–1325. DOI: 10.1021/ja02211a004.
65. Volkenstein Th. The Electron Theory of catalysis on semiconductors, p. 189–264. In: Eley D. D., Selwood P. W., Weisz B. P. (eds). Advances in Catalysis, v. 12, 1960, Academic Press, New York, London, 324 pages.
66. Anderson J. R., Boudart M. (eds) Catalysis: science and technology. Serial, v. 1–11, Springer-Verlag, 1981–1996.

67. Lunsford J. H. Role of surface generated gas-phase radicals in catalysis. Langmuir. 1989, v. 5, p. 12–16. DOI: 10.1021/la00085a003.
68. Boreskov G. K. Heterogeneous Catalysis, 2003, Nova Science Publishers, N. Y., 237 pages.
69. Polyakov M. V. Heterogeneous Catalysis in the Chemical Industry. 1955, Gos. Nauchno-Tekh. Izd. Khim. Lit., Moscow, p. 271.
70. Gorokhovatsky Ya B., Kornienko T. P., Shalya V. V. Heterogeneous–homogeneous Reactions. Tekhnika, Kiev, 1972.
71. Kono M. New Approaches to Controlling Combustion, p. 273–309. In: Someya T. (ed.). Advanced Combustion Science. Springer–Verlage, Tokyo, 1993, p. 301.
72. Pfefferle L. D., Pfefferle W. C. Catalysis in Combustion. Catalysis Reviews: Science and Engineering, 1987, 29, p. 219–267. http://dx.doi.org/10.1080/01614948708078071.
73. Reinke M., Mantzaras J., Schaeren R., Inauen A., Schenker S. High pressure catalytic combustion of methane over platinum: In situ experiments and detailed numerical predictions. Combustion and Flame, 2004, 136, p. 217–240. http://dx.doi.org/10.1016/j.combustflame.2003.10.003
74. Appel C., Mantzaras J., Schaeren R., Bombach R., Inauen A., Kaeppeli B., Hemmerling B., Stampanoni A. An experimental and numerical investigation of homogeneous ignition in catalytically stabilized combustion of hydrogen/air mixtures over platinum. Combust. Flame, 2002, v. 128, p. 340–368.
75. Garibyan T. A., Grigoryan R. R., Zeile L. A., Kurina L.N., Filicheva O. D. Effect of ethanol on the heterogeneous-homogeneous oxidation of methanol on silver. *Kinetika I Kataliz* (Kinetics and Catalysis, in Russian), 1990, v. 31, p. 376.
76. Vodyankina O. V., Kurina L. N., Izatulina G. A. Volume stages in the process of the ethylene glycol catalytic oxidation to glyoxal on silver. Reaction Kinetics and Catalysis Letters, 1998, v. 65, 2, p. 337–342.
77. Cybulski A., Moulijn J. A. Structured Catalysts and Reactors, 2006, Taylor and Francis, CRC Press, (USA), 856 pages, p. 223.
78. Jeleznyak A. S., Ioffe I. I. Methods of Calculation of Reactors of Multiphase Liquids. Leningrad, Khimia, 1974, 320 pages, p. 52–53.
79. Olsen R. J., Williams W. R., Song X., Schmidt L. D., Aris R. Dynamics of Homogeneous-Heterogeneous Reactors. Chemical Engineering Science, 1992, v. 47. 9–10, p. 2505–2510.
80. Krylov O. V., Shub R. B. Non-equilibrium Processes in Catalysis, 1994, CRC Press, Boca Raton, 299 pages.
81. Andrianova Z. S., Ivanova A. N., Barelko V. V. Nonlinear phenomena in heterogeneous catalytic reactions with a branched-chain mechanism of formation of active centers. Russian Journal of Physical Chemistry B. 2009, v. 3, 5, p. 764–769.
82. Barelko V. V., Andrianova Z. S., Ivanova A. N. Domain instability in heterogeneous catalytic reactions controlled by branched-chain kinetics. Doklady Physical Chemistry. 2008, v. 421, 1, p. 170–173. DOI:10.1134/S001250160807004X
83. Ivanova A. N., Andrianova Z. S., Barelko V. V. On the theory of nonlinear phenomena in catalytic combustion reactions. Doklady Physical Chemistry (in Russian), 2002, v. 386, 4, p. 257–261. DOI:10.1023/A:1020755100863
84. Garibyan T. A., Margolis L. Ya. Heterogeneous–Homogeneous Mechanism of Catalytic Oxidation. Catalysis Reviews Science and Engineering, 1989, v. 31, 4, p. 355–384.
85. Bakhchadjyan R. H., Vardanyan I. A. The heterogeneous propagation of the chains in reaction of the aliphatic Aldehydes oxidation. International Journal of Chemical Kinetics, 1994, v. 26, p. 595–603.
86. Jalali H. A., Vardanyan I. A. Modeling process of heterogeneous interaction of peroxy radicals with organic compound. Archivum Combustionis, 2010, v. 30, 4, p. 298–302.

87. Sinev M., Arutyunov V., Romanets A. Kinetic models of C_1-C_4 alkanes oxidation as applied to processing of hydrocarbon gases: principles, approaches and developments. In: Advances in Chemical Engineering, v. 32, 2007, Amsterdam, Elsevier, p. 167–258.
88. Astruc D., Lu F., Ruiz Aranzaes J. R. Nanoparticles as recyclable catalysts. The frontier between homogeneous and heterogeneous catalysis. Angew. Chem. Int. Ed. 2005, v. 44, p. 7852–7872.
89. Witham C. A., Huang W., Tsung C-K., Kuhn J. N., Somorjai G. A., Toste F. D. Converting homogeneous to heterogeneous in electrophilic catalysis using monodisperse metal nanoparticles. Nature Chemistry, 2010, v. 2, p. 36–41.

1.6

90. Rajaram J., Kuriacose J. C. Chemical Thermodynamics: Classical, Statistical and Irreversible. 2013, Pearson, Cainnai, Delhi, 676 pages.
91. Stanford Encyclopedia of Philosophy. Entry "Heraclitus", 2015 (Editor in chief E. N. Zalta). Website: www.plato.stanford.edu.

1.6.1

92. Shilov N. A. *O sopryazhennykh reaktsiyakh okisleniya* (On Conjugate Oxidation Reactions), Mamontov Publishers, Moscow, 1905, 304 pages.
93. Chernyi G. G., Losev S. A., Macheret S. O., Potapkin B. V. (eds). Progress in Astronautics and Aeronautics Physical and Chemical Processes in Gas Dynamics: Physical and Chemical Kinetics and Thermodynamics of Gases and Plasmas, Volume 197, 2004, American Institute of Aeronautics and Astronomic, Virginia, 334 pages. p. 135.
94. Bruk L. G., Temkin O. N. Conjugate reactions: New potentials of an old idea. *Kinetika i kataliz* (Kinetics and Catalysis, in Russian), 2016, v. 57, 3, p. 277–296.
95. Temkin O. N., Pozdeev P. P. Homogeneous Catalysis with Metal Complexes: Kinetic Aspects and Mechanisms, 2012, Wiley, 830 pages.
96. Deutschmann O., Knözinger H., Kochloefl K., Turek T. Heterogeneous Catalysis and Solid Catalysts. (Chapter) In: Ullmann's Encyclopedia of Industrial Chemistry, 2009. Wiley-VCH Verlag GmbH & Co. KGaA, Weinheim, p. 1–110, p. 17.
97. Boudart M. Kinetic coupling in and between catalytic cycles. Catalysis Letters, 1994, v. 29, 1, p. 7–13.

1.6.2

98. Prigogine I. Nobel Lecture: Time, Structure and Fluctuations. Nobelprize.org., Nobel Media, 2015.
99. Onsager L. Reciprocal Relations in Irreversible Processes. I. Phys. Rev., v. 37, 1931, 405–426.
100. De Donder T., Van Rysselberghe P. Thermodynamic Theory of Affinity: A Book of Principles. 1936, Oxford, England: Oxford University Press, 142 pages.
101. Boudart M. Some applications of the generalized De Donder equation to industrial reactions. Ind. Eng. Chem. Fundamen., 1986, v. 25, 1, p. 70–75.
102. Parmon V. Thermodynamics of Non-Equilibrium Processes for Chemists with a Particular Application in Catalysis. Elsevier, 2010, Oxford, UK, Amsterdam, The Netherlands, 340 pages.
103. Volkenstein M. V. Entropy and Information. Progress in Mathematical Physics. Vol. 57, Birkhäuser, 2009, ISBN 978-3-0346-0077-4.

104. Turing A. M. The chemical basis of morphogenesis. Philosophical Transactions of the Royal Society of London B: Biological Sciences, 1952, v. 237, 641, p. 37–72.

1.7

1.7.1

105. Laszlo S. I. Catalytic Activation of Dioxygen by Metal Complexes. Springer Science + Business Media. 1992, 396 pages, p. 1.
106. Bodenstein M. *Photochemische Kinetik des Chlorknallgases, Berichte der Bunsengesellschaft für physikalische Chemie*, 1913, v. 19, 2, p. 836–856.
107. Denisov E. T., Afanas'ev I. B. Oxidation and Antioxidants in Organic Chemistry and Biology. 2005, CRC Press, Taylor and Francis, Boca Raton, 1024 pages, p. 7.
108. McNesby J. R., Heller C. A. Jr. Oxidation of liquid aldehydes by molecular oxygen. Chem. Rev., 1954, v. 54, 2, p. 325–346. DOI: 10.1021/cr60168a004
109. Bolland J. L., Gee G. The kinetics of oxidation of unconjugated olefines. Trans. Faraday Soc., 1946, 42, 3–4, p. 236–243.
110. Misono M. Heterogeneous Catalysis of Mixed Oxides: Perovskite and Heteropoly Catalysts. Studies in Surface Science and Catalysis, Elsevier, Amsterdam, 2013, 181 pages, p. 13.
111. Sheldon R. A. Heterogeneous catalytic oxidation and fine chemicals. In: Guisnet M., Barrault J., Bouchoule C., Duprez D., Pérot G., Maurel P. R., Montassier C. (eds). Heterogeneous Catalysis and Fine Chemicals II. Elsevier, 1991, p. 33.
112. Sheldon R. A., Van Bekkum H. Fine Chemicals through Heterogeneous Catalysis, 2007, John Wiley and Sons, 636 pages, p. 475.
113. Koppenol W. H. The Haber-Weiss cycle –70 years later. Redox Report., 2001, v. 6, 4, p. 229–234.
114. Matar M. S., Mirbach M. J., Tayim H. A. Catalysis in Petrochemical Processes. 1989, Kluwer Academic Publishers, Dordreacht, 199 pages, p. 89.
115. Milas N. A. The hydroxylation of unsaturated substances. III. The use of vanadium pentoxide and chromium trioxide as catalysts of hydroxylation. J. Am. Chem. Soc., 1937, v. 59, 11, p. 2342–2344. DOI:10.1021/ja01298a065
116. Milas N. A., Sussman S., Mason H. S. The hydroxylation of unsaturated substances. V. The catalytic hydroxylation of certain unsaturated substances with functional groups J. Am. Chem. Soc., 1939, v. 61, 7, p. 1844–1847.
117. Arzoumanian H., Bakhtchadjian R. Oxo-atom transfer reactions of transition metal complexes in catalytic oxidation with O_2 on the light of some recent results in molybdenum-oxo chemistry (a review). Chemical Journal of Armenia, 2012, v. 65, 2, p. 168–188.

1.7.2

118. Henrici-Olive G., Olive S. The Chemistry of the Catalyzed Hydrogenation of Carbon Monoxide. Springer-Verlag, Berlin, 1984, p. 8.
119. Lazar M., Rychly J., Klimo V., Pelican P., Valko L. (eds). Free Radicals in Chemistry and Biology, CRC Press, 1989, Boca Raton, 312 pages, p. 9.
120. Arutyunyan A. Z., Gazaryan K. G., Garibyan T. A., Grigoryan G. L., Nalbandyan A. B. Formation of radicals on the surfaces of oxides in decomposition of H_2O_2. *Kinetika i Kataliz* (Kinetics and Catalysis, in Russian), 1988, v. 29, 4, p. 880–884.
121. Arutyunyan A. Z., Grigoryan G. L. Nalbandyan A. B. ESR study of the heterogeneous radical decomposition of hydrogen peroxide vapor on glass and silica. *Khim. Fizika* (Chem. Phys., in Russian), 1986, v. 5, 8, p. 1118.

122. Hinshelwood C. N., Prichard C. R. Two homogeneous gas phase reactions. Journal of the Chemical Society, Transactions, 1923, v. 23, p. 2725.
123. Ismagilov Z. R., Pak S. N., Krishtopa L. G., Yermolaev V. K. Role of free radicals in heterogeneous complete oxidation of organic compounds over IV period transition metal oxides. p. 240. In: Guczi L., Solymosi F., Tétényi P. (eds). New Frontiers in Catalysis, Parts A-C, 1993, Elsevier, 2859 pages.
124. Bakhchadjyan R. H., Vardanyan I. A., Nalbandyan A. B.: Low temperature oxidation of aldehydes in conditions excluding homogeneous propagation of chains. Dokl. AN Arm. SSR (in Russian), 1988, v. 86, 4, p. 170–173.

1.8

125. Grabow L. C. Computational Catalysis Screening, 1. 4., The Sabatier principle and the volcano-curve, p. 17. In: Asthagiri A., Janik M. (eds.). Computational Catalysis. RSC Publishing, Cambridge, 2014, 266 pages.
126. Van Santen R. A., Neurock M. Molecular Heterogeneous Catalysis: A Conceptual and Computational Approach. 2006, Wiley, Weinheim, 473 pages, p. 7.
127. Bhatia S. Zeolite Catalysts: Principles and Applications, 1990, CRC Press, Boca Raton, 293 pages, Chapter 6, p. 110.
128. Vasileva N. A., Buyanov R. A. Radical generation during pyrolysis of n-undecane on $BaCl_2$ and imperfect magnesium oxides. Chemistry of Sustainable Development. 2004, v. 12, p. 641–647.
129. Lunsford J. H. Formation and reactions of methyl radicals over metal oxide catalysts. p. 3. In: Wolf E E. (ed.). Methane Conversion by Oxidative Processes: Fundamental and Engineering Aspects, Springer Science, 1992, N.Y., 547 pages.
130. Golodets G. I. Specific catalyst activity in terms of the heterogeneous-homogeneous mechanism: Inversion of the Boreskov rule. *Kinetika i Kataliz* (Kinetics and Catalysis, in Russian), 1995, v. 36, 1, p. 27–30.
131. Califano S. Pathways to Modern Chemical Physics, 2012, Springer, N.Y., 288 pages, p. 48.
132. Medford A. J., Vojvodic, A., Hummelshoj J. S., Voss J., Abild-Pedersen F., Studt F., Bligaard T., Nilsson A., Nørskov J. K. From the Sabatier principle to a predictive theory of transition-metal heterogeneous catalysis. Journal of Catalysis, 2015, v. 328, p. 36–42.
133. Cheng J., Hu P., Ellis P., French S., Kelly G., Lok C. M. Brønsted–Evans–Polanyi relation of multistep reactions and volcano curve in heterogeneous catalysis. J. Phys. Chem. C, 2008, v. 112, 5, p. 1308–1311. DOI: 10.1021/jp711191j

1.9

134. Fierro J. L. G. Metal Oxides: Chemistry and Applications, 2006, CRC, Taylor and Francis, Boca Raton, 785 pages, p. 283.
135. Kung H. H. Transition Metal Oxides: Surface Chemistry and Catalysis, Studies in Surface Science and catalysis, 1989 and 1991, Elsevier Science, Amsterdam, 285 pages, p. 57.
136. Haber J. Molecular description of transition metal oxide catalysts. Chapter 11. p. 275–293. In: Carley A. F., Davies P., Hutchings G. J., Spencer M. S. (eds). Surface Chemistry and Catalysis, 2002, Springer Science, Amsterdam, 380 pages, p. 283.
137. *a*. Centi G., Cavani F., Trifiro F. Selective oxidation by heterogeneous catalysis. Chapter 8, p. 403–483, Control of the Surface Reactivity of Solid Catalysts, 2001, Springer Science, N.Y., 504 pages.
     b. ibid. 8.3; 1. Nature of the interaction between molecular oxygen and oxide surfaces and types of oxygen adspecies, p. 388.

138. Sashan K. Oxidative conversion of lower alkanes to olefins, p. 119–143. In: Spivey J. J., Dooley K. M. (ed.). Catalysis, v. 22, Royal Society of Chemistry, RSC Publishing, 2010, 317 pages, p. 123.

139. Védrine J. C. Heterogeneous partial (amm)oxidation and oxidative dehydrogenation catalysis on mixed metal oxides. Catalysts, 2016, v. 6, 2, p. 22; DOI:10.3390/catal6020022

140. Chiesa M., Giamello E., Che M. EPR characterization and reactivity of surface-localized inorganic radicals and radical ions. Chem. Rev., 2010, v. 110, 3, p. 1320–1347.

141. Ono Y., Hattori H. Solid Base Catalysis, 2011, Springer, Heidelberg, 419 pages, p. 5–10.

142. Védrine J. C. Heterogeneous catalysis on metal oxides (Review). Catalysts, 2017, v. 7, 11, p. 341–366. DOI:10.3390/catal7110341

2 Heterogeneous Generation and Reactions of Radicals. Heterogeneous–Homogeneous Reactions of Radical Decomposition

Among the different types of active intermediates mentioned in Chapter 1 (Section 1.2) are the free radicals and radical-like species that play a pivotal role in bimodal reaction sequences. Apparently, in a majority of the bimodal reactions at least, the homogeneous constituents are radical reactions. As regards the heterogeneous constituents, they may be radical or nonradical reactions, depending on the nature of the surface and reaction parameters. Heterogeneous reactions of radicals, in their turn, may be non-chain or chain processes with or without the transfer of part of the radicals to the fluid phases. This chapter includes a number of reactions of the mentioned types. Here the heterogeneous–homogeneous reactions are exemplified by the radical decomposition of hydrocarbons and peroxide compounds on the surfaces of the solid substances.

2.1 RADICALS AND RADICAL-LIKE SPECIES

The physical and chemical properties of the surface layer of the solid substances are completely different from those in the bulk (volume). Similarly, the properties of species in the adsorbed state are different from those in the gas or liquid phases. It is worth mentioning that the adsorbed atom of hydrogen on the surface of a solid substance is not the same atomic hydrogen in the gas phase, even if it saves its unpaired electron as a paramagnetic species. Therefore, in the adsorbed state, it may be named a radical-like species rather than a free radical. Evidently, its chemical properties, thermodynamic and other parameters, as well as paramagnetic properties, depend on the energy of the chemical bonding in the adsorbent–adsorbate system. Similarly, a large number of other intermediates, forming in the catalytic oxidation reactions on the surfaces of different solid substances, may be considered to be radical-like species rather than radicals (including anion or cation radicals) with one or two unpaired electron(s). The terms *radical-like species* and *radical-like intermediates* are ambiguous and may not be clearly defined. Radical-like species are species having one or two partially delocalized unpaired electron(s) that participate in the chemical bondings. In other words, in the case of the adsorbed radical-like species, the distribution of spin population (density) in these species is different from that in the free radicals, in the gas phase.

Presently, electron paramagnetic resonance (EPR) spectroscopy is one of the most powerful methods for complete characterization and identification of any paramagnetic species in chemistry. Therefore, analysis of the well-resolved EPR spectrum may indicate the differences between the free radicals and radical-like species. However, in certain conditions, the radicals or radical-like species may be nonobservable by the EPR method. Note that the absence of the EPR signal is not yet evidence of the absence of the paramagnetic properties of these species. The problem of the visibility of the EPR spectra in the adsorbed state of radical-like species was briefly discussed by Volodin et al.[1] They indicated at least four conditions that were necessary for detection of adsorbed radicals on the surfaces of solid substances using the EPR method.

The first of them is the condition requiring that the lifetime of the paramagnetic species must be greater than the recording time by the EPR-method (i). Then, the dipole-dipole interactions between paramagnetic centers, affecting to the EPR-linewidth, should not be considerable (the neighboring paramagnetic species would be far away), (ii). On the other hand, the relaxation properties of radicals should be suitable for the recording of the EPR spectra, in certain experimental conditions, and the "degeneracy" should be removed by interaction with the solid surface field (iii). Finally, the stabilization center on the metal oxide must be diamagnetic (iv).[1]

The presence of sustained evidences about the properties, similar to free radicals (isoelectronic structure and chemical reactivity), often permits to characterize the intermediate as a radical-like species. In other words, this is a specific state of atoms and molecules, which are radicals in the gas or liquid phase and partially save or partially lose their paramagnetic properties in combination or often in complexation with other molecules in the fluid phases or on the surface of the solids. However, not every radical species adsorbed on the surface may be considered a radical-like species. It is clear that in the case of the strong adsorption (chemisorption or complexation) with charge transfer, the radicals can completely lose their unpaired electron(s) and paramagnetic nature by creating a new chemical bond.

Seemingly, the formation of the adsorbed radical-like species is possible mainly in the weak adsorption (or complexation) of the radicals on the surface of the solid substances. This state of radical-like species opens the possibility for facilitated further transformations on the surface layer in heterogeneous catalytic reactions. For a sufficient reactivity, the energy of the bonding of the radical-like species with the surface will not be much higher, limiting their reactivity, and not much lower, making them only physically adsorbed on the surface. Willingly or unwillingly, this situation is reminiscent of Sabatier's famous principle in heterogeneous catalysis (for the Sabatier principle, see Section 1.8). Nearly the same fundamental notions were applied for intermediates in Volkenstein's theory of catalysis on semiconductor materials, known as the electronic theory of catalysis[2] (see Chapter 4).

Following are some examples of the radical-like species, known in the literature of chemistry. The existence of $CO_2^{\cdot-}$ in the gas phase was first evidenced by Paulson[3] in the reaction $O^- + CO_2 \rightarrow CO_2^{\cdot-} + O$, using mass-spectrometric analyses. This is an anion-radical, isoelectronic with NO_2. According to Aresta and Angelini,[4] the theoretical calculations of the structure of $CO_2^{\cdot-}$, using spin population (spin density) analysis, show a 68% delocalization of electrons on the C-atom and 16% on each O-atom. In this regard, $CO_2^{\cdot-}$ is a radical-like species at C-atom. Moreover, analysis of the hyperfine structure of EPR spectra[5] show that in the adsorbed $CO_2^{\cdot-}$ on MgO, the unpaired electron spin density was distributed over the entire radical anion. Thus, the adsorbed $CO_2^{\cdot-}$ on MgO may be considered a radical-like species rather than a free radical. Analogously, in certain conditions, azopropene on MoO_3 (bismuth-molybdate catalyst) generates π-allyl radicals,

$$H_2C \cdots\cdots CH \cdots\cdots CH_2$$

which are coordinated with surface Mo-atoms. The Mo-coordinated π-allyl radicals are radical-like species but not free radicals.[6] The presence of π-allyl radical-like species is well established in the selective oxidation of propylene on the bismuth-molybdate catalysts at 300–450°C. According to the mainly accepted mechanism suggestions,[6,7] the oxidation includes a stage of the α-H-atom abstraction from propylene on the surface, forming π-allylic intermediates bonded with the surface, on the coordinated metal-atom. Spectroscopic and chemical evidence, as well as the isotopic analysis experiments with D- (*deuterium*, ^2H) and ^{18}O-labeled products, are in agreement with the formation of this intermediate,[8,9] and it may not be considered as either a radical or an ion (anion or cation). It may be characterized only by an ambiguous term, as a radical-like species. There is yet other evidence for the primary formation of this species: the allyl radical-like species is a precursor for the formation of 1,5-hexadiene on Bi_2O_3 (above or about 400°C). The formation of π-allylic intermediates was established for many other systems, for instance, at the adsorption of propylene on the V_2O_5/Nb_2O_5 catalyst.[10]

Another more well-known example of the formation of radical-like species on the surfaces may be considered the weakly adsorbed atomic O^- ion-radical, on the metallic Ag, in epoxidation of ethylene.[11] Unfortunately, it was not detected in the reaction conditions (*in situ*) by the EPR method but its presence was confirmed on the basis of results obtained by other experimental methods, such as XPS (X-ray photoelectron spectrometry), AES, (Auger electron spectroscopy) and HREED (high-resolution electron energy loss spectroscopy).[12–14] Another form of the adsorbed oxygen, O_2^-, was detected by the EPR method on the Ag-catalyst.[15] However, the role of adsorbed O_2^- in oxidation of ethylene may not be very important, considering that its concentration was very low (only 3% of the chemisorbed O_2).[15] On the ZnO and V_2O_3, however, the adsorbed O^- was detected by the EPR method (see Section 3.4.2.1).

Dimerization of furan on Si(111)(7 × 7) surface, at 110 K, by recombination of two adsorbed radical-like precursor molecules located on two adjacent adatom sites, was investigated by the high-resolution electron energy loss spectroscopy (HREELS)

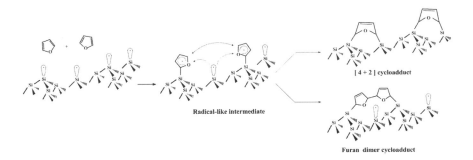

FIGURE 2.1 Radical-like intermediate in dimerization reaction of furan on Si(111). (Adapted from Ref.[17]).

and temperature programmed desorption (TPD) methods, combined with quantum-mechanical calculations.[16] It has been shown that the Si-dangling bond-mediated dimerization occurred through the radical-like species in neighboring adsorption sites[16, 17] (Figure 2.1).

Radical-like species, as intermediates, are also widespread in photocatalytic reactions. The formation and role of the radical-like species in photooxidation on a TiO_2[18] will be shown in Chapter 4.

Why, in general, do just the radical-like intermediates play an important role in some heterogeneous catalytic reactions? Seemingly, the reason is related to their "soft" chemical nature, expressed in the Fleming's words, "radicals are soft."[19] The majority of the adsorbed radicals or radical-like transient species are uncharged or very weakly charged species. Therefore, in interaction with other molecules, the Coulombic forces (interactions) play a less important role than the overlaps of the frontier molecular orbitals. As a result, the reaction through radical or radical-like species often has a lower activation barrier than other ways of product formation.

Valuable information about the motion of adsorbed radicals or radical-like species on the solid surfaces may be obtained by application of the muon spin spectroscopy, particularly the muon spine rotation (μSR) method,[20–22] employing spin-polarized positive muons in the heterogeneous catalytic reactions. One of the first investigations using this method, in the mid-1980s, permitted not only to detect the radicals, for instance, methyl, ethyl, 1,1,2-trimethylallyl, cyclohexanyl and others, on the surface of hydroxylated SiO_2 (Cabosil), but also to estimate the activation barriers of their motion (diffusion) on the surface at room temperatures, not available by "traditional" methods, such as EPR. For 1,1,2-trimethylallyl radicals on SiO_2 (in different coverages $\theta = 1 - 0.1$), it was shown that at room temperature, it might be considered "essentially free." Application of this method showed that the mobility of radicals on the surface was less than that in the liquid phase, but greater than that in the solid phase. At low coverage and at room temperature, the activation barrier of desorption were 16 kJ/mol ($\theta = 0.1$) and 13.6 kJ/mol ($\theta = 1$), respectively.

Alkyl radicals on the surface. As alkyl radicals play an important role in a number of heterogeneous–homogeneous reactions of oxidation and decomposition, it is important to know their activity on the surfaces of the solid substances. Are they conserved an unpaired electron in the adsorbed state, stabilizing on the surface, and how does their reactivity change in the adsorbed state? The investigations seeking to resolve these problems, for the first time, were carried out by V. B. Kazansky and co-workers in the 1960s, in the former Soviet Union.[23] The main experimental method was the EPR.

Alkyl radicals may be generated on the surfaces of different materials;[24–25] for instance, CH_3 radicals may be obtained by CH_3I photochemical decomposition under UV irradiation (6 h, 77 K) on silica gel, preliminarily degassed (573 K).[24] The recorded EPR spectrum consisted of four lines, with small linewidths (1–2 G), the amplitudes of which were in proportion 5:8.5:13:2.5 (g = 2.0001; note that binomial values for CH_3 are 1:3:3:1). The analysis and comparison of these spectra with the spectra obtained by the freezing of CH_3 radicals from the gas phase permits to conclude that in the adsorbed state the radical loses two rotational degrees of freedoms and remains only the third rotation degree of freedom, possessing a threefold symmetry axis. Thus, the rotational motion of the radical in the adsorbed state is limited. Nearly in analogous conditions, the C_2H_5 adsorbed radical was analyzed and compared with the same radicals obtained from the gas phase. The results showed that C_2H_5 "lies on the surface one its side." It is interesting that the CH_3 and CH_2 groups have a possibility of rotation relative to each other and "roll" on the surface. Other alkyl radicals on the surface also "lie one their sides." Thus, the unpaired electron makes only a small contribution to the chemical bonding between the radical and solid substance. By comparing hyperfine splitting of CH_3 and C_2H_5 radicals with the parameters of hyperfine splitting of radicals originated from the gas phase, another important conclusion was made: it was postulated that "the disturbing action of the surface on the unpaired electron cloud of the radicals is small" (3–5 kcal/mol).

The radicals CH_3 are highly stable on the Al_2O_3 surface (298–474 K).[24, 25] Their lifetime may even be months in the presence of molecular oxygen. The adsorption of CH_3 radicals on porous Vycor glass (96% SiO_2 and 3% B_2O_3) was studied in more detail by Gesser.[24] In general, hyperfine structure of the EPR spectra of radical-like species appears because of the hyperfine coupling (interaction between the spin of unpaired electron and nuclear spin).[25]

It is well known that the methyl radical, adsorbed on silica gel, exhibits reactivity toward H_2, O_2, NO, C_2H_4, etc.[23]

2.2 HETEROGENEOUS REACTIONS BY PARTICIPATION OF RADICALS

In general, the reactivity of the adsorbed free radicals or radical-like species on the surfaces of different substances is essentially different from that in the fluid phases. As a rule, the adsorbed radicals are less active in the reactions with the same species on the surface, in comparison with their activity in analogous reactions in the gas or liquid phase. In spite of low chemical activity, the reactions of the adsorbed radicals on the surfaces, in general, are more selective than those in the homogeneous conditions.

Below are examples of the heterogeneous reactions of radicals or radical-like species that demonstrate their specific reactivity on solid substances. They were taken from the "classical" organic chemistry and usually have primary application in chemical syntheses. Conditionally, here, they were divided into non-chain-radical and chain-radical reactions on the surfaces of solid substances.

2.2.1 EXAMPLES OF NON-CHAIN-RADICAL REACTIONS ON THE SURFACE OF SOLID SUBSTANCES

The history of the discovery of the heterogeneous reactions of radicals is related to the appearance of the first proof of the existence of free radicals altogether. As mentioned earlier (Section 1.2), after Gomberg's discovery of free radicals in 1900, for a long-time—the next two to three decades—their importance in chemical reactions was not recognized in the scientific world. Only in 1926 did Taylor and Jones obtain new, though indirect but convincing, proof of the existence of free radicals in the example of the thermal reaction of ethylene.[26] They observed that ethylene was stable in heating up to 260°C, but when it was mixed either with mercury or lead alkyls, a reaction forming "oil" was observed. They explained the formation of other hydrocarbons from ethylene by the generation of CH_3 radicals in thermal decomposition of lead alkyls and their further combination with ethylene giving new hydrocarbon radicals and, finally, "oil," a mixture of different hydrocarbons. One year later, in 1927, Paneth and Hofeditz[27] carried out the so-called lead-mirror experiment, the results of which may be considered as the first evidence for the heterogeneous reactions of radicals. The vapors of tetramethyllead $Pb(CH_3)_4$ in the hydrogen gas (used as an "inert" carrier gas) were passed through the quartz tube containing a cold mirror at the end of the reactor. The gaseous mixture was heated about 800°C. Tetramethyllead $Pb(CH_3)_4$ was decomposed, giving methyl radicals and lead. A part of methyl radicals was moved to the lead-mirror surface, where, as a result of their reaction on the surface, the mirror slowly disappeared. This experiment may also be considered as one of the first demonstrations of the existence of methyl radicals.

In general, it was not easy to identify and investigate the reactivity of the adsorbed radicals, as intermediates of the reactions on the solid substances. There are several reasons for this difficulty: firstly, the adsorbed radicals are a short-lifetime species; secondly, the detection of radical intermediates by a spectral technique, such as EPR, IR, Raman, LIF and other spectroscopies, mass spectrometry, and many others, may be directly (*in situ*) used in rare cases. Experimental evidence about their reactivity may be obtained mainly by indirect methods. In some cases, the trapping and scavenging of radical intermediates on the surface are more available than their direct detection by spectroscopic methods. There are not a great number of experimentally well-proven heterogeneous radical reactions of a catalytic or noncatalytic nature.

Synthesis of the Grignard reagent on magnesium. One of the classic examples of heterogeneous reactions that occur by radical intermediates is synthesis of the well-known Grignard reagent (RMgX) on the magnesium surface from organic halides RX (where R = alkyl, aryl, vinyl,..., and X a halogen atom) in organic solvents (ether or tetrahydrofuran). For discovering this reagent, which has great importance in synthetic organic chemistry, the French chemist F. A. V. Grignard, along with another famous French chemist, P. Sabatier, was awarded the Nobel Prize in Chemistry in 1912 (for discussion of the Sabatier principle, see Chapter 1).

According to the widely accepted suggestions, the $CH_3Br + Mg$ reaction occurs through formation of methyl radical intermediates.[28-31] In the primary stage, adsorption of CH_3Br on the surface of solid magnesium is accompanied by the single electron transfer from the surface of Mg to adsorbate, forming $(CH_3\text{-}Br)^{\bullet-}/Mg^{\bullet+}$. Mainly the latter decomposes into CH_3 and Br^- on the surface of Mg:

$$CH_3\text{-}Br + Mg \rightarrow CH_3\text{-}Br^{\bullet-} + Mg^{\bullet+}$$

$$CH_3Br^{\bullet-} \rightarrow CH_3^{\bullet} + Br^-$$

Then, the formation of the Grignard reagent occurs by the CH_3 radical intermediate:

$$CH_3^{\bullet} + Mg^{\bullet+} \rightarrow CH_3Mg^+$$

$$CH_3Mg^+ + Br^- \rightarrow CH_3MgBr$$

The detailed mechanism of the reaction has been the subject of long and continuous discussions in chemical literature. However, nearly all new suggestions regarding the mechanism are based on the radical (or ion-radical) nature of the intermediates, reacting on the surface.

Decomposition of methyl iodide on the surface of Cu(111).[8] Evidence regarding the formation and reactivity of adsorbed radicals on the solid surface in this reaction were obtained mainly by indirect methods. The reaction was investigated on the Cu(111) monocrystalline sample, at 140–165 K.[32] In decomposition of methyl iodide on the surface of Cu(111), a part of CH_3 radicals rests on the surface, and another part transfers to the gas phase. Methyl radicals, adsorbed on the surface, may be trapped using phenyl groups, generated thermally or by electron-induced dissociation from adsorbed iodobenzene. One product of reaction is methylbenzene, identified by the temperature-programmed reaction technique, involving mass-spectrometric measurements. Analogously, in decomposition of other alkyl iodides (C_2–C_5) on Cu(111), the adsorbed alkyl radicals may be trapped by surface phenyl species forming alkylbenzenes (T < 160 K). In this case, however, a transfer of alkyl radicals from the surface was not observed and, apparently, the products of reaction were mainly the result of the heterogeneous coupling and disproportionation of radicals.

Reaction of Kolbe on the electrode. Another example of the reaction occurring through formation of radical intermediates on the surface of the solid substances also may be taken from the "arsenal" of "classical" chemistry. The recognized reaction of Kolbe,[33] also named decarboxylative dimerization, initiated electrochemically or photochemically, occurs as follows:

$$RCO_2^{\bullet-} \xrightarrow{-e} \left[RCO_2^{\bullet}\right] \rightarrow R^{\bullet} + CO_2$$

$$2R^{\bullet} \rightarrow R\text{-}R \left(\text{or disproportionation}\right)$$

or

$$R^{\bullet} \xrightarrow{-e} R^+ \left(\text{carbonium ion products}\right)$$

The above reactions are heterogeneous processes, occurring through formation of radical intermediates on the surface of the electrode. The material of electrodes may be different noble metals. It is also known the photoassisted or "photo-Kolbe" decarboxylation reaction occurring upon illumination of a TiO_2 powder. This process is closely related to decarboxylation reactions in heterogeneous catalysis. Two carboxylic acids, acetic and triphenylacetic acid, were tested in the photo-Kolbe reaction on TiO$_2$.[34] The radical intermediates in a photo-Kolbe reaction were detected by registration of the EPR signals of the radical adducts: methyl radical with α-phenyl-n-tertbutylnitrone and triphenylmethyl radical with tetra-n-butylammonium triphenylacetate, respectively.

2.2.2 EXAMPLES OF HETEROGENEOUS CHAIN-RADICAL REACTIONS

The radical reaction of hydrosilylation of unsaturated compounds, initiated on the surface, is one of the best examples of heterogeneous chain-radical reactions.[35–42] The hydrosilylation reaction is well known in organic syntheses,[36, 37]

$$(S)\text{-Si-H} + CH_2 = CH\text{-R} \rightarrow (S)\text{-Si} - CH_2\text{-}CH_2\text{-R}$$

where R is the organic radical and (S) is the surface of the solid sample. The hydride-covered surface of Si(111) may be prepared by the wet-etch method.[39] The single silicon dangling bond site, surrounded with silicon hydride groups, may be generated using the tip of the STM (scanning tunneling microscopy) technique, in ultra-high vacuum (UHV) conditions.[36] The STM image of the sample Si(111) exposed to styrene (Ph-CH = CH$_2$) showed the formation of styrene islands located at these single silicon dangling bond sites. Creation of the surface silyl radical and propagation stage of reaction are presented in the scheme below (Figure 2.2).

Thus, a single site on the surface of Si(111) becomes an initiator for the bonding of a great number of styrene molecules. "Bunched-up styrene islands" were observed, evidencing the immobility of the chemically bonded molecules. According to the mechanism suggestions, this is a "self-propagating" and radical-chain reaction (Figure 2.2). As a result of the interaction of silicon with the dangling bond and styrene molecule, the chemically bonded organic radical ((S)-Si-CH$_2$-CH-R) forms on the surface site. It can abstract the neighboring hydrogen atom of the surface, producing a new adsorption site for the next adsorbate molecule. The process may continue

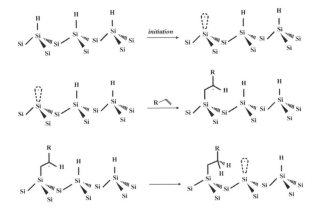

FIGURE 2.2 Scheme of the initiation and propagation stages of the hydrosilylation reaction of unsaturated organic compounds (Adapted from Ref.[36]).

as long as there are neighboring hydrogens. Hence, the radical-chain reaction propagates along the surface, producing the chemically bonded product of hydosilylation.

In a similar experiment on Si(100) 2 × 1 surface, Lopinsky et al.[37] observed the formation of the "molecular wire" whose length was 34 styrene molecules along the direction of the dimer row. Yates et al.[40] observed the formation of the linear molecular chain of dimethyldisulfide (CH_3SSCH_3) molecules, self-assembled on the surface of gold, Au(111) and Au(100) single crystal samples, at 70–200 K, in UHV-conditions, using the STM method. Reaction may be induced by electron injection from the tip to the surface. The charge transfer to the adsorbed molecule CH_3SSCH_3 from the surface leads to the dissociative adsorption, forming $CH_3S^{\bullet}$ radicals (radical-like species) on the twofold or threefold Au sites. Then, the adsorbed $CH_3S^{\bullet}$ radicals react with the neighboring adsorbed CH_3SSCH_3, resulting in cleavage of S-S. The formed radicals, attached on the surface, react with other molecules of CH_3SSCH_3, growing the chain of molecules in certain direction. A number of linearly "self-organized molecules" may be reached to 10–15. Note that the DFT (density functional theory) calculations predict a nearly zero activation barrier for this kind of polymerization. All described molecular events are clearly visible on the SMT images, including CH_3S radicals, as intermediates. At least two interesting behaviors of this system, from our viewpoint, are worthy of attention. First, the SMT images show the existence of the heterogeneous reaction with direct participation of the adsorbed CH_3S radicals, as reaction-active intermediates. Yates named it "a chain reaction molecule-by-molecule." In this regard, the mentioned reaction may be considered a "purely" heterogeneous-radical reaction, forming a product by the heterogeneously propagating chains. Here, the word "chain" has a double meaning. On the one hand, it indicates that the reaction forms a polymeric molecular chain, attached on the surface of gold; on the other hand, it characterizes the mechanism of reaction as a chain-process. In other words, the reaction is self-propagating, as every step of reaction growing the chain also "prepares" the next reaction, involving monomer. Principally, the reaction may be continued in one direction till the end of the crystalline plan or by the "random termination" of the chain. Second, here we obviously have the phenomenon of self-assembling (self-organization) in the chemical system via the initiation and recombination of radicals (cleavage and re-creation of the S–S bond), occurring as a heterogeneous chain process. This is also a process of the creation of new structure ("molecular wires" or "walking random," as expressed by the authors[36]). It is noteworthy that in the above reaction, as well as in other analogous reactions, the developing reaction itself creates islands of the product on the surface of the solid substance, exhibiting the self-assembling properties. The self-assembling of the adsorbed radicals on the surfaces of solid substances is well demonstrated in the system of the adsorbed radical α,γ-bisdiphenylene-β-phenylallyl (spin-1/2, BDPA) on Au(111) monocrystalline sample.[41] In this system, the impressive STM image of the sample shows the self-assembly, a stable and regular one- and two-dimensional radical composition on the surface.[41]

There are also some examples concerning the oxidation of hydrogen-terminated surfaces of H-Si(111) and H-Si(110) in the presence of O_2 and H_2O, in ordinary conditions, forming an oxidized monolayer on the surface.[43] In this case, the heterogeneous chain-radical reaction may be initiated photochemically through UV irradiation.

Note again that the above examples of reactions are "purely" heterogeneous-radical processes, where all stages of the reaction occur on the surface. Moreover, the active sites on the surface also have a paramagnetic nature. This kind of reaction is essentially different from the overall heterogeneous–homogeneous reactions by the participation of radicals, where some stages have a heterogeneous-radical, but others a homogeneous-radical nature. For example, radical pathways of the heterogeneous catalytic reactions were suggested for the oxidation of some hydrocarbons on the zeolite catalysts.[44] Specificity of the zeolite catalysts mainly relates to their porosity, as well as to the coexistence of the sites of a different chemical nature. Catalysis by zeolites with small pores may be considered a specific case of heterogeneous–homogeneous catalysis, when certain stages of the reaction occur on the surface sites, while others occur in the pores. The general schemes of the sequences of reactions correspond to the chain-radical reaction (initiation, propagation, and termination of chains). Among them, the important reaction of the decomposition of hydroperoxide ROOH (degenerate branching of chains, in the terminology of radical-chain reactions) usually takes place with the participation of metal–ion active centers of zeolite.

$$ROOH + M^{2+} \rightarrow RO + OH^- + M^{3+}$$
$$ROOH + M^{3+} \rightarrow RO_2 + H^- + M^{2+}$$

The thermal treatment of samples has an essential importance in catalysis on the zeolites. Dehydroxilation, according to van Santen and Neurock,[44] creates a three-fold coordinated Si and Al, as well as Si-O-Al (negatively charged) and AlO$^+$ sites from Si-OH-Al, on the proton-exchange zeolites. Leu and Roduner,[45] investigating the structure of the active centers by the EPR method, suggested the formation of so-called single-electron transfer sites (SETS), which involve molecular oxygen in the oxidation of organic compounds. The following scheme was proposed in the reaction of 2,5-dimethylhexa-2,4-diene (DMHD) with O_2 on the proton-exchange zeolite:[45]

$$DMHD + SETS \rightarrow DMHD^+ + SETS^-$$
$$SETS^- + O_2 \rightarrow SETS + O_2^-$$
$$DMHD^+ + O_2^- \rightarrow products$$

It was calculated that every active site (SETS) of the zeolite catalyst performed about 4000 cycles in the oxidation of hydrocarbon. However, being a radical reaction, here the overall process does not have a "purely" heterogeneous-radical nature.

2.3 HETEROGENEOUS–HOMOGENEOUS REACTIONS OF RADICAL DECOMPOSITION

Any covalent chemical bond may be dissociated if it receives and absorbs a required quantity of energy in any form (heat, energy of irradiation, etc.), which is more than the energy of the chemical bond. In heating, the energy driving the thermal decomposition reactions of chemical compounds has the Boltzmann distribution. In being dependent on the temperature, the chemical bond may be dissociated due to the thermal energy of molecule. Obviously, the dissociation energy is dependent

Table 2.1 Standard dissociation energies of some chemical bonds[47]

Chemical Bond	Compound	Bond Energy (*kcal/mol*)
H-H	H_2	104
C-C	CH_3-CH_3	83
C-H	CH_4	105
O-H	hydroxyl	110
O=O	O_2	119
O-O	RO-OR	35
N≡N	N_2	226
N=N	RN=NR	109
N-N	R_2N-NR_2	38.4

on the strength of the chemical bond. The homolytic dissociation of the covalent chemical bond may even occur at room temperature in the gas or liquid phases if the chemical bond is relatively weak, being not more than 25–35 kcal/mol.[46] Table 2.1 presents some bond energies.[47] Though incomplete, this table shows that the energies of the chemical bond for peroxo and azo compounds have relatively low values. Therefore, it is expected that the chemical compounds containing these bonds will be a convenient source of radicals in thermal or photochemical decomposition. The activation barrier of decomposition reaction is usually close to the bond energy. However, often, the dissociation of the mentioned bonds may also occur at low temperatures, even at room temperature, as in the case of some peroxides

$$ROOR \rightarrow RO + RO$$

The thermal or photochemical decomposition of azoalkanes is also a convenient means of generating free radicals

$$R-N=N-R \rightarrow R + N_2 + R$$

The homolytic decompositions of peroxides are exothermic reactions.[48]

$$H_2O_2 \rightarrow OH + OH \qquad\qquad 49.9 \text{ kcal/mol}$$

$$CH_3C(O)OOH \rightarrow CH_3C(O)O + OH \qquad\qquad 45.6 \text{ kcal/mol}$$

The thermal dissociation reactions, even in the case of the compounds, containing relatively weak chemical bonds, are complex processes. Many factors are involved: solvent effects; the presence of other compounds in the reaction medium, even in very small amounts; and secondary reactions of the intermediate species, including their heterogeneous reactions, which do not always permit direct use of these reactions, in the aims of the preparative chemistry of radicals. In the majority of cases, the radicals obtained by decomposition of these compounds undergo further decomposition, isomerization, and disproportionation, often in combination with dioxygen, nearly always presenting in the reaction medium, especially in the case of peroxides. In other words, the radicals formed in the primary dissociation of the chemical bond, particularly O-O in ROOR or ROOH, indispensably undergo further stabilization in any way.

The main ways of generating radicals in the chemical literature are thermal (including creaking or pyrolysis), photochemical, high-energy radiation (as γ-rays), and redox (by ions of transition metals) pathways. Usually, the majority of researchers bypass the feasibility of the heterogeneous or heterogeneous-catalytic generation of radicals. As the generation of radicals by the above-mentioned ways is well documented in the chemistry of free radicals, here we will focus solely on the problems of the heterogeneous or, more precisely, heterogeneous-homogeneous pathways of the generation of free radicals.

The heterogeneous generation of radicals plays an important role not only in complex chain-radical (branched or degenerate-branched), but also in heterogeneous catalytic reactions of oxidation. As mentioned earlier, the role of the heterogeneous reactions of radicals, particularly in photocatalytic reactions on the surfaces of materials, such as TiO_2, SiO_2, and zeolites, is increasingly being recognized (see Chapter 4).

As examples of the chemical systems generating radicals by heterogeneous pathways, the reactions of two classes of the chemical compounds will be briefly discussed in this chapter. One of these is the generation of radicals through the decomposition of peroxide compounds involving H_2O_2 and organic peroxide compounds (peroxides, hydroperoxides, suprooxides, etc.) and the second is the pyrolysis of hydrocarbons, both on the solid surfaces of different nature (mainly, metal oxides and salts). Some reactions of the heterogeneous decomposition of ozone will be presented in Chapter 5.

As this chapter will show, both classes of the organic compounds, from the point of view of the general kinetic behaviors in decomposition reaction, are analogous. Note that the examples of the heterogeneous radical decomposition reactions given below concern the gas-phase processes. The same phenomenon was also observed in certain liquid-phase decomposition processes.[107 in Ch.1]

Presently, the generation of radicals by heterogeneous decomposition is a wide class of reactions involving a great number of organic, as well as inorganic compounds, and all diversity of their structural analogs.

2.3.1 HETEROGENEOUS–HOMOGENEOUS PYROLYSIS OF HYDROCARBONS

Pyrolysis of hydrocarbons[49–56, 128 in Ch.1] is a thermal decomposition reaction in the absence of oxygen. In the thermal decomposition reactions of higher hydrocarbons (mainly in the liquid phase), the chemical term *pyrolysis* is replaced by the term *creaking*. At first glance, it seems that pyrolysis has nothing in common with the oxidation processes. In reality, knowledge of the reaction mechanisms of pyrolysis is an important cornerstone in studies of oxidation reactions, as "the pyrolysis, combustion and oxidation are intrinsically similar classes of chemical processes" (Parmon[49]). Moreover, pyrolysis is often one of the stages of combustion or oxidation processes. In chemical practice, especially on the industrial scale, oxidation and pyrolysis accompany each other to some extent.

The temperature of pyrolysis depends on the chemical structure of hydrocarbons. In this regard, the temperatures of pyrolysis of acyclic hydrocarbons and hydrocarbons, containing carbonic rings ($C_n \geq 3$), are very different.[49] Methane, being a thermodynamically more stable compound than other hydrocarbons, undergoes pyrolysis at about 1000–1100°C, compared to ethane at 800–900°C. The decomposition of hydrocarbons usually is a nonbranched chain-radical process and is strong dependent on the wall effects. At not very high temperatures, the rate of the breaking of C–C bond in hydrocarbons prevails over the C–H bond breaking ($E_{C-C} < E_{C-H}$); hence, in the absence of the catalyst, the main primary reaction in pyrolysis, for instance, for ethane, is the formation of two methyl radicals:

$$C_2H_6 \rightarrow 2CH_3$$

but not

$$C_2H_6 \rightarrow C_2H_5 + H$$

Then, however, the H-abstraction reaction from ethane may take place:

$$CH_3 + C_2H_6 \rightarrow CH_4 + C_2H_5$$

The propagation of the homogeneous chains also includes

$$C_2H_5 \rightarrow C_2H_4 + H$$

$$H + C_2H_6 \rightarrow C_2H_5 + H_2$$

The termination of chains involves at least four reactions of recombination of the H, CH_3, and C_2H_5 radicals:

$$H + CH_3 \rightarrow CH_4$$

$$CH_3 + CH_3 \rightarrow C_2H_6$$

$$C_2H_5 + C_2H_5 \rightarrow C_4H_{10}$$

$$C_2H_5 + H \rightarrow C_2H_6$$

Two other chain termination reactions are relatively slow processes:

$$H + H \rightarrow H_2$$

$$CH_3 + C_2H_5 \rightarrow C_4H_8$$

This is a very simplified scheme of the pyrolysis of ethane, as an example of the chain-radical nonbranched reaction. The length of chains reaches about 100–200, depending on the conditions and material of the reactor used.[52, p.54]

In pyrolysis, the concentration of the leading radicals is usually lower than their thermodynamically equilibrium concentrations, due to the fast recombination of a part of the leading active centers. In pyrolysis of higher alkanes, the radicals formed in the primary reaction, as propyl radical formed in pyrolysis of propane, may undergo isomerization and/or decomposition

$$CH_3\text{-}CH_2\text{-}CH_2 \rightarrow CH_3\text{-}CH\text{-}CH_3$$

$$CH_3\text{-}CH_2\text{-}CH_2 \rightarrow CH_3 + CH_2 = CH_2$$

Often, it may also undergo substitution by other radicals:

$$R' + RH \rightarrow R'H + R$$

C_5–C_6 hydrocarbons usually form cyclic intermediates.

As a rule, the further creaking of higher hydrocarbons occurs by the breaking of C–C in β-position, where the bond is weaker than in the α-position.[52] All these and many other features are well investigated and documented in the chemical literature.[49]

The presence of the catalyst in the reaction zone not only effectively lowers the temperature of pyrolysis, but also often affects the selectivity of reaction, which may be used to obtain desirable products by pyrolysis.[50] For a great number of pyrolytic processes, it is known that in the primary process, the hydrocarbons, adsorbed on the surface of the solid substance, undergo decomposition, forming radicals, a part of which may desorb from the surface and diffuse to the gas phase. Thus, the pyrolysis of hydrocarbons in the presence of solid-phase catalysts can be a typically heterogeneous–homogeneous process.

As an example of the pyrolysis of hydrocarbons, having a bimodal reaction sequence, let us discuss the thermal reaction of n-undecane (C_{11} alkane) in the presence of solid substances in the reaction zone.[128 in Ch.1] This reaction was investigated in the presence of $BaCl_2$ and MgO, as well as so-called imperfect magnesium oxide (defect-bearing MgO). The solid substances in the form of granules were placed in the reactor. Pyrolysis in the presence of these solid substances occurred as a heterogeneous–homogeneous reaction. The EPR method, combined with the freezing of radicals, was applied for the detection of radicals in the gas phase, near the surface. Ethyl radicals were detected in the gas phase, in decomposition of undecane at 873–953 K. It was observed that the activation barrier of the generation of radicals was 130 kJ/mol lower in the presence of $BaCl_2$ than in its absence (322 kJ/mol). The authors considered that the used solid surfaces had different functionality in their action on the process, accelerating either the heterogeneous ($BaCl_2$) or homogeneous constituent (defect-bearing MgO) of the reaction. They posited that the active centers on both surfaces have a radical nature, and, in the case of $BaCl_2$, the radical reactions occur on the surface, analogous to propagation of chains in the gas phase. On the other hand, it was shown that the "defected" MgO samples also exhibited activity in the generation of free radicals, which, unlike $BaCl_2$ surface, were able to desorb to the gas phase and accelerate the homogeneous reactions, leading to pyrolysis. In the case of the "defected" MgO samples, at a certain critical temperature, the so-called ignition phenomenon was observed—a drastic acceleration of the homogeneous process due to intensification of the generation of radicals on the surface and their accumulation at the near-surface layer. This observation was named the phenomenon of the "catalytic sphere." The activation barrier of the radical generation at a certain temperature (for instance, 903 K) for n-undecane on MgO, suddenly decreased from 322 to 63 kJ/mol and 4 kJ/mol for "defected" MgO. Analogous critical phenomena were observed in the pyrolysis of other hydrocarbons and dichloromethane (the experimental results were reviewed in Ref.[51]). How to explain these observations? According to the authors of these works, at the pyrolysis of $CH_3(CH_2)_9CH_3$, the strength of the chemical bond between the radical and surface active center (D_{R-S}) correlates with the activation energy (E_a) of the radical generation by the surface:

	E_a (kJ/mol)	D_{R-S} (kJ/mol)
Quartz	335	–
$BaCl_2$	192	200
MgO	63	360
MgO ("def.")	4	470

As these data show, it follows that the "escape" of radicals from the surface—therefore, the probability of the heterogeneous–homogeneous reaction—increases with the increase in the bonding energy of hydrocarbon radicals with the surface active center (see also Section 1.7). The authors also considered that "a substantial energy, equal to D_{R-S}, is released as a result of the radical adsorption on the active center of the surface" and the "oxide crystal can accumulate substantial part of this energy," creating electron-hole centers. At the moment of "ignition," the active centers of radical nature with high-energy potential, in nonequilibrium state, form on the surface. Moreover, the surface reactions energetically enhance the activity of catalyst to generate the radicals to the gas phase. This attractive explanation, however, involves some hypotheses, which must be proven experimentally.

As mentioned above, the homogeneous pyrolysis of alkanes can usually be presented by the schemes of the nonbranched chain reaction. The following generalized scheme of pyrolytic processes was suggested by Buyanov and Vasileva and presented in a number of works:[53–55]

1. Initiation of chains

$$R\text{-}R_{sh} \rightarrow R_{sh} + R$$

2. Propagation of chain

$$R \rightarrow R_{sh} + \alpha\text{-olefin}$$

$$RH + R_{sh} \rightarrow R_{sh}H + R$$

3. Termination of chain

$$R_{sh} + R_{sh} \rightarrow R'_{sh} + R'_{sh} \text{ (molecular products)}$$

R_{sh} is a short-chain radical (CH_3, C_2H_5, H), R is a long-chain radical, and correspondingly, $R\text{-}R_{sh}$, RH, and $R_{sh}H$ are hydrocarbons. In the presence of solid-phase substances, playing the role of catalyst,[54–56] according to these authors, the scheme of the overall heterogeneous–homogeneous reaction must be completed by the stage of the heterogeneous propagation of chain by the following reaction:

$$S + R\text{-}R_{sh} \rightarrow S\text{-}R_{sh} + R$$

and also, obviously, by the termination reaction of chains:

$$S - R_{sh} + R \rightarrow S + R_{sh} - R_{sh}$$

where S is the active site on the surface. It was considered that in this scheme, the active centers on the surface might be born during the reaction, propagating the chains on the surface. According to this scheme, the surface always participates in the generation of radicals to the gas phase. Thus, the radicals are simultaneously active centers both on the surface and in the gas phase.

These very brief notes on the pyrolytic processes of hydrocarbons demonstrate the typically heterogeneous–homogeneous nature of the reaction in the presence of a number of solid-phase substances, playing the role of catalysts, and in the absence of oxygen. One of the peculiarities of such systems, in certain conditions, may be the emergence of the critical phenomena mentioned above. From our point of view, these phenomena may also be considered as a specific exhibition of the self-organization phenomena in complex systems with pyrolytic reactions.

2.3.2   HETEROGENEOUS–HOMOGENEOUS DECOMPOSITION
OF PEROXIDE COMPOUNDS

Hydrogen peroxide and organic peroxide compounds are the main intermediate and often final products in the thermal and photochemical oxidation of organic substances with dioxygen.[57, 58] They are metastable compounds, which may be separated from other products and intermediates of the oxidation and analyzed as individual substances.

For the majority of the noncatalytic and catalytic reactions of oxidation of the organic compounds with oxygen, the formation and decomposition reactions of peroxide compounds determine the main kinetic peculiarities and composition of the final products. As a rule, both in the gas and liquid phases the peroxide compounds are the main intermediates of the autocatalytic chain-radical reactions, responsible for the degenerate branching of chains[34 in Ch. 1] in the oxidation of hydrocarbons and certain classes of oxygenated compounds. As will be shown in Section 3.4.2, in particular cases, the heterogeneous–homogeneous character of the complex reactions of oxidation also is related to the reactions of the organic peroxide compounds or hydrogen peroxide.

The main groups of the organic peroxides are ROOH (hydroperoxides) and ROOR, as well as ROOR′, where R or R′ are alkyl aryl, acyl, *tert*-buthyl or other radicals. The O–O bond in organic peroxide compounds is relatively weak (in comparison with the O–O bond in dioxygen), being not more than about 45–50 kcal/mol (190–210 kJ/mol); for instance, in dialkyl ROOR, it is 37 ± 1 kcal/mol and is not strongly dependent on the nature of R.[34 in Ch. 1, 58] In hydrogen peroxide, it is about 142 kJ/mol. The decomposition of peroxides is an exothermic reaction, and the released energy is approximately equal to the dissociation energy of the O–O bond.[58]

2.3.2.1   Heterogeneous–Homogeneous Decomposition
of Hydrogen Peroxide

The decomposition reaction of hydrogen peroxide has a pivotal importance in a great number of complex reactions investigated from the point of view of the different problems of chemistry, biology, atmospheric science, etc.

Usually, the vapors of hydrogen peroxide thermally decompose by noncatalytic homogeneous reaction above 400–450°C.[59, 60] In the presence of the solid substances in reaction media, at relatively low temperatures, the rate of the heterogeneous decomposition of H_2O_2 prevails over the homogeneous one.[56] The rate of the decomposition of hydrogen peroxide strongly depends on the nature and phase state of the solid substances, including the materials of the walls of the reaction vessel.[61–64] This also applies to the liquid-phase decomposition of hydrogen peroxide in water solutions.[63, 65] The history of chemistry reveals that the experiment of the Belgian physical chemist W. Spring with metallic Pt in decomposition of H_2O_2, demonstrated that a small but fresh scratch on the glanced surface of Pt, was able noticeably to lower the decomposition temperature, for instance, from 60° to 12°C, for a 38% solution of H_2O_2 in water.[66] In the gas phase, the presence of metals in the reaction zone, as a rule, accelerates decomposition of hydrogen peroxide in comparison with the empty reactor or reactors with the walls, made from the "soft" glasses, borosilicate and quartz. The rate of the decomposition of H_2O_2 on metallic Pt is about 10^4 order of magnitude higher than in noncatalytic decomposition.[59] In comparison with the noncatalytic decomposition, the rate of enzymatic decomposition of H_2O_2 may be considered exceptional, as, for example, it is about 10^{11} order of magnitude higher in the presence of catalase (with a low activation barrier about 7 kJ/mol) than in noncatalytic decomposition (activation barrier is about 70 kJ/mol).[67] Among a great number of heterogeneous catalysts for decomposition of H_2O_2, Ag, Pt, Au, Mn_2O_3, dusts are known as more effective in their action.[68]

The catalytic activity of metal oxides in the decomposition of the gas phase H_2O_2, in more or less comparable conditions, changes by the following sequence: $Mn_2O_3 >$ PbO > Ag_2O > CoO/Co_2O_3 > Cu_2O > NiO (green) > CuO > Fe_2O_3 > CdO > ZnO=MgO > SnO > Glass (Pyrex), investigated at 34–184°C and at the partial pressure of H_2O_2 about 1 mm Hg.[69] The first order of the rate in respect to the concentration of H_2O_2 usually was observed in the mentioned and in other analogous experiments of catalytic decomposition.[68–70]

In the liquid phase, investigations of the mechanism of hydrogen peroxide decomposition, in the presence of metal ions (for instance, Fe(III), Cu(II)), in homogeneous conditions, were started, beginning with the discovery of the Fenton reagent (the Haber–Weiss mechanism; see Ref.[67] in Section 1.7.1). Both presently and formerly, the mechanism suggestions in homogeneous catalytic conditions, mainly, are based on the oxidative/reductive role of metal ions, producing free radicals,[68] which consequently may develop radical reactions or chain-radical catalytic cycles.

In this section, attention is paid to the low-temperature heterogeneous–homogeneous reactions of decomposition of H_2O_2 on solid surfaces, which are relatively more frequently investigated in the gas phase.

The heterogeneous catalytic decomposition of hydrogen peroxide has some noticeable differences from homogeneous catalytic decomposition, although the reaction products (H_2O and O_2) are usually the same in both cases, as well as in noncatalytic reactions.[71] Obviously, in heterogeneous catalytic decomposition, the interaction begins by adsorption of hydrogen peroxide on the surface of the solid substance, forming an adsorption complex. Since the early kinetic investigations of the heterogeneous decomposition of H_2O_2,[72] the mechanism suggestions were mainly related to the electron transfer between the adsorbed peroxide and active centers of the solid, in the primary reaction. For instance, on the metallic surfaces, the primary reaction might occur by the following scheme:[72]

$$H_2O_2 + S \rightarrow S^+ + OH + OH^-$$
$$H_2O_2 + OH \rightarrow H_2O + HO_2$$
$$S + HO_2 \rightarrow S^+ + HO_2^-$$
$$S^+ + HO_2^- \rightarrow S + HO_2$$

where S is surface of the solid substances.

Evidence of charge transfer in the heterogeneous decomposition of H_2O_2 on the surfaces of some metal oxides (NiO, ZnO) was also obtained by electroconductivity measurements.[73] Moreover, the existing linear correlation between the decomposed quantity of peroxide and electroconductivity of the solid sample often permits us to use it for analytical purposes.[73 and within]

Nearly all kinetic investigations, by the application of different experimental methods, show that the primary reaction of hydrogen peroxide on the surfaces of metals, metal oxides, other solid materials, including different glasses, and salts (for instance, alkali metal halides) is rupture of the O–O bond, giving two hydroxyl species (in the above scheme, hydroxyl radical and hydroxyl ion pair). The detection of the OH radicals on the surface by such direct method as EPR is complicated (a number of reasons are indicated in Ref.[74, 75]). The further reactions of the adsorbed species are responsible for formation of other radical-like species, such as HO_2, O_2^-, O^- on the surface, mainly, formed by hydrogen transfer reactions. For example, the EPR investigation of the decomposition of H_2O_2 on rutile TiO_2 indicates the formation not only of O_2^-, but also of the interacting O^-... O^- radical pair (or radical-like species in triplet state, with g = 2.0104 and g = 2.00644, correspondingly[76]).

Presently, there is enough experimental evidence indicating that, in certain conditions, the low-temperature thermal decomposition of hydrogen peroxide on the surfaces of solid substances is the typically heterogeneous–homogeneous reaction.[77] The radicals generated on the solid surface are able to desorb from the surface and diffuse to the fluid phases. The radicals, which appear in the gas phase, may be detected, using the matrix isolation technique of the EPR spectroscopic method. For the first time, Grigoryan and Nalbandyan[78] showed the appearance of the hydroperoxy radicals in the gas phase, generated on the surfaces of a number of solid substances, such as glass materials, SiO_2, γ-Al_2O_3 (473–623 K). The further kinetic investigations of the decomposition of H_2O_2 at low temperatures, on a great number of solid substances, clearly proved these conclusions.[79–82, 121 in Ch.1] The heterogeneous decomposition of hydrogen peroxide and the formation of peroxy radicals, based on analysis of the EPR spectra and kinetic data may be represented by the following simplified sequence of heterogeneous reactions:

$$H_2O_2 \rightarrow 2OH \tag{1}$$

$$OH + H_2O_2 \rightarrow HO_2 + H_2O \tag{2}$$

$$HO_2 \rightarrow (HO_2)_{gas} \tag{3}$$

The EPR spectra obtained contain a component assigned to the presence of OH radicals that disappears with the increase of temperature or at pumping. This is also evidence of the existence of the equilibrium between two stages (1) and (2). However, it was practically impossible to obtain the EPR signal of "pure" OH generated on the surface and, then, accumulated in the gas phase, by freezing of the reaction mixture.

Later, the HO_2 radicals, formed in heterogeneous decomposition of H_2O_2 (333–393 K) on SiO_2 and γ-Al_2O_3, even at room temperature, were successfully detected by the EPR method (*in situ* experiments).[83] The radicals on the surface were stable in heating below about 400 K. The activation barrier of their decay was 31 kJ/mol. The EPR signal obtained in the case of ZnO was assigned to the O_2^- adsorbed radical.[83]

The formation of OH radicals in the decomposition of H_2O_2 in the liquid phase, on a SiO_2 surface, at low temperature was confirmed by application of radical scavengers, as 5,5-dimethyl-1-pyrroline N-oxide (DMPO).[84] The active centers arise in the grinding of SiO_2 or in mechanical treatment. Two types of active centers formed in this case by homolytic or heterolytic cleavage of Si-O-Si are $Si^{\bullet}$ + $^{\bullet}O$-Si and Si^{+} + ^{-}O-Si, respectively

$$Si^{\bullet} + H_2O_2 \rightarrow SiOH + OH^{\bullet}$$

$$Si^{\bullet} + O_2 \rightarrow SiOO^{\bullet}$$

and

$$Si^{+} + SiO^{-} + HO_2^{\bullet} \rightarrow Si^{+}O_2^{-} + SiOH$$

$$HO_2^{\bullet} \rightleftarrows H^{+} + O_2^{\bullet -}$$

Other important reactions were considered, such as

$$Si^{\bullet} + {}^{\bullet}OH \rightarrow SiOH$$

$$SiO^{\bullet} + {}^{\bullet}OH \rightarrow SiOOH$$

On the other hand, the superoxide anion-radical $(O_2^{\bullet -})_{ads}$ was identified on the surface of SiO_2. Analogously, the formation of OH radicals on MgO and TiO_2, in their simple contact with H_2O_2 was reported by Amorelli et al.[85]

The kinetic studies and theoretical calculations (DFT) of the decomposition of H_2O_2 on noble metal catalysts,[61] at around room temperature, showed that on SiO_2 supported Pd-metal, the Pd(111) and Pd(100) facets of metal participate in decomposition of H_2O_2, forming OH and O surface species; then, H_2O_2 and HO_2 reduce them to H_2O by H-transfer reactions.[61] It was considered that on Pd(111) the OH species occupied a bridge-tilted, OOH bent-top, O fcc and products H_2O and H_2O_2, the top positions on the surface structure of metal, respectively. Thus, the main intermediates on the surface are O, OH, and HO_2, often also H-adsorbed species.

It has already been mentioned that early investigations showed no direct spectroscopic evidence of the formation of OH radical-like species on the solid substances in heterogeneous decomposition of H_2O_2. In the most recent decade, development of the so-called single-molecule detection spectroscopic technique yields valuable information not only about the intermediates, but also the surface sites in heterogeneous reactions. Tachikawa and Majima[86–89] successfully detected hydroxyl radical and singlet oxygen "diffused from the surface," applying the single-molecule fluorescence imaging technique for TiO_2-based (nanotubes) photocatalytic reactions. The single-molecule fluorescence imaging experiments were described in full details by Tachikawa and Majima.[88] In general, this method is reminiscent of the Kautsky experiment (see Section 3.4.2.7). On the other hand, it is similar to the Nosaka experiment (using the chemiluminescent probe method).[90] OH radicals or singlet oxygen were produced under UV irradiation, on the nanotube of TiO_2 (attached on the glass), in contact with the gas- or liquid-phase reaction mixture. Dye molecules were on the other surface. Reactive oxygen species, generated on TiO_2, reaching to the surface of the supported dye molecules, reacted with them under laser irradiation (Figure 2.3).

The fluorescence emission from the molecules, produced in interaction of the excited dye molecules and intermediates, was detected and analyzed by the fluorescence signal detector. Fluorescence imaging of molecules may be obtained. The applied technique permits obtaining the signals from the individual molecules (10–20 molecules on the area 10×10 μm^2, according to the work of Tachikawa et al.[86, 87]). Scanning over the TiO_2 surface, it may be seen that the information obtained from single-molecule fluorescence imaging also permits detecting inhomogeneity of the solid surface and

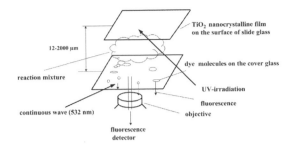

FIGURE 2.3   Simplified scheme of the Tachikawa and Majima experiment (Adapted from Ref.[87, 88]).

obtaining the spatial and temporal distribution of intermediates, for instance, of the so-called airborne singlet oxygen molecules (Figure 2.4)[87] or radicals OH,[89] diffused from the surface. In analogous experiments, the same authors showed that a correlation exists between the number of catalytic sites and fluorescence intensity.[88] It was concluded that the distribution of active sites on the nanotube surface was related to the inhomogeneity of the surface defects.[88] According to these authors, the spatial distribution of $^1O_2$ in Figure 2.4 was the first example in the single-molecule detection experiments demonstrating "traveling such a long distance in ambient air" of the active intermediate species, generated on the surface.

The single-molecule fluorescence microscopy provides more and more valuable information. In 2014, the Nobel Prize in Chemistry was awarded to three researchers, E. Betzig, S. W. Hell, and W. E. Moerner, for the development of super-resolved fluorescence microscopy. For single-molecule detection, they applied an ultrasensitive microscopy that is not limited by the diffraction of light. This technique permits researchers to "observe the behavior of individual molecules," obtain the image of three-dimensional (3D) structures, and record dynamic processes, even in living cells at the nanometer scale.[91]

The above-mentioned experiments clearly approve the generation of radicals on the solid surface and their diffusion to the gas phase, on a TiO_2 and some other metal oxide materials. Although this method was not yet applied in study of the

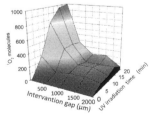

FIGURE 2.4   Spatial and temporal distributions of airborne $^1O_2$ molecules, photogenerated on the surface of TiO_2 film and diffused to the volume, obtained by single-molecule fluorescence microscopy (70 × 70 μm²). (Reproduced from the work of Tachikawa T., Majima T. (Ref.[86]), by the permission of the Research Center of Formatex).

heterogeneous decomposition of hydrogen peroxide on different solid surfaces, however, nearly all the reactive oxygen species, including even singlet oxygen, as well as H_2O_2, were successfully detected and identified. In this regard, single-molecule spectroscopy opens up unprecedented possibilities in investigations of heterogeneous–homogeneous reaction systems.

As an important conclusion of this section, devoted to some heterogeneous–homogeneous reactions of the decomposition of hydrogen peroxide, note that the heterogeneous constituent of reaction is the supplier of radicals, ion-radicals, and perhaps also other intermediates, including electronically excited species, to the fluid phases. The integrity of the mentioned data, obtained by different methods, once again indicates the heterogeneous–homogeneous nature of this decomposition reaction in the presence of solid substances, as a well-established experimental fact. However, the details of the formation of radicals on the surfaces and the chemical nature and structure of active sites, therefore, the detailed mechanism of their generation are not clear in many aspects.

2.3.2.2 Heterogeneous–Homogeneous Decomposition of Organic Peroxides

In early investigations, it was considered that, in the gas phase, the homogeneous thermal decomposition (usually at temperatures T > 400–500°C) of ROOH or ROOR mainly occurred by the homolytic breaking of the O–O bond with the formation of radicals.[92] The radicals (primarily species of the RO_2 type), which usually are products of the secondary reactions, may be detected by the EPR method using the matrix isolation technique or by trapping with the scavengers (radical traps). In the recent decades, development of the single-molecule spectroscopic methods has opened up new possibilities for detecting intermediates in the decomposition reactions of peroxides (see Section 2.3.2.1).

In homogeneous catalytic decomposition of the organic peroxides in the liquid phase by the transition metal ions, the redox mechanism suggestions of reaction, like mechanism suggestions in the decomposition of hydrogen peroxide, were based on radical and ion-radical reactions.[34 in Ch. 1]

In the case of the heterogeneous catalytic decomposition of peroxides, there are some important differences between gas-phase and liquid-phase reactions. Obviously, first of all, they are mainly related to the peculiarities of solid–liquid interactions in heterogeneous decomposition reactions. In particular for reactions on silica surfaces, Mello et al.[95] mentioned at least two important features in the case of the liquid-phase substrate interaction with the surface. One of these features is polarizability, and the other is the ability of the surface to form hydrogen bonds with the intermediate species in the liquid-phase reaction. For example, the active sites on the surface of SiO_2 have a high polar character and acidic properties, and "they promote polar reaction pathways in the decomposition" of peroxides.[93] As a result, in the liquid phase, on a SiO_2, the heterolytic decomposition of peroxides may become a major reaction channel. Moreover, the chain-radical pathway of the surface reaction is hindered, as the adsorbed radicals, which were formed in decomposition, have "restricted transitional mobility."[93] As the reaction rate depends on the formation and decomposition of the hydrogen bonds with the surface, here the hydroxyl groups, which are more than one type on the surface, play an important role. In the case of γ-Al$_2$O$_3$, the preheating at 400–800°C may gradually remove 50 to 100% of the surface hydroxyl groups. It is known that two important types of Al-ions (tetrahedral and octahedral coordinated Al-ions) on the surface of γ-Al$_2$O are differently bonded with OHs and form five types of surface sites by different reactivity.[93]

The possibility of the heterogeneous decomposition of the organic peroxides in gas-phase reactions was a subject of interest, first of all, in investigations of the combustion and low–temperature slow oxidation of hydrocarbons. Early experimental investigations in this field in the 1930s have proved that the yield of organic peroxide compounds depended on the nature of solid substances in the reaction medium. For instance, Pease and Munro[94] found that in the slow oxidation of propane, some materials, such as silica and Pyrex, were "favorable" for the accumulation of peroxides, while the alkali-washed and soda-glass surfaces reduced their quantities in comparable conditions. They explained the different rates of decomposition by the differences of the nature and surface area of the reaction vessels.[95] Later, in 1939, Harris[96, 97] suggested that "alkyl hydrogen peroxides (*alkyl hydroperoxides in modern terminology*) decomposes heterogeneously." It was observed that the major products of the heterogeneous decomposition of methyl hydroperoxide, in a packed reactor, were formaldehyde and water, while oxygen and methanol were minor products. Unlike heterogeneous decomposition, in homogeneous decomposition, the main products were formaldehyde and methanol, in equal quantities. The radical pathway in the mechanisms, in both cases, were evidenced by the so-called toluene carrier flow experiments (374°C) performed by Kirk and Knox.[98] In a series of experiments, they observed the formation of dibenzyl, in the presence of toluene in gaseous mixture. In analogous experiments on the decomposition of ethyl hydroperoxide, the formation of diphenyl was observed.[99] It is hardly possible to imagine, in both cases, a nonradical mechanism resulting in the formation of a product as dibenzyl. That is why, in these reactions, the radical pathway of heterogeneous decomposition was considered predominant, although the experimental approvals were not enough. In early investigations, some observations, approving the heterogeneous nature of the decomposition of peroxides, were made on oxidation of aldehydes, the intermediates of which were peroxyacids (some examples are presented in Section 3.5.4, devoted to the oxidation of aldehydes).

Thus, already, in the mid-1970s, there were certain preconditions for an important finding in the mechanism of the oxidation of organic compounds, at low temperatures, related to the role of heterogeneous reactions of hydrogen peroxide and organic peroxides in fluid-phase reactions. Systematic investigations in this area revealed a phenomenon characterized as heterogeneous radical decomposition of organic compounds on the surfaces of solid substances, accompanied by the generation and desorption of radicals from the surface of solid-phase substances to the fluid phases. This important conclusion was first experimentally proven by Nalbandyan and Vardanyan,[100] using as examples the decomposition of organic peroxide compounds: as methyl hydroperoxide decomposition on Pt, at 398 K,[101] as well as on Ag, Au, and other metals,[102] latter, also on the examples of the decomposition of peracids RCO$_3$H (where R is H; CH$_3$; C$_2$H$_5$) on the surfaces precoated with halogenids of alkali metals.[38 in Ch. 1, 100]

It is now clear that this class of reactions is more widespread than was formerly thought and includes the decomposition not only of hydrogen peroxide and a large number of organic peroxides, but also other classes of chemical compounds such as hydrocarbons (example: undecane[128 in Ch.1]), nitrogen-containing compounds (example: HNO_3,[103] amines),[68 in Ch.1] etc. Moreover, the action of the heterogenizied Fenton-like catalysts,[104–106] in different radical-initiated reactions of oxidation of the organic compounds, is based on the heterogeneous radical decomposition of H_2O_2 by the further desorption and diffusion of part of the radicals (OH and HO_2) to the bulk of the liquid phase. The radicals, appearing in the liquid phase, are able to oxidize organic substances in the reaction media. Immobilized on the solid support, the metal or metal-organic compounds (mainly of the elements Fe or Cu) are known as heterogenizied Fenton-like catalysts. In this regard, the heterogeneous Fenton reaction, applied for the oxidation of many substrates, is in fact a heterogeneous–homogeneous process, the homogeneous constituent of which is a radical reaction.

Initially, in the 1970–1980s, the decomposition of organic peroxides, including peroxyacides, were investigated on the surfaces of different solid substances, such as glasses or their surfaces, precoated by boric, nitric, and fluoric acids, alkali salts, as well as some metals and metal oxides.[38 in Ch. 1] The typical experiments on the heterogeneous decomposition of the vapors of organic peroxide compounds were carried out at relatively low temperatures, usually above the room temperature, but dozens, even hundreds of degrees below the temperature of their decomposition in homogeneous conditions, in the absence of solid substances.[38 in Ch. 1] The experiments were usually carried out on the diluted mixtures of peroxide with nitrogen or inert gases, containing small amounts of dioxygen. In nearly all cases, the decomposition was a first-order reaction with regard to the concentration of peroxide compounds. The characteristic activation barrier was not more than 20–30 kcal/mol, indirectly evidencing the heterogeneous nature of the decomposition of the peroxide compounds. The main evidences of the heterogeneous–homogeneous character of these reactions were obtained by detecting radicals in the gas phase, applying the matrix-isolation technique with the EPR method. In all the above-mentioned cases, the peroxy radicals (by a general formula RO_2, where R is H, CH_3, C_2H_5, ... or CH_3CO ...) were detected in the gas phase. Kinetic investigations of the overall reaction, as well as the behaviors of the detected radicals and analysis of the final products, clearly showed that the decomposition of the peroxide radical-like species, formed in the primary and secondary surface reactions, desorb from the surface and diffuse to the gaseous reaction mixture. In general form, the primary surface reactions involve the following acts

$$RC(=O)OOH \rightarrow RC(=O)O + OH$$
$$RC(=O)O \rightarrow R + CO_{2\,gas}$$
$$R + O_2 \rightarrow RO_2 \rightarrow RO_{2\,gas}$$

Subsequently, the heterogeneous character of decomposition of the peroxide compounds on the surfaces were proven by many other experimental facts. For instance, in the heterogeneous decomposition of methyl hydroproxide on Pt catalyst,[101] it was shown that at temperatures lower than the temperature of the decomposition in homogeneous conditions, even, in the conditions fully excluding the homogeneous proceeding of the reaction, the reaction might occur by the heterogeneous generation of radicals.[101, 107] In another case, the peroxy radicals formed on the surface in the decomposition of diacetyl on the SiO_2 were detected by the EPR method, not only in the gas phase (using the matrix-isolation method), but also directly on the surface.[108] In decomposition of tert-butyl hydroperoxide, the partial charge transfer from the peroxide to the NiO surface in the adsorption complex was also observed.[109] As a result of the further decomposition of the adsorption complex and the generation of radicals from the surface, in the same system, simultaneously the peroxy radicals were detected in the gas phase.

The chemical nature of the surface plays a determining role in these decomposition reactions. For example, for peroxide compounds of the type RC(O)O-OH (peroxyacides), in more or less comparable conditions, the activity of the surfaces in the generation of radicals for noble metals had the following sequence: Pt > Au > Ag, for alkali halides LiCl > NaCl > KCl > RbCl > CsCl and in decomposition of $(CH_3)_3CO-OH$ on metal oxides NiO > SiO_2 > ZnO.[38 in Ch. 1] Of special note, not every reaction of heterogeneous decomposition (catalyzed or inhibited) of the organic or inorganic compounds is analogous to the described phenomenon. A great number of decomposition reactions may be heterogeneous, but not radical processes. On the other hand, not every heterogeneous decomposition process, producing adsorbed radicals or radical-like species on the surface, corresponds to the above-described sequence of the reaction stages, as the desorption of species from the surface is not always possible. An example is the decomposition of peroxyacetic acid on Pt; in certain conditions, the decomposition occurs without the desorption of radicals to the gas phase.[107]

Related to these investigations, the following question arises: How significant is the role of the heterogeneous decomposition with the "escape" of radicals from the surface in overall oxidations? Already, as mentioned earlier, the appearance of radicals in the fluid phases is not yet evidence of their determining role in the homogeneous reaction, and it must be proven by kinetic and spectroscopic investigations in more detail. The key role and reactivity of radicals in the gas phase, generated by the heterogeneous decomposition of peroxides, in oxidation reactions of organic compounds are briefly presented in Sections 3.4–3.5.

These findings permit the elucidation of some interesting observations in the chemistry of oxidation reactions, which previously were unclear in the framework of the "purely homogeneous" or "purely heterogeneous and nonradical" schemes of the oxidation mechanisms suggestions. Let us give one fitting example. In 1950, investigating the gas-phase oxidation reaction of acetaldehyde, Emanuel[110] made an interesting observation. The reaction of the ignition of acetaldehyde, in oxidation with dioxygen, suddenly stopped by rapid freezing of the reaction mixture, was then continued by the rapid heating of the mixture. It appeared that the ignition temperature of the mixture became noticeably lower (at least by a few dozen degrees) than that for the initial reaction. The cause of this effect was not clear at that time. In a number of investigations, Nalbandyan and co-workers[111–112] showed that this observation may be explained by taking into account the fact that the organic peroxides, particularly peracetic acid, formed in the initial reaction, adsorbs on the surface of the vessel in a rapid freezing of the reacting mixture. In heating of the reaction vessel, the adsorbed peracetic acid undergoes heterogeneous decomposition by the formation of radicals, which are able "to escape" from the surface to the gas phase, initiating the ignition of the mixture at relatively low temperatures. The reaction was accompanied by the propagating blue flame in the reactor that may be detected and analyzed. Subsequently, this effect was investigated in more detail. The dependence of the ignition temperature on the nature of the surface and decomposed organic peroxide compound, as well as the length of homogeneous chains and other important parameters of the process, were determined in these investigations.[111-113]

A number of later investigations revealed that the heterogeneous-radical decomposition of peroxide compounds also was directly related to phenomena such as cool flames,[114] oscillations,[115] and certain critical phenomena,[93] in the oxidation of hydrocarbons and oxygenated compounds at relatively low temperatures.

REFERENCES

2.1

1. Volodin A. M., Malykhin S. E., Zhidomirov G. M. O⁻ radical anions on oxide catalysts: Formation, properties, and reactions. *Kinetika i Kataliz*, 2011, v. 52, 4, p. 615–629 (in Russian). Kinetics and Catalysis; 2011, v. 52, 4, p. 605–619 (in English).
2. Royter V. A., Golodets G. I. Introduction to Theory of Kinetics and Catalysis (in Russian). *Naukova Dumka*, 2nd ed., 1971, p. 100.
3. Paulson J. F. Some negative-ion reactions with CO_2. The Journal of Chemical Physics, 1970, v. 52, p. 963. DOI: http://dx.doi.org/10.1063/1.1673083
4. Aresta M. Angelini A. The carbon dioxide molecule and the effect of its interaction with electrophiles and nucleophiles, p. 1–38 (see also ref. 30 within). In: Xiao-Bing Lu (ed.). Topics in Organometallic Chemistry, 2016, 53. Springer, Heidelberg, 308 pages.
5. Chiesa M., Giamello E. Carbon dioxide activation by surface excess electrons: an EPR study of the CO_2^- radical-ion adsorbed on the surface of MgO. Chemistry. 2007, v. 13, 4, p. 1261–1267.
6. Grasselli R. K. Surface reaction mechanism of selective olefine oxidation over mixed metal oxides. In: Che M., Bond G. C. (eds) Adsorption and Catalysis on Oxide Surfaces, 1985, Elsevier Science Publishers, Amsterdam, 442 pages, p. 277.
7. Arora N., Deo G., Wachs I. E., Hirt A. M. Surface aspects of bismuth–metal oxide catalysts. J. Catalysis, 1996, v. 159, p. 1–13.
8. Burrington J. D., Kartisek C. T., Grasselli R. K. Infrared study of sulfided Co-Mo/Al_2O_3 catalysts: The nature of surface hydroxyl groups. J. Catal., 1980, v. 63, p. 235–238.
9. Grzybowska B., Haber J., Janus J. Interaction of allyl iodide with molybdate catalysts for the selective oxidation of hydrocarbons. J. Catal., 1977, v. 49, p. 150–163.
10. Zhao C., Wachs I. E. Selective oxidation of propylene to acrolein over supported V_2O_5/Nb_2O_5 catalysts: An *in situ* Raman, IR, TPSR and kinetic study. Catalysis Today, 2006, v. 118, p. 332–343.
11. Kobayashi H., Nakashiro K., Iwakura T. Density Functional Study of Ethylene Oxidation on Ag(111) Surface. Mechanism of Ethylene–Oxide Formation and Complete Oxidation with Influence of Subsurface Oxygen, Internet Electron. J. Mol. Design. 2002, v. 1, p. 620–635. http://www.biochempress.com.

12. Campbell C. T., Paffett M. T. Model studies of ethylene epoxidation catalyzed by the Ag(110) surface. Surf. Sci., 1984, v. 139, p. 396–416.

13. Campbell C. T. The selective epoxidation of ethylene catalyzed by Ag(111): A comparison with Ag(110). J. Catal., 1985, v. 94, p. 436–444.

14. Grant R. B., Lambert R. M. A single crystal study of the silver–catalyzed selective oxidation and total oxidation of ethylene. J. Catal., 1985, v. 92, p. 364–375.

15. Clarkson R. B., Cirillo A. C. Jr. The formation and reactivity of oxygen as O_2^- on a supported silver surface. J. Catal., 1974, v. 33, p. 392–401.

16. Cao Y., Wang Z., Deng J-F., Xu G. Q. Evidence for dangling bond-mediated dimerization of furan on the silicon Si(111)-(7 × 7) surface. Angew. Chem., 2000, v. 39, 15, p. 2740–2743.

17. Tao F., Bernasek S. L. Chemical bonding five-membered and six-membered aromatic molecules, 5.2 - Five-membered aromatic molecules containing one heteroatom, p. 89–104. In: Tao F., Bernasek S. L. (eds). Functionalization of Semiconductor Surfaces. Wiley, Hoboken, N.J., 2012, 443 pages, p. 91.

18. Minero C. Surface modified photocatalysts, p. 23–44. In: Bahnemann D. W., Robertson P. K. (eds). Environmental Photochemistry, Part 3, Springer, 2015, 345 pages, p. 40.

19. Fleming I. Molecular Orbitals and Organic Chemical Reactions. Chapter 7.1, Molecular Orbitals. 2009. University of Cambridge, UK. John Wiley, 378 pages.

20. Yaouanc A., de Réotier P. D. Muon Spin Rotation, Relaxation, and Resonance: Applications to Condensed Matter. Oxford University Press, 2011, 504 pages, p. 28.

21. Rhodes C. J. Spectroscopic characterization of molecules adsorbed at zeolite surfaces, Annu. Rep. Prog. Chem., Sect. C: Phys. Chem., 2010, v. 106, p. 36–76. DOI: 10.1039/B903505M

22. Heming M., Roduner E. Muon. Formation and dynamics on a SiO_2-adsorbed observed by muon spin rotation. Surface Science, 189/190, 1987, North-Holland, Amsterdam, p. 534.

23. Kazansky V. B., Pariisky G. B., Voevodsky V. V. Radiation-induced processes on the silica-gel surface. Discuss. Faraday Soc., 1961, v. 31, p. 203–208. DOI: 10.1039/DF9613100203, as well as in Russian: Pariskii G. B., Zhidomirov G. M., Kazanski V. B. *Zhurnal Structurnoy Khimii* (Journal of Structural Chemistry, in Russian), 1963, 4, p. 364; Kazanski V. B., Pariskii G. B. *Kinetika i Kataliz* (Kinetics and Catalysis, in Russian) 1961, 2, p. 507; Pariskii G. B., Kazanski V. B. *Kinetika i Kataliz* (Kinetics and Catalysis, in Russian), 1964, 5, p. 96.

24. Gesser H. D. ESR study of alkyl radicals adsorbed on porous Vycor glass. p. 168 In: Matsuura T., Anpo M. Photochemistry on Solid Surfaces. 1989, Elsevier, 624 pages.

25. Roberts M. W., Thomas J. M. (eds). Surface and Defect Properties of Solids, v. 2, Royal Chem. Soc., London. 1972. p. 83.

2.2

2.2.1

26. Rice F. O. The genesis of free radical chemistry, p. 3–6. In: Wall L. A. (ed) The Mechanisms of Pyrolysis, Oxidation, and Burning of Organic Materials, v. 13, 1972, National Bureau of Standards, publication 357, Washington, USA, 192 pages.

27. Paneth F., Hofeditz W. *Über die Darstellung von freiem Methyl. Berichte der deutschen chemischen Gesellschaft* (A and B Series), 1929, v. 62, 5, p. 1335–1347. DOI:10.1002/cber.19290620537

28. Kochi J. K. Organometallic Mechanisms and Catalysis. The Role of Reactive Intermediates in Organic Processes. 1978, Academic Press, New York, 623 pages.

29. Walborski H. M. Mechanism of Grignard reactive formation. The surface nature of reaction. Acc. Chem. Res. 1990, v. 23, p. 286–293.

30. Garst G. F. Grignard reagent formation and freely diffusing radical intermediates. Acc. Chem. Res., 1991, v. 24, 4, p. 95–97.

31. Péralez E., Négrel J-Cl., Goursot A., Chanon M. Mechanism of the Grignard reagent formation - Part 1, Theoretical investigations of the Mg_n and RMg_n participation in the mechanism. In: Main Group Metal Chemistry. 1999, v. 22, 3, p. 185–200. DOI: 10.1515/MGMC.10.1515/MGMC.1999.22.3.185

32. Xi M., Bent B. E. Adsorbed phenyl groups as traps for radical intermediates in reactions on copper surfaces. Langmuir, 1994, v. 10, 2, p. 505–509. DOI: 10.1021/la00014a027

33. Vijh A. K., Conway B. E. Electrode kinetic aspects of the Kolbe reaction, Chem. Rev., 1967, v. 67, 6, p. 623–664. DOI: 10.1021/cr60250a003

34. Kraeutler B., Jaeger C. D., Bard A. J. Direct observation of radical intermediates in the photo-Kolbe reaction-heterogeneous photocatalytic radical formation by electron spin resonance. Journal of the American Chemical Society, 1978, v. 100, 15, p. 4903–4905.

2.2.2

35. Linford M. R., Chidsey C. E. D. Alkyl monolayers covalently bonded to silicon surfaces, J. Am. Chem. Soc., 1993, v. 115, 26, p. 12631–12632.

36. Wayner D. D. M., Wolkow R. A. Organic modification of hydrogen terminated silicon surfaces. J. Chem. Soc., Perkin Trans. 2, 2002, p. 23–34. DOI: 10.1039/B100704L (Review Article).

37. Eves B. J., Lopinski G. P. Formation of organic monolayers on silicon via gas-phase photochemical reactions. Langmuir, 2006, v. 22, p. 3180–3185.

38. Nemanich E. J., Hurley P. T., Brushwig B. E., Lewis N. S. Chemical and electrical passivation of Silicon (111) surfaces through functionalization with sterically hindered alkyl groups. J. Phys. Chem. B, 2006, v. 110, p. 14800–14808.

39. Cicero R. L., Linford M. R., Chidsey C. E. D. Photo reactivity of unsaturated compounds with hydrogen-terminated Silicon(111). Langmuir, 2000, v. 16, 13, p. 5688–5695.

40. Maksymovych P., Sorescu D. C., Jordan K. D., Yates J. T. Collective reactivity of molecular chains self-assembled on a surface, Science, 12 December 2008, v. 322, p. 1664–1667. DOI: www.sciencemag.org.

41. Müllegger S., Rashidi M., Fattinger M., Koch R. Interactions and self-assembly of stable hydrocarbon radicals on a metal support. Nanomaterials and Interfaces. J. Phys. Chem. C, Nanomater. Interfaces. 2012, v. 116, 42, p. 22587–22594. DOI: 10.1021/jp3068409

42. Rijksen B., Caipa Campos M. A., Paulusse J. M. J., Zuilhof H. Silicon radical surface chemistry. Polymers and materials, 68, p. 2081. In: Chatgilialoglu C., Studer A. (eds). Encyclopedia of Radicals in Chemistry, Biology and Materials, 2012, Wiley, 2324 pages.

43. Chatgilialoglu C., Timokhin I. V. Silyl radicals in chemical synthesis, p. 117–182, p. 173. In: Hill A. F., Fink M. J. (ed). Advances in Organometallic Chemistry, v. 57, 2008, Elsevier, Amsterdam, 466 pages.

44. Van Santen R. A., Neurock M. Molecular Heterogeneous Catalysis: A Conceptual and Computational Approach. 2006, Wiley VCH Verlag, Weinheim, 488 pages, p. 189.

45. Leu T. M., Roduner E. Oxidation catalysis of unsaturated hydrocarbons with molecular oxygen via single-electron transfer in thermally treated H zeolites. Journal of Catalysis, 2004, v. 228, 2, p. 397–404.

2.3

46. Walling C. Free radical reactions. p. 1–21; Francisco J. S., Montgomery J. A. Jr. Theoretical Studies of the Energetics of Radicals, p. 110–149. In: Martinho Simoes J. A., Greenberg A., Liebman J. F. (eds). Energetics of Organic Free Radicals. 1996, Springer, Netherlands. DOI 10.1007/978-94-009-0099-8

47. Frenking, G. Shaik S. (eds). The Chemical Bond: Chemical Bonding Across the Periodic Table. 2014, Wiley, Weinheim, Germany, 566 pages.

48. Radzig V. A. Point defects on the silica surface. In; Trakhtenberg L., Lin S., Ilegbusi O. (eds). Physico-Chemical Phenomena in Thin Films and at Solid Surfaces (Series: Thin Films and Nanostructures), v. 34, Elsevier, Amsterdam, 2007, 800 pages, p. 308.

2.3.1

49. Berlin A. A., Novakov I. A., Khalturinsk N. A., Zaikov G. E. (eds). Chemical Physics of Pyrolysis, Combustion and Oxidation. Nova Science Publishers, New York, 2005, 722 pages, p. 13.

50. Odegova G. V., Vasil'eva N. A., Plyasova L. M., Kriger T. A., Zaikovskii V. I. Defective magnesium oxide: synthesis and studies of structure formation. *Kinetika i kataliz* (Kinetics and Catalysis, in Russian), 1997, v. 38, 6, p. 848–854.

51. Vasil'eva N. A., Buyanov R. A. Mechanism and principles of organization of catalytic radical heterogeneous-homogeneous processes *Kinetika i kataliz* (Kinetics and Catalysis, in Russian), 2011, v. 1, 4, p. 334–349; English version in: Review Journal of Chemistry, 2011, v. 1, 4, p. 344–358.

52. Raseev S. Thermal and Catalytic Processes in Petroleum Refining. 2003, Marcel Dekker, New York, 925 pages, p. 13; 26.

53. Vasileva N. A., Buyanov R. A. The mechanism of action of heterogeneous catalysts in the radical-chain processes of hydrocarbon pyrolysis, *Kinetika i kataliz* (Kinetics and Catalysis, in Russian), 1993, v. 34, 5, p. 748–755.

54. Vasileva N. A., Buyanov R. A. Role of the catalysis sphere in the chain-radical process of hydrocarbon pyrolysis. *Kinetika i kataliz* (Kinetics and Catalysis, in Russian), 1996, v. 37, 3, p. 409–411.

55. Vasileva N. A., Buyanov R. A. Catalytic pyrolysis of dichloroethane—calculation of the catalysis sphere. *Kinetika i kataliz* (Kinetics and Catalysis, in Russian), 1998, v. 39, 4, p. 584–587.

56. Tsyganova E. I., Shekunova V. M., Aleksandrov Yu. A., Filofeev S. V., Lelekov V. E. Effect of the VI Group metals on the catalytic pyrolysis of lower alkanes. Bulletin of the South Ural State University. Ser. Chemistry.2016, v. 8, 3, p. 19–27. DOI: 10.14529/chem160303

2.3.2

57. Bach R. D. General and theoretical aspects of the peroxide group (Chapter 1). In: Rappoport Z. (ed). The Chemistry of Peroxides, Volume 2, Parts 1 and 2. Chemistry of Functional Groups Series, Wiley, 2006, 1518 pages.

58. Liebman J. F., Greer A. (eds), Rappoport Z., Marek I. (series editors). The Chemistry of Peroxides, volume 3, part 1, 2014, Wiley, 1120 pages.

2.3.2.1

59. Homann K. H., Haas A. Kinetics of the Homogeneous Decomposition of Hydrides. p. 6. In: Bamford C. H., Compton R. G., Tipper C. F. H. (eds). Comprehensive Chemical Kinetics. Decomposition of Inorganic and Organometallic Compounds. V. 4. 1972, Elsevier Publishing Company, Amsterdam, 272 pages.

60. Satterfield C., Stein T. Decomposition of hydrogen peroxide vapor on relatively inert surfaces. Ind. Eng. Chem., 1957, 49, 7, p. 1173–1180. DOI: 10.1021/ie50571a042

61. Plaucka A., Stanglandb E. E., Dumesica J. A., Mavrikakisa M. Active sites and mechanisms for H_2O_2 decomposition over Pd catalysts. Proc Nat. Acad. Sci. USA. 2016, v. 113, 14, E1973-82. DOI: 10.1073/pnas.1602172113. Epub. 2016 Mar 22.

62. Lousada C. M., Johansson A. J., Brinck T., Jonsson M. Mechanism of H_2O_2 decomposition on transition metal oxide surfaces. J. Phys. Chem. C, 2012, v. 116, 17, p. 9533–9543.

63. Hiroki A, Laverne J. A. Decomposition of hydrogen peroxide at water-ceramic oxide interfaces. J. Phys. Chem. B, 2005, v. 109, 8, p. 3364–3370.

64. Lin S. S., Gurol M. D. Catalytic decomposition of hydrogen peroxide on iron oxide: Kinetics, mechanism, and implications. Environ. Sci. Technol. 1998, v. 32, 10, p. 1417–1423.

65. Rice F. O., Reiff O. M. The thermal decomposition of hydrogen peroxide. J. Phys. Chem., 1927, v. 31, 9, p. 1352–1356. DOI: 10.1021/j150279a006

66. Lionetti F., Mager M. Walter Spring, an early physical chemist. J. Chem. Educ., 1951, v. 28, 11, p. 604. DOI: 10.1021/ed028p604

67. Heinz H. J. Notes on enzymes and their importance (Chapter 11, B), p. 301. In: Shapton D. A., Shapton N. F. (eds). Principles and Practices for the Safe Processing of Foods, Butterworth Heinemann, Oxford, London, 1991, 456 pages.

68. Wiberg E., Wiberg N. Inorganic Chemistry. Academic Press, San Diego, 2001, 1795 pages, p. 499.

69. Hart A. B., McFadyen J., Ross R. A. Solid-oxide-catalyzed decomposition of hydrogen peroxide vapour. Trans. Faraday Soc., 1963, v. 59, p. 1458–1469. DOI: 10.1039/TF9635901458

70. Lousada C. M., Yang M., Nilsson K., Jonsson M. Catalytic decomposition of hydrogen peroxide on transition metal and lanthanide oxides Journal of Molecular Catalysis A: Chemical, 2013, v. 379, pages 178–184.

71. Murzin, D. Y., Salmi T. Catalytic Kinetics: Chemistry and Engineering. 2016 (second edition), and 2005 (first edition), Elsevier, Amsterdam, 740 pages, p. 449.

72. Weiss J. The free radical mechanism in the reactions of hydrogen peroxide. p. 333, In. Frankenbourg W. G. Komarewsky V. I., Rideal E. K. (eds). Advances in Catalysis. Volume 4, 1952, Academic Press. Pub. New York, 457 pages.

73. Bakhchadjyan R. H., Vardanyan I. A. Studies of the radical decomposition of peroxy compounds on the surface of the semiconductor structure by the method of EPR and electroconductivity. *Khimicheskaya Fizika* (Chemical Physics, in Russian), 1983, v. 11, p. 1536–1540.

74. Volodin A. M., Malykhin S. E., Zhidomirov G. M. O^- radical anion on oxide catalysys: Formation, properties, and reactions. *Kinetika i kataliz* (Kinetics and Catalysis, in Russian), 2011, v. 52, 4, p. 615–629.

75. Wang Z., Ma W., Chen C., Ji H., Zhao J. Probing paramagnetic species in titania-based heterogeneous photocatalysis by electron spin resonance (ESR) spectroscopy (A mini review). Chemical Engineering Journal, 2011, v. 170, p. 353–362.

76. Murphy D. M., Griffiths E. W., Rowlands C. C., Hancock F. E., Giamello E. EPR study of the H_2O_2 interaction with TiO_2; evidence for a novel $S = 1$ surface radical pair. Chem. Commun., 1997, v. 22, p. 2177–2178. DOI: 10.1039/A705856J

77. Shteinberg A. S. Fast Reactions in Energetic Materials: High-Temperature Decomposition of Rocket Propellants and Explosives. Springer-Verlag, Berlin, 2006, 200 pages, p. 183.

78. Grigoryan G. L., Nalbandyan A. B. Radical decomposition of hydrogen peroxide on the solid surfaces. *Docladi* of the Academy of Sciences of the USSR (in Russian), 1977, v. 235, 2. p. 381–383.

79. Sarkissyan E. G., Grigoryan G. L., Nalbandyan A. B. About the nature of radicals cooled from the gas phase in heterogeneous decomposition of hydrogen peroxide. *Docladi* of the Academy of Sciences of the USSR (in Russian), 1980. v. 253, 3. p. 648–650.

80. Vartikyan L. A., Grigoryan G. L., Nalbandyan A. B. Kinetic peculiarities of the formation of radicals HO_2 in the heterogeneous decomposition of H_2O_2. *Docladi* of the Academy of Sciences of the USSR (in Russian), 1980. v. 254, 4. p. 914–917.

81. Arutyunyan A. Z., Grigoryan G. L., Nalbandyan A. B. A study of the heterogeneous radical decomposition of vapors of hydrogen peroxide on the glass and silica by the method of EPR. *Kinetika i kataliz* (Kinetics and Catalysis, in Russian), 1985, v. 26, 4. p. 785–789.

82. Grigoryan G. L., Gukasyan P. S., Martiryan A. I., Beglaryan A. A., Grigoryan G. S. The nature of the intermediate compound formed in the $ZnO + H_2O_2$ reaction. 2007, *Zhurnal Fizicheskoi Khimii* (J. Phys. Chem., in Russian). 2007, v. 81, 8, p. 1379–1384.

83. Arutyunyan A. Z., Gazaryan K. G., Garibyan T. A., Grigoryan G. L., Nalbandyan A. B. Formation of the radicals on the surfaces of oxides in decomposition of H_2O_2. *Kinetika i kataliz* (Kinetics and Catalysis, in Russian), 1988, 29, 4, p. 880–884.

84. Giamello E., Fubini B., Volante M., Costa D. Surface Oxygen Radicals Originating via Redox Reactions during the Mechanical Activation of Crystalline SiO_2 in Hydrogen Peroxide. Colloids and Surfaces, 1990, v. 45, p. 155–165.

85. Amorelli A., Evans J. C., Rowlands C. C., Egerton T. A. An electron spin resonance study of rutile and anatase titanium dioxide polycrystalline powders treated with transition-metal ions. J. Chem. Soc. Faraday Trans. 1, 1987, v. 83, 12, p. 3541–3548.

86. Tachikawa T., Majima T. Single molecule fluorescence technic for the detection of reactive oxygen species. p. 651–659. In: Méndez-Vilas A., Diaz J. (eds). Modern Research and Educational Topics in Microscopy. 2007 Edition, Formatex, (Microscopy Series, volume 2 (3), Badajoz, Spain, 1033 pages (http://www.formatex.org).

87. Tachikawa T., Majima T. Single-Molecule, single-particle approaches for exploring the structure and kinetics of nanocatalysts. Langmuir, 2012, v. 28, 24, p. 8933–8943. DOI: 10.1021/la300177h

88. Tachikawa T., Majima T. Single molecule reactive oxygen species detection in photocatalytic reactions. Chapter 10, p. 421–435. In: S. Patai (ed.). The Chemistry of Peroxides, Volume 3. 1995, Wiley, p. 430.

89. Naito K., Tachikawa T., Fujitsuka M., Majima T. Real-time single-molecule imaging of the spatial and temporal distribution of reactive oxygen species with fluorescent probes: Applications to TiO_2 Photocatalysts. J. Phys. Chem. C, 2008, v. 112, 4, p. 1048–1059.

90. Nosaka Y., Yamashita Y., Fukuyama H. Application of chemiluminescent probe to monitoring superoxide radicals and hydrogen peroxide in TiO_2 photocatalysis. J. Phys. Chem. B, 1997, v. 101, 30, p. 5822–5827, DOI: 10.1021/jp970400h

91. The Nobel Prize in Chemistry 2014. Nobelprize.org., Nobel Media, AB, 2014. Web. 18 Jul. 2018. http://www.nobelprize.org/nobel_prizes/chemistry/laureates/2014.

2.3.2.2

92. Richardson W. H., O'Neal H. E. Decomposition and isomerization of organic compounds. In: Compton R. G., Bamford C. H., Tipper C. F. H. (eds). Comprehensive Chemical Kinetics. Volume 5, 1st Edition. 1971, Elsevier Science, 771 pages, p. 483.

93. Mello R. G., Nuez M. S. Reactions of peroxides on the surfaces. In: Liebman J. F., Greer A. (eds). The Chemistry of Peroxides. Volume 3, part 1, 2014, John Wiley and Sons; 1120 pages, p. 437.

94. Pease R. N., Munro W. P. The slow oxidation of propane. Journal of the American Chemical Society. 1933, v. 56, p. 2034–2038.

95. Harris E. J., Egerton A. Observation on the oxidation of propane. Chem. Rev., 1937, v. 21, 2, p. 287–297.

96. Harris E. J. The Decomposition of alkyl peroxides: dipropyl peroxide, ethyl hydrogen peroxide and propyl hydrogen peroxide. Royal Society Proceedings; 1939, p. 128.

97. Harris E. J. Note on the decomposition of hydrogen peroxide and its formation during the slow combustion of hydrocarbons Trans. Faraday Soc., 1948, v. 44, p. 764–766. DOI: 10.1039/TF9484400764

98. Kirk A. D., Knox J. H. The pyrolysis of alkyl hydroperoxides in the gas phase. Trans. Faraday Soc., 1960, v. 56, p. 1296–1303.

99. Cubbon R. C. P. The kinetics of thermal decomposition of peroxides. p. 29–112. In: G. Porter (ed). Progress in Reaction Kinetics, Volume 5, Pergamon Press, Oxford, 1970, 467 pages.

100. Nalbandyan A. B., Oganessyan E. A., Vardanyan I. A., Griffiths J. F. J. Radicals generated by the heterogeneous decomposition of peracetic acid studied by electron spin resonance spectroscopy. Faraday Trans. 1, 1975, v. 71, p. 1203–1210.

101. Bagdassaryan G. O., Vardanyan I. A., Nalbandyan A. B. Study of the radical decomposition of peracetic acid. *Docladi AN SSSR* (in Russian), 1975, v. 244, 2, p. 359–362.

102. Bagdassaryan G. O., Vardanyan I. A., Nalbandyan A. B. Catalytically radical decomposition of organic peroxides on platina. *Docladi AN SSSR* (in Russian), 1976, v. 231, 2, p. 362–365.

103. Ballod A. P., Titarchuk T. A., Ticker G. S., Rozovskii A. Ya. Heterogeneous-homogeneous decomposition of HNO_3. Reaction Kinetics and Catalysis Letters, 1989, v. 40, 1, p. 95–100.

104. Arshadi M., Abdolmaleki M. K., Khalafi-Nezhad A., Firouzabadi H., Gil A. Degradation of methyl orange by heterogeneous Fenton-like oxidation on a nano-organometallic compound in the presence of multi-walled carbon nanotubes. Chemical Engineering Research and Design, 2016, v. 112, p. 113–121.

105. Kwan W. P., Voelker B. M. Rates of hydroxyl radical generation and organic compound oxidation in mineral-catalyzed Fenton-like systems. Environ. Sci. Technol., 2003, v. 37, 6, p. 1150–1158.

106. Araujo F. V. F., Yokoyama L., Teixeira L. A. C., Campos J. C. Heterogeneous Fenton process using the mineral hematite for the discolouration of a reactive dye solution. Brazilian Journal of Chemical Engineering, 2011, v. 28, 4, p. 605–616.

107. Vardanyan I. A., Nalbandyan A. B. Advances in mechanism of the gas phase oxidation of aldehydes, *Uspekhi khimii* (Russian Chemical Review Journal), 1985, v. 54, 6, p. 903–921.

108. Bakhchadjian R. H., Gazaryan K. G., Vardanyan I. A. Formation and reactions of acetylperoxy radicals on the solid surface Chemical Physics, 1991, v. 10, 5, p. 659–663.

109. Bakhchadjyan R. H., Vardanyan I. A. Studies of the radical decomposition of peroxy compounds on the surface of the semiconductor structure by the methods of EPR and electroconductivity, *Khimicheskaya Fizika* (Chemical Physics, in Russian), 1983, v. 11, p. 1536–1540.

110. Emanuel N. M. Kinetics of aldehydes oxidation in the gas phase. Kinetics of oxidation chain reactions collection of works Moscow, Leningrad, 1950, p. 185–232.

111. Nalbadyan A. B., Vardanyan I. A., Arustamyan A. M., Dorunts A. G. Low-temperature ignition of acetaldehyde oxygen mixtures initiated by organic peroxides adsorbed on a reaction vessel surface. p. 58–63, In: Kuhl A. L., Bowen J. R., Borisov A., Leyer J.-C. (eds). Dynamics of Reactive Systems Part I: Flames; Part II: Heterogeneous Combustion and Applications, Progress in Astronautics and Aeronautics, 1988, 113, Washington, 432 pages. http://dx.DOI.org/10.2514/5.9781600865879.0058.0063

112. Doruntz A. G., Arustamyan A. M., Nalbandyan A. B. Low temperature ignition of acetaldehyde/oxygen mixtures initiated by organic peroxides adsorbed on the surface of a reaction vessel. Combustion and Flame, 1987, v. 69, p. 251–255. DOI: 10.1016/0010-2180(87)90118-0

113. Vardanyan I. A. On the mechanism of radical initiation of acetaldehyde-oxygen gas mixtures ignition. Reports of the National Academy of Sciences of Armenia, 2017, v. 117, 2, p. 162–165.

114. Sargsyan G. A., Vardanyan I. A. Modeling of cool flame combustion of gas mixtures in flow conditions. The Millennium 9th Intern. Symp. on Flow Visualisation, Edinburgh, Proceedings, 2000, p. 266.1–266.10.

115. Sargsyan, G. N., Yessayan R. S., Vardanyan I. A. Analysis of heterogeneous–homogeneous model, describing oscillations in system $CH_3CHO + O_2$. Applied Catalysis A: General, 2000, v. 203, 2, p. 285–291.

3 Bimodal Reaction Sequences in Oxidation of Hydrogen and Organic Compounds with Dioxygen

This chapter covers some general problems of the oxidation process and focuses on the bimodal (heterogeneous–homogeneous or homogeneous–heterogeneous) reaction sequences in the interaction of hydrogen and organic compounds of some major classes, such as hydrocarbons, alcohols, and aldehydes with dioxygen in the presence of different solid substances. Much attention is given, on the one hand, to the discussion of the reactivity of active oxygen species (radicals, ion–radical, radical-like species, their adsorbed forms) formed from dioxygen and, on the other hand, to the other intermediates formed from substrates in the fluid phases and on the surfaces of solid substances. In addition, this chapter discusses the peculiarities of the mechanism and synergistic effects in bimodal oxidation by coupling homogeneous and heterogeneous constituents in a unified process.

3.1 OXIDATION WITH DIOXYGEN

Dioxygen is cheaper than any other oxidant agent, for it exists in nature in mega-ton amounts. Obviously, use of dioxygen as a raw material is advantageous. However, the engineering and design of oxidation processes must correspond to several other requirements. The process must be environmentally benign, having minimal consumption of energy, without side products. Both in the homogeneous and heterogeneous oxidation processes, the catalyst used must be cheap, stable, durable, easily separable from products, and its regeneration must be realizable by easy procedures.

The difficulties in the gas-phase oxidation processes of dioxygen relate to the explosivity of certain gaseous mixtures. The material and construction of reactors must be specially chosen in this case. If the oxidation reaction occurs in the liquid phase, the solubility of dioxygen in the given solvent must be sufficient (often, it is available only at high pressures), and the boiling point of the solvent must be appropriate to the thermal regime of reaction. In laboratory synthesis, as well as in chemical industries, in most cases, the separation of final products from catalysts is not a simple problem in homogeneous oxidation reactions. These problems sometimes require replacement of dioxygen with other oxidants, mainly with hydrogen peroxide, which after dioxygen is the second most available oxidant agent.

Oxidation is usually considered as half the reaction of the redox (reduction–oxidation) process. Obviously, the second half is reduction reaction. Based on the redox concept, IUPAC's Gold Book gives the following basic definition of the oxidation reaction:[8 in Ch. 1]

1. The complete, net removal of one or more electrons from a molecular entity (also called deelectronation)
2. An increase in the oxidation number of any atom within any substrate
3. Gain of oxygen and/or loss of hydrogen of an organic substrate.

Accordingly, a reduction is the gain of electrons from another chemical entity and a decrease of its oxidation number.

The application of these definitions, especially in organic chemistry, sometimes meets with difficulties. According to Breslow,[1] "oxidation is not a well-defined concept in organic chemistry." What is oxidation in organic chemistry? The majority of the reactions of organic substances occur without net or complete transfer of the electron(s) from one "molecular entity" to the other. Therefore, this definition has no physical meaning in the case of the formation of a covalent chemical bonds. For example, the formation of H_2O from H_2 and O_2 is considered a redox reaction. According to the concept of redox reactions, hydrogen is oxidized and dioxygen is reduced. In reality, there are neither electron transfer, nor ions in the formation of water. It is well known that in the gas phase, this reaction occurs by the branched-chain-radical mechanism and the intermediates are not ions but free radicals. Thus, when one considers the concept of redox reactions, assigning the charge to atoms, taking into account the electronegativities of atoms, it is a formality. Moreover, not every reaction of the insertion of the oxygen atom in organic compounds is oxidation. In other words, not every reaction yielding a product containing oxygen atoms (or any other more electronegative atom than carbon) is oxidation. For example, the reaction of ethylene with water forming ethyl alcohol is not an oxidation reaction, while the dehydrogenation of ethane to ethylene is oxidation:

$$CH_2=CH_2 + H_2O \rightarrow CH_3\text{-}CH_2\text{-}OH \quad \text{no oxidation/reduction}$$

formal oxidation number of carbon -2 -2 $\qquad$ -3 -1 $\qquad$ $\Sigma = 0$

$$CH_3\text{-}CH_3 \rightarrow CH_2=CH_2 + H_2 \qquad \text{oxidation}$$

formal oxidation number of carbon -3 -3 -2 -2 $\qquad$ $\Sigma = -2$

In the first reaction, the organic compound neither gains nor loses electrons in overall reactions (here, a rearrangement of the electrons within the molecule takes place). Therefore, there is no oxidation (or reduction). In contrast, the second reaction is a true oxidation of the organic compound. Analogously, the hydrogenation of ethylene with hydrogen to ethane is a reduction reaction.

Often, in organic chemistry, especially, in organic synthesis, the formation of oxygenated products is a multistep process. For example, methanol from methane may be obtained through the chlorination reaction followed by hydrolysis.

$$CH_4 + Cl_2 \rightarrow CH_3Cl + HCl$$
$$CH_3Cl + H_2O \rightarrow CH_3OH + HCl$$

Here, only the first step is an oxidation–reduction reaction, but the second step, the formation of the oxygenated compound methanol is a result of the hydrolysis of an already oxidized organic compound.[1] A more accurate presentation of the oxidation reaction requires the precision of the oxidation step.

Let us examine the case of the oxidation of a chemical bond. Using the terminology of the redox concept, it may be considered that the oxidation of the C-H bond takes place when hydrogen is replaced by a more electronegative atom, formally able to show that the carbon atom becomes more "positively charged." Analogously, oxidation for the C-C bond is the formation of a new chemical bond between carbon and a more electronegative atom. In CH_3CH_2OH, the carbon atom, combined with OH, has a higher oxidized state relative to the same carbon atoms in ethane; hence, we say that it is oxidized. The point is that the oxidation of organic compounds we observe is relative to the initial state of the carbon atom.

Here is another example: the direct insertion of electronegative atom in the bond C-C (in a carbonic chain) is oxidation:

$$2CH_2 = CH_2 + O_2 \rightarrow 2\,(H_2C \overset{O}{\underset{}{\diagup \diagdown}} CH_2)$$

Breslow[1] exemplified the difficulties in knowledge of oxidation within the framework of the redox conception by the following situation related to the formation of a new C-C bond in C-H bond oxidation. According to the redox concept, in organic syntheses, the insertion of the carbonyl group into the C-H bond of the secondary carbon is an oxidation, giving C-[C(O)]-H. The latter may undergo dehydration (which is not oxidation), giving an unsaturated carbon (toward the original carbon). When carbonyl insertion takes place in the C-H bond of tertiary carbon, the dehydration becomes impossible. Indeed, it is "hard to imagine" that the formation of unsaturated carbon by preliminary insertion of the carbonyl group in the case of a secondary carbon is oxidation, while in the case of a tertiary carbon its formation via carbonyl insertion, as an oxidation, and so further dehydration may not take place at all.

In spite of the gaps in knowledge regarding reactions, the redox concept, based on the electronegativity of atoms, remains the only way for diversification of oxidation–reduction reactions to be distinguished from other reactions, such as acid–base and hydrolysis reactions.

3.2   ELECTRONIC STRUCTURE OF DIOXYGEN

Understanding of some of the physical and chemical properties of molecular oxygen and oxygenated radicals became possible only after development of the quantum-mechanical theory in the 1920s–1930s.[2] American physicist and chemist Robert Mulliken (1896–1986) applied the molecular orbital (MO) method to determine the electronic structure and properties of small molecules, including dioxygen.[3, 4] Mulliken was the first to explain the paramagnetic properties of gaseous dioxygen in the ground-state electronic configuration on the basis of his own theoretical perceptions. In 1966, he received the Nobel Prize in Chemistry for his "fundamental work concerning chemical bonds and the electronic structure of molecules."

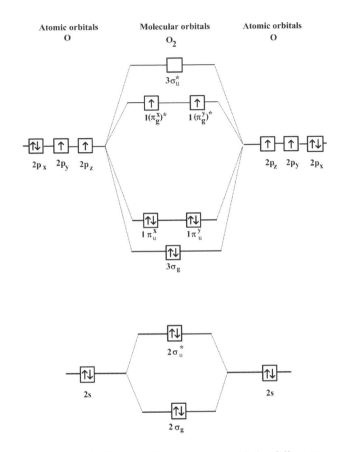

FIGURE 3.1 Energetic diagram of the molecular orbitals of dioxygen.

This discovery was a triumph not only for the molecular orbital method, but also for the quantum-mechanical theory. For homonuclear diatomic molecules, Mulliken gave the following energetic sequence of molecular orbitals (σ, π …) in dioxygen:

$$1\sigma_g < 1\sigma_u < 2\sigma_g < 2\sigma_u < 3\sigma_g < 1\pi_u < 1\pi_g < 3\sigma_u$$

Electronic configuration diagram of molecular orbitals of dioxygen, in the ground state, is presented in Figure 3.1 (The asterisk (*) in the figure indicates the antibonding molecular orbital; index g (*gerade*) is applied if the molecule has a center of symmetry, indicating the inversion of the wave function through the center of symmetry; and u (*ungerade*) indicates that the wave function is antisymmetric for inversion through the center).

The molecule of dioxygen (O_2) has the following electronic configuration

$$(1\sigma_g)^2(1\sigma_u^*)^2(2\sigma_g)^2(2\sigma_u^*)^2(3\sigma_g)^2(1\pi_u)^4(1\pi_g^*)^2$$

Due to two unpaired electrons in antibonding $1\pi_g^*$ molecular orbitals, dioxygen has paramagnetic properties (two π molecular orbitals may also be differentiated as π^+ and π^-).

According to the theory, this configuration corresponds to the molecular term[5] of dioxygen in its ground state:

$$^3\Sigma_g^- \text{ triplet (ground state)}$$

Molecular term symbols characterize the quantum state of molecules.[2, 5, 6] Electrons in atoms, as well as in molecules, have their own angular and spin momenta along the internuclear axis (z), and their couple determines the molecular term:

$$^{2S+1}\Lambda_{g/u}^{(+/-)}$$

For molecules, the values of the total angular momentum projections on the intermolecular axis are known as molecular terms (similar to atomic terms), denoted by the following Greek letters:

$$\Lambda = 0, \ 1, \ 2 \ldots$$
$$\Sigma, \ \Pi, \ \Delta \ldots$$

S is the total spin quantum number; g/u (gerate/ungerate) indicates a parity; and g-symmetric and u-antisymmetric represent inversion through a center. The notation + or – (right singe, applicable only for Σ states) indicates the symmetry in reflection of the wave function through a plane, containing an internuclear axis (+ for symmetric and – for antisymmetric reflection).

For O_2, six electrons may be placed in π_u and π_g molecular orbitals in different ways, represented in Figure 3.2. Taking into consideration two Hund rules,[3, 6] we see that the molecular term corresponding to the energetically lowest electronic state of dioxygen is the triplet state $^3\Sigma_g^-$.

Two low-lying excited states of molecular oxygen have the following terms:

$$^1\Delta_g \text{ singlet (first excited state)}$$

$$^1\Sigma_g^+ \text{ singlet (second excited state)}$$

The existence of the ground and low-lying excited states of dioxygen were predicted by Mulliken's calculations in 1928.[2, 3] All electronic transitions between the ground and the excited energetic states of molecular oxygen have been confirmed by spectroscopic data.

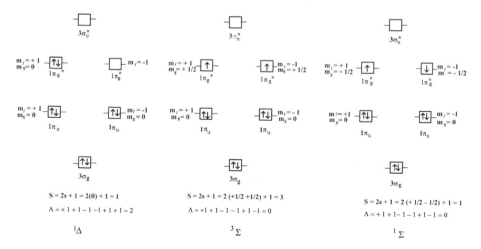

FIGURE 3.2 Three different ways of arranging six electrons in π molecular orbitals of dioxygen and corresponding molecular terms. m_l is magnetic quantum number ($m_l = -l, \ldots, 0, \ldots, +l$), l is the angular quantum number of the atomic orbitals of oxygen, and m_s is the spin quantum number ($m_s = +\frac{1}{2}$ or $-\frac{1}{2}$).

Above the ground state of triplet oxygen, the low-lying excited states have energies of 95 and 158 kJ/mol, respectively. The electronic transition

$$^3\Sigma_g^-(0) \rightleftarrows\, ^1\Sigma_g^+(0)$$

where (0) indicates the vibrational level, corresponding to the 762 nm wavelength, was known long before the discovery of singlet oxygen as a dark red Fraunhofer line.[7] In 1814, it was observed by J. Fraunhofer (1787–1826) in the solar spectrum (the earliest mention was made by W. H. Wollaston), and it was later assigned to atmospheric oxygen. Unfortunately, the electronic transition between the ground state and the first low-lying excited energetic state

$$^3\Sigma_g^-(0) \rightleftarrows\, ^1\Sigma_g^+(0)$$

is forbidden by selection rules known in quantum mechanics. However, this weak band 1268 nm experimentally was observed in Earth's atmosphere and liquid oxygen.[5, 8, 9] Finally, the band 1910 nm, corresponding to

$$^1\Delta_g(0) \rightleftarrows\, ^1\Sigma_g^+(0)$$

was also observed experimentally as a weak fluorescent emission of singlet-triplet transition.[10]

A diagram of the energetic levels of O_2 is presented in Figure 3.3, according to the most accurate data of Tachikawa and Majima.[11]

The existence of molecules in a triplet ground state, like dioxygen, is a rare phenomenon in nature. The majority of organic and inorganic compounds in the ground state are in the singlet electronic state. This specificity of the electronic structure of molecular oxygen is exhibited in its reactivity toward different substances.

As it will be shown in the next sections, the electronically excited dioxygen molecules $^1\Delta_g$, exhibit chemical properties different from those for dioxygen in the ground state. The following question naturally arises: "Is dioxygen in the excited energetic state the same individual chemical species in comparison with dioxygen in the ground energetic state?" Many investigations have reported that the length of chemical bond in homonuclear molecules increases in electronically excited states $^1\Delta_g$ and Σ_g^+.[11–14] For molecular oxygen in the ground state $^3\Sigma_g^-$, the bond length is

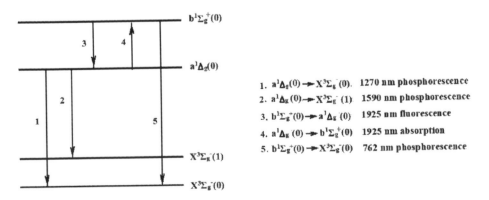

1. $a^1\Delta_g(0) \to X^3\Sigma_g^-(0)$.   1270 nm phosphorescence
2. $a^1\Delta_g(0) \to X^3\Sigma_g^-(1)$ 1590 nm phosphorescence
3. $b^1\Sigma_g^+(0) \to a^1\Delta_g(0)$ 1925 nm fluorescence
4. $a^1\Delta_g(0) \to b^1\Sigma_g^+(0)$ 1925 nm absorption
5. $b^1\Sigma_g^+(0) \to X^3\Sigma_g^-(0)$ 762 nm phosphorescence

FIGURE 3.3 Diagram of some transitions between the energetic levels in dioxygen and their spectroscopic notations, adapted from the work of Tachikawa and Majima T. (Ref.[11]).

1.207 Å, while for $^1\Delta_g$ it is 1.216 Å and 1.228 Å for the $^1\Sigma_g^+$ state. In agreement with this fact, in the excited state, the bond energy is lower than in the ground state, being 402 kJ/mol and 498 kJ/mol, respectively.[13]

Geometric changes in the chemical structures of a number of molecules in electronically excited states, in comparison with those in the ground state, were known beginning from the classical spectroscopic investigations of G. Herzberg.[14] For example, the molecule HCN that is linear in the ground state becomes non-linear in the electronically excited state. For three different excited states of this molecule, the determined angles HCN are 125°, 114.5° and 141°.[15] Another example is formaldehyde, which in the ground state is planar but in the excited state is pyramidal.

These facts indicate that the molecules in electronically excited and ground states have different electronic configurations, structural properties, and geometric forms, and therefore different chemical reactivities. These differences are reminiscent of the well-known phenomenon of the existence of isomeric compounds in chemistry. From this point of view, the molecules in the different electronic states are electromers or electroisomers.

The problems of the reactivity of singlet oxygen in the fluid phases were a subject of a long and intensive investigation. Relatively few investigations are known in the field of reactions of singlet oxygen on the surfaces of solid substances. Some heterogeneous reactions of the singlet oxygen and their role in heterogeneous catalysis will be briefly discussed in Section 3.4.2.7.

3.3 SYMMETRY CONSIDERATIONS. APPLICATION OF THE WIGNER–WITMER RULES IN OXIDATION WITH DIOXYGEN

Expression of the symmetry in natural phenomena has a universal character.[16, 17] A number of great discoveries in the natural sciences in the twentieth century were based on or linked to considerations of symmetry. The German mathematician Emmy Noether (1882–1935) showed that symmetry considerations led directly to the different conservation laws in nature.[17] Symmetry considerations were first applied to chemical reactions by E. P. Wigner and E. E. Witmer, in their work published in 1928.[18]

According to the IUPAC Gold Book, Wigner's conservation rules state that "for both radiative and radiationless transitions, the principle that transitions between terms of the same multiplicity are spin-allowed, while transitions between terms of different multiplicity are spin-forbidden."[8 in Ch. 1] As a chemical reaction is a rearrangement of electrons and nucleus, it includes the changes in position of the nucleus and in electronic structure. The symmetry laws govern all these changes in the micro world. The Wigner–Witmer rules (also known as correlations rules), essentially based on symmetry considerations, state that the orbital angular momentum and spine are conserved in the reactions of diatomic molecules. Application of the so-called correlation diagrams permits prediction of the feasibility of certain chemical reactions. Like spectroscopic transitions, the electronic transitions in the chemical reactions may be "allowed" or "forbidden."

As an example of the application of conservation rules, let us consider the oxidation reactions of organic compounds with molecular oxygen in ordinary conditions. From the viewpoint of chemical thermodynamics, the reactions between dioxygen and most organic compounds, in the majority of cases, are "allowed" because, in general, they are exergonic reactions. However, as a rule, organic matter is relatively stable toward dioxygen in ordinary conditions, in the absence of catalysts. The majority of the organic compounds in ordinary conditions, at room temperature, are singlets ($\uparrow\downarrow$) in their ground states, while dioxygen in the ground energetic state, in the same conditions, is triplet ($\uparrow\uparrow$). Chemical reaction between oxygen and organic substance, giving a new organic compound, also in singlet state, is spin forbidden by the Wigner–Witmer rules, as it violates the spin conservation rule. The majority of the oxidation reactions of organic compounds with dioxygen have a high energetic barrier and may not occur without heating, application of a catalyst, external irradiation, or other means of initiation. This is one of the most important reasons, explaining the relative stability of life on Earth, in an atmosphere of molecular oxygen.

Application of the Wigner conservation rules in chemical reactions has certain limitations.[19] The rules are applicable only for concerted reactions, occurring by one elementary step. Note that nature often circumvents the restrictions imposed by these rules, choosing energetically more accessible pathway of the oxidation reaction—for instance, a stepwise reaction pathway, via the formation of intermediates. On the other hand, symmetry considerations are not the only factor determining the feasibility of the chemical reaction; other factors, for example, steric (geometric) considerations, also have an important role in the theoretical prediction of chemical transformations. However, the historical significance of the first application of symmetry considerations in chemical reactions was great. In 1963, Wigner was awarded the Nobel Prize in Physics "for his contributions to the theory of the atomic nucleus and the elementary particles, particularly through the discovery and application of fundamental symmetry." Subsequently, Wigner's discovery became the theoretical basis for R. Woodward, R. Hoffmann, K. Fukui, and others in their investigations of the role of symmetry considerations in theoretical analyses of chemical reactions. Woodward and Hoffmann formulated more general rules on symmetry considerations in chemical reactions.[20] In 1965, Woodward received the Nobel Prize in Chemistry, but in another area of chemistry—"for his works in organic synthesis." He synthesized many complex natural products such as quinine, cholesterol, chlorophyll, and vitamin B_{12}. During his investigations of the synthesis of vitamin B_{12}, in collaboration with Hoffmann, Woodward discovered the principle of conservation of orbital symmetry in so-called pericyclic reactions. However, he did not receive a second Nobel Prize for this discovery. The Nobel Prize in Chemistry for development of the theory of symmetry applied in chemical reactions was instead attributed jointly to K. Fukui and R. Hoffmann, in 1981, two years after Woodward's death.[5]

The Woodward–Hoffmann rules of orbital symmetry state that "in the course of concerted reaction, the molecular orbitals of reactant molecules are transformed into the molecular orbitals of the product by a continuous pathway."[7] According to other formulations, given by the IUPAC definition: "Conservation of orbital symmetry

requires the transformation of the molecular orbitals of reactants into those of products to proceed continuously by following a reaction path along which the symmetry
of these orbitals remains unchanged."[8 in Ch. 1]

Fukui developed another approach to symmetry considerations in the chemical reaction, building his conclusions on the interactions of the frontier molecular
orbitals of the initial reactants, in terms of the higher occupied molecular orbitals
(HOMOs) and lower unoccupied molecular orbitals (LUMOs). The main conclusion,
however, did not differ from that in Woodward and Hoffmann's theory. In a number
of chemical reactions, when a new product is formed by the participation of the frontier MOs of the initial reactants, the orbital symmetry is conserved.[21]

As a classical example of the application of the spin and orbital symmetry conservation rules in oxidation reactions, let us briefly observe the following oxidation
reaction

$$N_2 + O_2 \rightarrow 2NO \qquad \Delta H = 180 \text{ kJ/mol}$$

This reaction is not thermodynamically favorable, as it is an endothermic process.
Moreover, it is forbidden from the viewpoint of the symmetry considerations. From
the simple comparison of HOMO and LUMO of oxygen and nitrogen molecules
(Figures 3.1 and 3.4a), it is obvious that due to symmetry restrictions, the electrons
of $1\pi_u$ orbitals of N_2 cannot flow to the $1\pi_g^*$ orbital of oxygen. Hypothetical electronic
flow is possible only from the oxygen molecule to N_2, as N_2 has $1\pi_g$ empty orbitals (Figures 3.4b and 3.5). However, it is impossible in reality, as oxygen is a more
electronegative element than nitrogen.[22] As a result of these factors, the energetic
barrier of this reaction will be very high. It is about 390 kJ/mol. The reverse reaction,

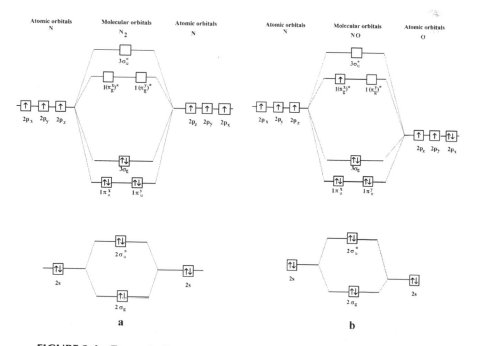

FIGURE 3.4 Energetic diagrams of the molecular orbitals: (a) N_2 and (b) NO.

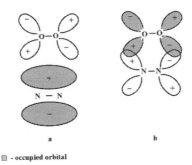

☐ - occupied orbital

FIGURE 3.5 Molecular orbitals in the reaction $N_2 + O_2$: (a) absence of overlap of the frontier MOs of $N_2 + O_2$; (b) hypothetical overlap of the frontier MO by flow of electrons from N_2 to O_2 that is chemically unacceptable.[22, 23]

decomposition of NO into N_2 and O_2, is an exothermic process, but, surprisingly, it has a high energetic barrier (210 kJ/mol). Therefore, it is a very slow process in ordinary conditions. From the viewpoint of symmetry considerations, it is also a forbidden reaction, as the same arguments mentioned above for forward reaction also apply for reverse reaction.

Although the oxidation of NO by molecular oxygen

$$2NO + O_2 \rightarrow 2NO_2$$

kinetically shows the third-order reaction behaviors, experimental data indicate that it may be described as a process of successive bimolecular reactions.[22, 24] The suggested mechanisms of this reaction include the following reactions

$$NO + NO \rightleftarrows N_2O_2 \tag{1}$$

$$N_2O_2 + O_2 \rightarrow 2NO_2 \tag{2}$$

The positive overlap of MO of N_2O_2 and oxygen is shown in Figure 3.6a. The reaction is "allowed" from the point of view of symmetry considerations, as the electronic flux takes place from molecular orbitals of N_2O_2 to the half-filled $1\pi_g^*$ orbital of oxygen. This results in breaking the O-O bond and forming two new N-O bonds. The product is N_2O_4, but not NO_2 formation by the reaction (2), which is spin-forbidden.

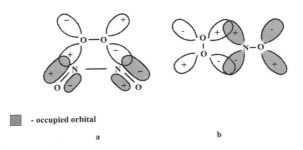

☐ - occupied orbital

a   b

FIGURE 3.6 Frontier molecular orbitals: (a) in the reaction $N_2O_2 + O_2$ producing N_2O_4, (b) in the reaction $NO + O_2$ producing NO_3.[22]

Another version of the mechanism suggestions includes the reactions

$$NO + O_2 \rightleftarrows 2NO_3 \tag{3}$$

$$NO_3 + NO \rightarrow 2NO_2 \tag{4}$$

Both reactions (3) (Figure 3.6b) and (4) are "allowed" by spin and orbital symmetry conservation rules. Here, according to the experimental kinetic data, the faster reaction (3) is in equilibrium with the slow reaction (4). In addition, note that the two above mechanisms for NO oxidation by dioxygen are nearly equally accepted in the scientific literature.[24]

3.4 REACTIVITY OF DIOXYGEN AND ACTIVE INTERMEDIATE SPECIES OF OXYGEN IN HETEROGENEOUS–HOMOGENEOUS OXIDATION

3.4.1 REACTIVITY OF DIOXYGEN IN OXIDATION

3.4.1.1 Reactivity of Dioxygen

The limited reactivity of molecular oxygen in ordinary conditions related to its electronic structure was shown briefly in the above sections. There are various ways of activating molecular oxygen (thermal, photochemical, catalytic, mechanical, enzymatic). One of them is the generation of electronically excited molecules, as singlet oxygen, by different physical and chemical methods. Active oxygen species,[25] other than singlet oxygen, such as radicals, ions, and ion-radicals, radical-like species on the surfaces of solid surfaces, may be generated in homogeneous noncatalytic and catalytic systems, as well as heterogeneous catalytic systems. Principally, another way is biological or enzymatic activation of dioxygen, which is not the subject of the present work.

One of the first interpretations of the overall reactivity of molecular oxygen, based on the quantum-mechanical perceptions of its electronic structure, was given by H. Taube, the Nobel Prize winner in Chemistry in 1983 "for his work on the mechanisms of electron transfer reactions, especially in metal complexes." In his paper[26] published in 1965, Taube showed why the restrictions imposed by spin conservation rules might be circumvented in the reactions of molecular oxygen with some transition metal ions. These principal elucidations were useful for understanding dioxygen activation in homogeneous catalysis by transition metals. Vast experience is accumulated in the field of oxygen activation in homogeneous catalytic systems by transition metal ions, during the last decades.[27]

One of the chemical ways of activating dioxygen is heterogeneous catalysis. Two main classes of heterogeneous systems will be discussed in this context: surfaces of metals and metal oxides. There are essential differences in the adsorption of molecular oxygen on the surfaces for these two classes of solid substances.

3.4.1.2 Activation of Dioxygen on the Surfaces of Noble Metals

On the majority of noble metal surfaces, dioxygen is adsorbed molecularly, at low temperatures below 100–120 K, and dissociative, at high temperatures. The adsorption characteristics of dioxygen on the metals depend on the nature and crystal face

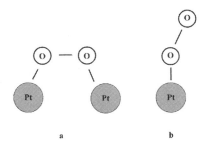

FIGURE 3.7   Peroxo-like (a) and superoxo-like (b) molecular oxygen adsorbed on
Pt(111).

of adsorbents, as well as the dispersion degree of the catalyst, in the case of supported metal catalysis and a number of other factors.

One of the most investigated systems in heterogeneous catalysis is dioxygen
adsorption on Pt(111).[28–30] Investigations by different methods of surface science
involving quantum-mechanical calculations indicate that at low temperatures (below
100 K) molecular oxygen on the surface of Pt(111) of the "clean" monocrystalline
samples is adsorbed nondissociatively and may exist in three forms. At very low
temperatures, up to about 30–40 K, dioxygen predominantly adsorbs physically.
With the rise of temperature (approximately in the range $30 < T < 100$ K), two other
forms of adsorbed O_2 appear on this surface. One of them is the so-called peroxo-
like form O_2^{-2} (Figure 3.7a), occupying threefold sites on top-hollow(fcc)-bridge and
top-hollow(hcp)-bridge configurations (fcc: *face-centered cubic*; hcp: *hexagonal
close-packed*) on the Pt(111) surface (Figure 3.8).

The second structure is a superoxo-like form of the adsorbed O_2^- (Figure 3.7b).
Evidently, it is a paramagnetic species, occupying twofold sites in top-bridge-top
configuration, the O–O bond being perpendicular to the surface of Pt(111) and
slightly canted (Figures 3.7b and 3.8).

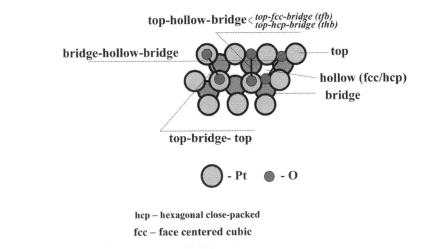

FIGURE 3.8   Top view of the adsorption of oxygen on Pt(111).

The adsorbed O-atom (in a threefold position) appears in nonadjacent sites separated by two lattice constants. O may migrate to a more stable fcc site.

In adsorption of O_2 on Pt(111) (below 120 K), practically no energetic barrier exists. Heat of adsorption is in strong dependence on coverage of the surface. In temperature-dependent desorption experiments (TDD), below 150 K, the main desorption product in heating is molecular oxygen, while at 150–500 K it is atomic oxygen, indicating the activated dissociation of oxygen from the precursor, adsorbed molecular oxygen, forming a p(2 × 2) overlayer on the surface.[29, 30] Thus, the main surface reaction is

$$(O_2)_{ads} \rightarrow 2(O)_{ads}$$

Maximum coverage of Pt(111) with atomic oxygen is 0.25 ML (monolayer); therefore, the atom of oxygen occupies every fourth fcc site.

Above 500 K and up to 1250 K, the so-called subsurface oxygen has been observed as a result of penetration of atomic oxygen into the internal surface layer of metal. Some authors report the α-PtO_2 phase formation in adsorption of oxygen below 650 K, at the ultra-vacuum experiments at low pressure of oxygen.[31] At 1 atm oxygen pressure, the appearance of "subsurface" oxygen is possible even in 400–650 K.[32] Adsorption of dioxygen on the surfaces of other crystalline faces of Pt is not very different.

The reactivity of the adsorbed oxygen species on the Pt in hydrogen + oxygen reaction will be shown in Section 3.5.1.

On the Ag (111) surface at a wide range of temperatures, up to 500 K, oxygen adsorbs dissociative and adsorbed O-atoms form p(4 × 4) structure on the surface.[33, 34]

In nearly all cases, the dissociative adsorption of dioxygen at a certain temperature range is predominant on the surfaces of Pt and Pt-group metal monocrystalline samples.[30, 34]

The above described pattern of adsorbed oxygen on Pt(111) was clearly proved by the low-energy electron diffraction spectroscopy (LEED),[33] high-resolution electron energy loss spectroscopy (HREELS),[34] scanning tunneling microscopy (STM),[29] and other methods. Seemingly, the most powerful method permitting to obtain a direct view of atoms on the surface is STM.[29] An attractive image was obtained by the STM method in adsorption of dioxygen on Pt(111) at 165 K. In 2007, Ertl presented it in his Nobel lecture.[35] The adsorbed oxygen atoms on the surface were clearly visible.

Atomic oxygen on the surface of platinum and palladium is very reactive and able to activate C-H, N-H, or O-H bonds of organic compounds. Platinum and palladium are known as catalysts for complete oxidation of the majority of organic compounds.

3.4.1.3   Activation of Dioxygen on the Surface of the Metal Oxides

A great number of metal oxides are known as heterogeneous catalysts in various reactions of oxidation with molecular oxygen.[135, Ch. 1; 36–40; 142, Ch. 1] Different classifications of them exist as catalysts; for instance, the transition metals oxides and nonreducible metal oxides (alkaline and earth metal oxides; oxides of rare earth metal oxides) compose different groups of the oxidation catalysts. On the other hand, the semiconductors (such as TiO_2) occupy a special place among metal oxide catalysts due to their wide applications in different fields.[38]

The heterogeneous reactions of oxidation with molecular oxygen of the organic or inorganic compounds on metal oxides involve nearly all varieties of the active oxygen species (see Section 3.5.4.3) as intermediates,[38, 39] indeed, in dependence on the nature of the metal oxide and thermodynamic parameters of the chemical system.

In adsorption of dioxygen, the electronic configurations of both transition metals ion and oxygen considerably change in the crystal field of metal oxide.[38–40] For the majority of transition metal ions at the octahedral symmetry, due to crystal field energy, the electronic configuration of d orbitals (d_{xy}, d_{xz}, d_{yz}, d_{x2-y2}, d_{z2}) is transformed, giving t_{2g} triply degenerated (d_{xy}, d_{xz}, d_{yz}) and double-degenerated e_g (d_{x2-y2}, d_{z2}) orbital sets. When the energy difference between these two new electronic levels is small, unpaired electrons occupy the e_g orbital, and, when the difference is greater than the energy for formation of a pair, the t_{2g} are occupied by electrons.

It was shown that π^* antibonding molecular orbital of dioxygen (composed from components π_x and π_y) is double degenerated (Figure 3.1). In the strong crystal field, this degeneracy may be "removed" (i.e., the energy differences between π_x and π_y becomes larger than the energy of the pairing of electrons) and electrons will be paired, occupying the low energetic level π_y. As a result of these changes in the electronic configuration of oxygen, six sp^2 hybrid molecular orbitals will be formed. Two of them lie on the internuclear axis, and others have angles of 120° with the axis (Figure 3.9).

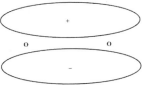

π_x - bonding orbital

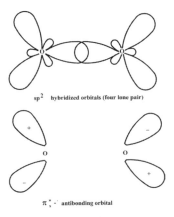

sp^2 hybridized orbitals (four lone pair)

π^*_x - antibonding orbital

FIGURE 3.9   Some molecular orbitals of dioxygen.

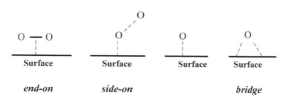

FIGURE 3.10 Different forms of the adsorption of dioxygen and oxygen-atoms on the surface.

The different forms of adsorption, which geometrically are either close to the parallel or near to the perpendicular in respect to the axis of the O-O bond, were named end-on and side-on oxygen species (Figures 3.10–3.12).

Adsorbed atomic oxygen may also be in various positions on the surface: near to the perpendicular or bridge structures. These two types of interactions are discussed in the monograph by Centi et al.[39, p. 390]

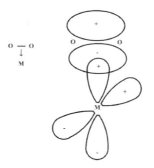

FIGURE 3.11 Overlap of d_{x2-y2} metal orbital and π_x orbital of O_2. The metal atom is perpendicular to the axis of the O-O bond. (Adapted from Ref.[39])

FIGURE 3.12 Overlap of metal atom orbitals with O_2 orbitals for "side-on" adsorption configuration of O_2: (a) overlap of empty e_g metal orbital by occupied lone pair hybrid orbital of O_2 with back donation from filled t_{2g} metal orbital with empty π_x^*; (b) Overlap of metal atom orbitals with antibonding π^* orbital of O_2 (Adapted from Ref.[39, p.388]).

The back donations stabilize the M-O bond, thereby facilitating rupture of O-O and formation of the adsorbed O-atoms on the surface.[39]

In general, the chemical bond of the oxygen intermediate species with metal oxide may be more or less polarized. Obviously, their nucleophilic or electrophilic properties[36] depend on the coordination state of the surface ions. On combination of the metal with oxygen, the accepted nomenclature in the coordination chemistry is presented in Figure 3.13.

Due to the different electron affinities of metal ions in the oxides, the complete or partial electron transfer to dioxygen or oxygen atom takes place from the surface through the formation of different species: O_2^-, O^-, or O^{2-}. The general scheme of the transformations of oxygen species in adsorption on the metal oxides was given by Bielanski and Haber.[34]

$$(O_2)_g \rightarrow (O_2)_{ads} \rightarrow (O_2^-)_{ads} \rightarrow 2(O^-)_{ads} \rightarrow 2(O^{2-})_{ads} \rightarrow 2(O^{2-})_{subsurface}$$

O_2^- and O^- are paramagnetic species, and their adsorbed forms are often detectable by the EPR method. Formation of O_2^- and O^- in the adsorption of dioxygen may be considered well established for the majority of metal oxides such as TiO_2, ZnO, NiO, SnO_2, V_2O_5 WO_3, MgO, Al_2O_3, MoO_3, and As_2O_5. In addition, many works have also shown the formation of radicals O_3^- and CO_2^- on the metal oxide surfaces.[37]

All experimental data indicate that the rate of the dissociative adsorption of molecular oxygen is much lower on the metal oxides than on the metals. According

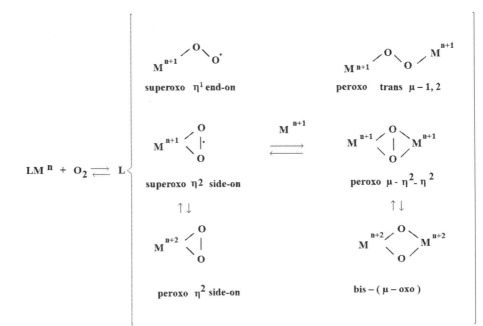

FIGURE 3.13   Structures and nomenclature of the coordination complexes in combination with dioxygen. (Adapted from Ref.[142 in Ch. 1] where this scheme is presented for Cu^{2+} ions.)

to Kung,[135 in Ch. 1] one of the reasons for this observation is the different availability of a pair of neighboring metal-ion sites for dioxygen to give atomic oxygen.

Preliminary notions about the adsorption of dioxygen on the surfaces of metal oxides may be obtained by the temperature programmed desorption experiments (TPD). These experiments on metal oxides show three peaks, corresponding to molecular oxygen (usually desorbed at low temperatures, below 300°C); atomic oxygen in relatively small amounts; and "subsurface" oxygen species above about 600°C.[135 in Ch. 1] However, this is a rough, generalized pattern showing the diversity of the adsorbed species of oxygen on the metal oxide. More detailed information on the adsorption forms of oxygen was obtained by the EPR method; in some cases, by LIF (laser induced fluorescence) methods; and then, in general, by surface science methods.[135 in Ch. 1; 37; 40; 142 in Ch. 1]

In the case of the transition metal oxides, partially reduced and dehydroxylated samples are more active in adsorption of oxygen.[135 in Ch. 1, p.116] This feature is exhibited in catalytic properties of the transition metal oxides and in oxidation of organic compounds with oxygen.

3.4.2 REACTIVITY OF OXYGEN SPECIES ON SOLID SURFACES AND IN THE FLUID PHASES

In relation to the heterogeneous–homogeneous reactions of oxidation with oxygen, a number of major types of active intermediates—free radicals, ions, excited species, their adsorbed forms on the surfaces of the solid substances, as well as a number of other surface intermediate complexes—are one of the main subjects of the present work. The diversity of the active intermediate species in oxidation processes is not limited to the frameworks of oxygen species. Particularly in the case of catalytic oxidation, both oxygen and oxidizing substances may be activated in interactions with the catalyst, in primary reactions, giving a wide diversity of intermediate species.

The interaction of molecular oxygen with the catalyst (or inhibitor) produces active intermediates which are called active (or reactive) oxygen species. They include not only species of the oxygen element, but also a great number of species where oxygen is combined with the atoms of other elements (for example, OH, HO_2, NO, CO_2^-).

There exist a wide diversity of active oxygen species, and they may be classified by the principles presented in Table 3.1.[41]

This table is incomplete and may be developed on the basis of other properties of oxygen species. For example, at least, six different types of OH species by different reactivity may be differentiated on the metal oxide surfaces of TiO_2 (Section 3.4.2.3). Therefore, this table also may be completed including classification by types and strengths of the chemical bonds of active oxygen species with the surface.

The majority of the above listed active intermediates of oxygen, as chemically individual species, have definite lifetimes in the fluid phases and on the surfaces of solid substances. On the solid surfaces, as a rule, they exhibit chemical properties and reactivity different from those in the fluid phases. From the point of view of heterogeneous catalysis or heterogeneous–homogeneous reactions, they will be observed to be in dependence on the nature and chemical properties of the solid substances.

Table 3.1 Classification of active oxygen species

1. *By number of oxygen atoms in species:*

    mono, di, and three atomic oxygen species

    monoatomic – O (atomic oxygen); O^- (anion–radical of atomic oxygen); O^{2-}
      (oxygen anion)

    biatomic – O_2 (molecular); O_2^- (superoxide anion–radical), O_2^{2-} (peroxy anion); O_2^+
      (cation–radical of oxygen)

    triatomic – O_3^- (ozonid anion)

2. *By charge*

    ions – O^-; O_2^{2-}; O^{2-}; O_2^+; O_3^-

    neutral species – O_2; O; O_3

3. *By magnetic properties*

    paramagnetic – O_2; O; O^-; O_2^-; O_3^-; O_2^+

    diamagnetic – O^{2-}; O_2^{2-}; O_3

4. *By electronic state*

    excited state – ΔO_2 (singlet)

    ground state – ΣO_2 (triplet)

5. *By combination with other atoms*

    combined with hydrogen – HO_2 (radical), HO_2^- (radical anion), OH (radical),
      OH^- (anion), H_2O_2 (peroxide, neutral molecule)

    combined with carbon – CO_2^- (anion–radical), RO_2 (peroxy radical), RO (oxy radical)

    combined with nitrogen – NO (neutral molecule, radical), NO_2 (neutral molecule, radical),
      NO_3^- (ion)

    combined with other elements.

6. *By oxidation number*

$-1/3$ - O_3	0 - O_2	-2 - O^{2-}	$+2$ - O^{2+}
$-1/2$ - O_2^-	-1 - O_2^{2-}	$+1$ - O_2^+	

7. *By phase state*

    lattice oxygen (as O^{2-} in the solid substances)

    adsorbed (as O_{ads}^- on the solid substances)

    in the gas or liquid phases (as O).

   The surfaces of solid substances, as catalysts or inhibitors in different oxidation reactions with dioxygen, may be divided into groups.[42] The surfaces of the metals and semiconductors, ionic structures (metal oxides, salts), covalent structures (including macromolecular 2- and 3-D networks, as graphite, polymers), as well as their various compositions (as zeolites), constitute different groups of heterogeneous oxidation catalysts or inhibitors. Metals and metal oxides, in their turn, may be divided into subgroups of the noble and transition metals and their oxides, as well as subgroups of the different metallic elements.

   *As shown in Table 3.1-(2), a number of oxygen active intermediates in oxidation reactions were charged species. **What is a charge?** Although this fundamental problem does not directly relate to the problems of oxidation reactions, it is present in our efforts to understand the nature of chemical phenomena. Every chemical reaction occurs by the rearrangement of the electronic density and nuclear configuration of reactants. A notable part of oxidation reactions occurs by the formation and destruction of* *charged species through partial or full electron transfer. Electrons possess a definite negative charge and mass. However, what is a charge? We don't know. And what is the difference between positive and negative charges? We don't know the answer to that either, nor do we know why a negative charge attracts a positive charge and repels other negative charges.*

Modern fundamental physics teaches that the charge is an intrinsic property of matter or, more precisely, of subatomic particles. An electron possesses the smallest negative charge in nature. We successfully use this property of matter to explain many physical and chemical phenomena, without understanding its nature. Nearly all academic publications circumvent this fundamental question, stating only the usefulness of the conception of the existence of charge as a fundamental property of matter. Analogously, we accept other fundamental properties of matter, such as mass, gravity, and energy, without any other explanation.

In modern physics, however, the problem of the origin of charge is a subject of both experimental and theoretical investigations, as an intrinsic property of the subatomic particles. Positron is a subatomic particle that has the same mass and spin as an electron but a positive charge. The positron, also named the antielectron, is the first antimatter particle discovered. The existence of antimatter was predicted by the Dirac equation, formulated in 1932 by Paul Dirac who as a result was awarded the Nobel Prize in Physics in 1936. In annihilation at the low-energy collision, an electron and positron pair produce two or more γ photons by the reaction.

$$e^- + e^+ \rightarrow \gamma + \gamma$$

In 1934, Landau, in the former Soviet Union, and Breit and Wheeler in the United Kingdom, predicted this reaction. The reverse process, the production of an electron and positron pair, when γ photons have enough energy (1.02 MeV) also is known:

$$\gamma + \gamma \rightarrow e^- + e^+$$

In 2014, the English-language press released the following sensational information: "Scientists discover how to turn light into matter after 80-year quest."[43] Rose and co-workers used the polarized short-pulsed electromagnetic field for creation of an electron–positron pair. According to their report, about 105 electron–positron pairs may be created using the experimental pathway.[43] The photon has not mass and has energy and momentum; thus, "the light creates matter." However, this remains the limit of our knowledge about the origin of matter and charge creation. Unfortunately, the birth of the electric charge remains inexplicable and mysterious in the physics of subatomic particles.

3.4.2.1 Oxygen Atom and Ion–Radical O⁻

Atomic oxygen may be considered one of the allotropic modifications of natural oxygen. Although it doesn't exist naturally on Earth's surface, it is the major atmospheric component at high altitudes (160 km >).[44] Atomic oxygen plays an essential role not only in a great number of photochemical reactions at the atmosphere of Earth, but also in combustion processes, plasma and radiation chemistry, and heterogeneous catalytic reactions of oxidation.

Knowledge of the overall properties and reactivity of atomic oxygen in the gas phase is useful in understanding its chemical reactivity on the surfaces of solid substances. Atomic oxygen, having an electronic configuration of $1s^2 2s^2 2p^4$, like molecular oxygen, is biradical in the ground state (atomic term ^3P). The first excited state, ^1D, as well as the second ^1S, are singlets (Figure 3.14). The energetically excited state

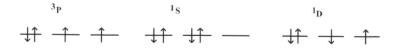

FIGURE 3.14 Energetic configurations of atomic oxygen in the ground ^3P and low-lying excited ^1D and ^1S states.

^{1}D is laid 189.6 kJ/atom higher than the ground state ^{3}P, and the second excited state, about 404.4 kJ/atom.[45] For comparison, the dissociation energy of the O-O bond in the ground state of dioxygen is 497 kJ/mol.[46]

Atomic oxygen in different energetic states in the gas phase can be generated by the photolysis of dioxygen (a–c):[46, 47]

$$240 \text{ nm} < \lambda < 242.4 \text{ nm} \quad O_2\left(^3\Sigma_g^-\right) + h\nu \rightarrow O\left(^3P\right) + O\left(^3P\right) \quad \Delta H_0 = 493.56 \text{ kJ/mol} \quad \text{(a)}$$

$$137 \text{ nm} < \lambda < 175 \text{ nm} \quad O_2\left(^3\Sigma_g^-\right) + h\nu \rightarrow O\left(^3P\right) + O\left(^1D\right) \quad \Delta H_0 = 683.38 \text{ kJ/mol} \quad \text{(b)}$$

$$110.6 \text{ nm} < \lambda < 134.2 \text{ nm} \quad O_2\left(^3\Sigma_g^-\right) + h\nu \rightarrow O\left(^3P\right) + O\left(^1S\right) \quad \Delta H_0 = 897.8 \text{ kJ/mol} \quad \text{(c)}$$

Like molecular oxygen, the reactions of triplet atomic oxygen with the atoms and molecules, which are in the singlet electronic state, are forbidden by spin conservation rules of Wigner (see Section 3.3). Therefore, in general, the $O(^1D)$ is more reactive than the triplet atom of oxygen in the reactions with organic compounds; for instance, the triplet O (^{3}P) oxygen atoms are unreactive with methane and methyl alcohol, while singlet oxygen atoms O (^{1}D) directly inserts into CH bond to give CH_3OH and $HOCH_2OH$ (methanediol, finally giving CH_2O and H_2O).[48]

Atomic oxygen plays an important role in atmospheric chemistry, particularly in the regeneration of ozone in Earth's stratosphere. This process is called the Chapman cycle.[49, 50]

In the stratosphere, dioxygen under UV radiation (240 nm < radiation < 310 nm) dissociates, forming atomic oxygen

$$O_2 \rightarrow O + O$$

In the low stratosphere (15–30 km altitude), atomic oxygen readily reacts with dioxygen giving ozone

$$O_2 + O + M \rightarrow O_3 + M$$

where M is any neutral particle, more probably N_2 or O_2 (since they comprise 99% of atmospheric gases).[50] It is noteworthy that, according to the atmospheric measurements, the lifetime for atomic oxygen, which depends on the altitude, is very short, for example, its lifetime is about 0.002 second in 30 km, while the lifetime of ozone is approximately 1000 seconds.

On the other hand, the oxygen atom reacts with ozone

$$O_3 + O \rightarrow 2O_2$$

Simultaneously, the decomposition of ozone under UV irradiation occurs according to the reaction

$$O_3 + h\nu\left(240 \text{ nm} < \text{radiation} < 310 \text{ nm}\right) \rightarrow O_2 + O$$

Finally, two atoms of oxygen may recombine

$$2O \rightarrow O_2$$

Thus, the concentration of ozone in the stratosphere depends on the ratios of the rates of the above reactions. In Chapman's cycle, the loss of ozone is balanced with its formation via the photo dissociation of dioxygen. Chapman's scheme can explain about a 20% loss of ozone. The equilibrium between the loss and formation of ozone in the atmosphere can be unbalanced not only naturally, but also because of other ozone-loss reactions, such as free radical reactions with Cl and Br containing compounds, azote oxides, and many other chemical compounds emitted into the atmosphere as a result of human activity. The emission of chlorofluorocarbons (CFCs) was considered one of the main causes of the depletion of the ozone:

$$CCl_2F_2 \, (CFC\text{-}12) + h\nu \rightarrow CClF_2 + Cl$$

Cl atoms participating in the reactions of the depletion of ozone play the role of catalyst:

$$Cl + O_3 \rightarrow ClO + O_2$$
$$\underline{ClO + O \rightarrow Cl + O_2}$$
$$O_3 + O \rightarrow 2O_2$$

As the reaction of Cl is a chain-radical process, in principle one atom of Cl permits to destroy hundreds or even thousands of ozone molecules.

Analogously, the nitric oxides NO_x ($x = 1/2; 1; 2$) participate in the catalytic cycles that destroy ozone.

$$NO + O_3 \rightarrow NO_2 + O_2$$
$$\underline{NO_2 + O \rightarrow NO + O_2}$$
$$O_3 + O \rightarrow 2O_2$$

Finally, OH radicals are also an important cause for the depletion of ozone:

$$OH + O_3 \rightarrow HO_2 + O_2$$
$$\underline{HO_2 + O_3 \rightarrow HO + 2O_2}$$
$$2O_3 \rightarrow 3O_2$$

The depletion of atmospheric ozone and the role of atomic oxygen in the heterogeneous–homogeneous process will be discussed in Chapter 5.

The first electron affinity of oxygen atom is -142 kJ/mol[51]; therefore, the formation of the O^- anion–radical from the O atom-radical is an "allowed" process. Although the electron transfer to adsorbed oxygen strongly depends on the nature of the metal oxide, it takes place in nearly any interaction of oxygen atoms with solid-phase metal ions. This charge transfer is accompanied by further stabilization of O-atoms on surface sites, which may be examined through analysis of the EPR spectra of the adsorbed O^- anion–radical.[52]

Adsorbed O-atom radical-anions on the surfaces of metal oxides may be generated by thermal or photochemical dissociation of adsorbed dioxygen, as well as decomposition of N_2O.[52 and within] It may be detected directly by the EPR method, mainly at cryogenic temperatures, on the dielectrics, such as CaO, MgO; semiconductors, such as ZnO, TiO_2; supported catalysts such as V/SiO_2, etc. However, the detection of O^- on the surfaces of metal oxides requires the existence of certain conditions. It may not be detected by the EPR method on the clean surfaces of metals such as platinum and palladium, or on the metal oxides, when the adsorption site has intrinsic paramagnetic properties, as in the case of Fe^{2+}, Fe^{3+}, Mn^{2+}, and Cu^{2+} cations.[52] Kazansky and co-workers, in a long series of experimental works, applying the EPR method, demonstrated the generation of adsorbed active oxygen species, including O^- anion-radicals on the surfaces of different metal oxides.[53]

Detailed analysis of EPR spectra, as well as theoretical calculations, give evidence of the existence of two types of $(O^-)_{ads}$ by different energetic configurations (named σ-type and π-type). They appear as a result of the splitting of the degenerate energetic level of oxygen atom $O^-(^2P)$ into singlet and doublet states, in the axial crystal field of the solid. This is well exhibited in differences of g-tensors of the EPR spectra for $(O^-)_{ads}$, in various crystalline structures. The reader can find the main references for this subject in the work of Volodin et al.[52]

The high reactivity of anion-radical O^- in homogeneous conditions is well documented in the chemical literature.[61] Similarly, in heterogeneous reactions, according to a very great number of former and present investigations on the surfaces of metals and metal oxides, anion-radical O^- exhibits relatively higher reactivity[52] than other active intermediate species of oxygen listed in Table 3.1. As mentioned above, it can activate the C-H, O-H, N-H bonds in "mild" conditions and often at very low temperatures, depending on the nature of the surface.

One of the characteristic chemical reactions of $(O^-)_{ads}$ on different surfaces of metal oxides (ZnO, V_2O_5, MoO_3, WO_3) is abstraction of the H-atom from hydrocarbons. For example, it readily reacts with methane at room temperature on the surfaces of metal oxides such as ZnO.

$$(O^{-\bullet})_{ads} + CH_4 \rightarrow \left(OH^-\right)_{ads} + (CH_3^{\bullet})_{ads}$$

For the silica-supported catalysts V_2O_5, MoO_3, and WO_3, the reactivity of $(O^{-\bullet})_{ads}$ decreases in H-abstraction from CH_4 and H_2.[135, Ch.1, p.118] Analogous reactions of H-abstraction by $(O^{-\bullet})_{ads}$ are also known for alkenes, although further events may be different in this case (Section 3.5.2). From the viewpoint of the heterogeneous–homogeneous processes, this reaction on ZnO is interesting. The EPR spectra of radicals $(CH_3^{\bullet})_{ads}$ on ZnO are fully identified at 90 K. It is noteworthy that part of the formed CH_3 radicals escapes from the surface to the gas phase and continues to react in the gaseous mixture.[52]

Analysis of the EPR spectra of anion-radical $O^{-\bullet}$ adsorbed on MgO and CaO shows two types of species having a different coordination environment.[52] They are named 3C and 4C $(O^{-\bullet})_{ads}$ (three and four coordinated), depending on their position on the surface. It has been suggested that in the presence of molecular oxygen, two types of intermediate complexes might be formed, each being in an equilibrium state with others on the surface

$$\left[O_{3C}^- \ldots O_2\right] \rightleftarrows \left[O_{4C}^- \ldots O_2\right]$$

g-tensors of these $(O^{-\bullet})_{ads}$ are clearly different. It has also been shown that $[O_{3C}^- \ldots O_2]$ would be a more stable form than $[O_{4C}^- \ldots O_2]$, having higher heat of formation. In short-term pumping at 180–200 K, the complex $[O_{4C}^- \ldots O_2]$ disappears

$$\left[O_{4C}^- \ldots O_2\right] \rightleftarrows \left[O_{4C}^- + O_2\right]$$

while $[O_{3C}^- \ldots O_2]$ remains on the surface. Accordingly, the $[O_{4C}^- \ldots O_2]$ type of adsorbed anion-radical is more reactive than $[O_{3C}^- \ldots O_2]$. This was proven in interactions with ^{13}CO investigated by the EPR method. The reaction gives an adsorbed CO_2^- anion-radical on the surface. In general, $[O_{4C}^- \ldots O_2]$ exhibits high reactivity at 163 K in H-abstraction reactions with H_2, CH_4, and C_2H_4, while other types $[O_{3C}^- \ldots O_2]$ remain unreactive in the same conditions. Similar observations on CaO also give analogous results. The generation of $(O^{-\bullet})_{ads}$ on the mentioned surfaces relates to the dissociative adsorption of water on the partially dehydroxylated surfaces. $(O^{-\bullet})_{ads}$ on oxide surfaces also exhibit high reactivity toward O_2, forming adsorbed anion-radicals O_3^-.

$$O^- + O_2 \rightarrow O_3^-$$

The existence of anion-radical ozonide O_3^- on the MgO is proven by the EPR method at low pressure of O_2.[52]

3.4.2.2 Superoxide Ion-Radicals

Like $(O^-)_{ads}$, the superoxide anion-radicals $(O_2^-)_{ads}$ are paramagnetic species.[1 in Ch. 2; 55–59] The generation of $(O_2^-)_{ads}$ on the surfaces of metal oxides may occur via the formation of surface complexes simultaneously with the formation of O^-.[1 in Ch. 2] In general, presently, the scheme for the consecutive reactions of oxygen intermediate species on the surface of metal oxides (Section 3.4.1.3) is questionable. In this regard the step

$$\left(O_2^-\right)_{ads} \xrightarrow{e} 2\left(O^-\right)_{ads}$$

seems doubtful, in consideration of the analysis of the corresponding EPR data[1 in Ch. 2] in the adsorption of oxygen. Hence, other ways to explain the formation $(O_2^-)_{ads}$ and $(O^-)_{ads}$ have been proposed. As the experimental data show, in adsorption of dioxygen on the V_2O_5/SiO_2 catalyst at 430 K, some EPR spectrum lines of $(O_2^-)_{ads}$ may be eliminated by short-term pumping of the sample, due to the shifting of the equilibrium

$$\left(O_2^-\right)_{ads} \underset{}{\overset{-e}{\rightleftarrows}} (O_2)_{gas}$$

while the anion-radical O^- is stable in the same conditions. It has been suggested formation of the charge-transfer complex, simultaneously forming two oxygen species[51 in Ch. 1, 52]

$$\left[V^{5+} .... O^{2-}\right] \rightleftarrows \left[V^{4+} .... O^-\right]^*$$

where $[V^{4+} .... O^-]^*$ is charge-transfer complex

$$\left[V^{4+} ... O^-\right]^* \xrightarrow{O_2} \left[V^{5+} ... O^-\right] + \left[V^{5+} ... O_2^-\right]$$

On the polycrystalline samples of a TiO_2, three different types of $(O_2^-)_{ads}$ may be identified at 140–250 K by three different g-factors, 2.019, 2023, and 2026, respectively.[58]

The mobility of the adsorbed O_2^- radical on the different surfaces was studied by the temperatures dependencies of their EPR spectra by Che and Giamello.[60] For O_2^- on a Ti^+ site (at 4–400 K), it was found that the radical had highly anisotropic motion (g_{xx} and g_{zz} were changed, and only g_{yy} remained unchanged), indicating the perpendicular attachment of molecules on the surface. It was also found that the rotation was even observed at 57.4 K, and the correlation time of rotation was 10^{-5} s at 14 K and 10^{-9} s at 263 K. The activation barrier was about 0.5 kcal/mol estimated above 100 K.

Like the O^- atom radical, the superoxide anion-radical adsorbed on the nonreducible metal oxide exhibits high reactivity toward organic compounds in hydrogen abstraction reactions that apparently occur through the radical mechanism. However, as the kinetic investigations using the EPR method show, for instance, on the MgO, it is less active than the O^- atom anion-radical.[61] Modern notions of the mechanism involved in the heterogeneous catalytic oxidation of organic compounds with dioxygen, especially on the reducible metal oxides, are based on the Mars–van Krevelen type of mechanism suggestions, where nucleophilic lattice oxygen O^{-2} plays a central role. It implies the transformation of adsorbed superoxide anion-radical to lattice oxygen. As mentioned above, the mechanism of the transformation of extrafacial species to interfacial species (see the Haber–Bielanski cycle[34]) was not proven in a number of cases.[1 in Ch. 2] Moreover, there is no clear differentiation in the mechanism suggestions for transformation of superoxide anion-radicals on the reducible and nonreducible metal oxides in the oxidation of organic compounds. In this regard, certain mechanism suggestions attempt to elucidate the participation of the adsorbed superoxide radicals in the oxidation of some hydrocarbons by the Mars–van Krevelen type of mechanism applied for reactions on nonreducible metal oxides.[57] For example, the heterogeneous catalytic oxidation on MgO of propene, at room temperature, forming acetates, formates, and carbonates, according to Ampo et al.,[57] may proceed by the following scheme:

$$R\text{–}H + O_{(surf)}^{2-} \rightarrow R_{(ads)}^{-} + OH_{(surf)}^{-}$$

$$R_{(ads)}^{-} + O_{2(g)} \rightarrow \left(O_2^{\bullet -}\right)_{ads} + R_{(ads)}^{\bullet}$$

According to this scheme, the R–H bond may be heterolytically dissociated by low coordinated O^{2-} in metal oxides of the basic nature, as MgO. Moreover, this scheme is based on the concept of the surface intermolecular electron transfer (SIET) from carbanion to adsorbed dioxygen from the gas phase, by the formation of superoxide anion-radicals on the surface. Carbanion is detectable by UV spectrometry and adsorbed superoxide anion-radical or radical HO_2 is spectroscopically visible by the EPR method. Here, direct activation of dioxygen coming from the gas phase and corresponding to the Haber–Bielanski[34] scheme does not take place.

There are also other evidences obtained by the experiments with the isotopically labeled atoms of dioxygen in the gas phase.[62, 63] They showed that the main products formed on the surface were not a result of the insertion of the labeled atoms of oxygen into the structure of organic compounds. Clearly, in this case, oxidation occurs through participation of the nucleophilic (lattice) oxygen in the reaction.

According to Puigdollers et al.,[62] the transformation of adsorbed oxygen into lattice is in dependence on the "cost of formation" of the oxygen vacancies ("reducibility") of metal oxides.

3.4.2.3 Hydroxyl Radicals

In general, hydroxyl radicals were considered highly reactive species in various reactions of oxidation, playing a determining role in chemical transformations. They form on the surfaces of solid substances in the thermal, as well as in photochemical oxidation of hydrogen and hydrocarbons with dioxygen.[64] Thermal and photochemical heterogeneous catalytic decomposition of the peroxide compounds (H_2O_2, ROOH) may also serve as sources of OH radicals on the surfaces of metals and metal oxides. However, often, particularly in photocatalytic oxidation reactions, their role is overestimated.[65] There exists a wide diversity of the generation of different types of the adsorbed OH species on the surfaces of the solid substances.[66]

In the gas or liquid phase, the chemical identity of OH species is more or less definite as individual chemical compounds (molecules, entities) in the form either of hydroxyl radical or hydroxyl ion. On the surface of the solid substances, the formation of the different types of OH species may be expected in dependence on the nature of the adsorption site.[66] Each of them has its own characteristics and, often, different reactivities in dependence on the strengh of the chemical bonding with the surface sites, as well as other factors (p^H, nature of the solvent). How many types of OH species were observed on the surfaces of different solid substances? For example: there are four different types of hydroxyls with different reactivities on the surface of α-FeOOH.[66] At least, six types of hydroxyl species were mentioned on the illuminated surface of TiO_2 (particularly, P25) in photooxidation reactions.[67] Three of them are the so-called linear hydroxyls, and three others are bridging hydroxyls. They were detected by different spectroscopic methods and were visible by the STM method.[67]

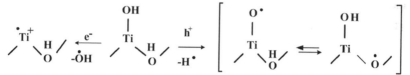

This schematic presentation of the formation of different types of hydroxyls by trapped holes and electrons on an illuminated TiO_2 was given by Nosaka et al.[67] Hence, in discussing the role of OH species in heterogeneous reactions, it is important to indicate the type of OH on the surface.

The existence of OH radicals (radical-like species), adsorbed on the surface of Pt(111), was first proven by Fisher and Sexton.[68] Applying ultraviolet photoelectron spectroscopy (UPS) and electron energy loss spectroscopy (EELS) in co-adsorption of O_2 and H_2O, the adsorbed OH, above 150 K, was detected.[68] In addition, they found that the O–H axis appears to be bent relative to the surface normal. Talley and Lin[69] detected the OH radicals by the LIF method in the gas phase, desorbed at 820°C from the Pt polycrystalline sample, in water decomposition on the surface. They determined the activation barrier of desorption, 30 kcal/mol, which was in agreement with the values reported previously for the Pt-catalyzed oxidation of H_2 by O_2 and N_2O.[69]

Adsorption characteristics (heat of adsorption, geometry of structure and the lifetime) of OH on metals strongly depend not only on the nature of metal, but also on the crystalline face and adsorption sites.[67, 70] For instance, the calculated heats of adsorption of OH decrease in order Ag(110)>Ag(100)>Ag(111), being 118.3 kcal mol^{-1} at the short bridge site on Ag(110), 108.6 kcal mol^{-1} at the fourfold hollow site on Ag(100), and 97.3 kcal mol^{-1} at the threefold hollow site on Ag(111). These data indicate a strong chemical bonding of OH with the surface sites. The adsorption geometries also are different; in respect to the surface plan, the OH axis varies from perpendicular up to 50° tilted. Characteristics of the adsorption of OH radical-like species on the metal surfaces Ag, Cu, Ni, Fe, Pd, Ru, Pt, Rh were investigated in a number of works.[70 and ref. 1–16 within] Evidently, in all cases, the adsorption of OH on the surface of the clean metal surfaces occurs by the partial electron transfer from the surface to OH species. For example, according to calculations,[70] when OH is adsorbed on Ag, the net charge transfer is 0.78e, 0.64e, and 0.43e, respectively, at the on-top, bridge, and hollow sites on Ag(100).

In general, the direct detection of OH radicals in heterogeneous reaction and determination of their type, as well as concentration, is difficult for certain reasons. Certain types of OH (such as the so-called free OH radicals) are reactive species and possess high-oxidation potential. Single-molecule fluorescence imaging technique applied for detection of OH, generated in heterogeneous reactions was described in Section 2.3.2.1, and the Nosaca experiment will be briefly described in Section 4.4.3. Talley et al.,[71–73] also applied LIF method in investigations of the reactivity of adsorbed OH on Pt(111), including experiments with OH co-adsorption with oxygen, water, and hydrogen on the same surfaces (see also Section 3.5.1.1).

From the viewpoint of the present book, the problem of the reactivity of OH species in heterogeneous–homogeneous reactions relates not only to their surface reactivity, but also to the possibility of their desorption to the fluid phases. In the case of Pt or other noble metals, the desorption of OH radicals from the surface takes place only beginning from the relatively high temperatures (above about 800–1000 K). The behaviors and internal energy of OH radicals, formed in the hydrogen–oxygen reaction, was investigated by the laser-induced fluorescence method at 1300 K, in relation to their escape from the surface of a Pt-polycrystalline sample.[73] It was found that the rate of energy randomization in interaction of OH with the surface was greater than the rate of its desorption. This also means that OHs are strongly bonded species on the solid surfaces.

3.4.2.4 Hydroperoxyl Radicals

Marshall predicted the existence of HO_2 radicals in 1926, while investigating the photooxidation of hydrogen with oxygen in order to explain the formation of hydrogen peroxide in this reaction.[74] Applying the mass spectroscopic technique, Foner and Hudson[75] obtained direct experimental evidence about the radicals HO_2 by the reaction of hydrogen atoms with molecular oxygen, which occurs in the presence of the third body (M):

$$H + O_2 + M \rightarrow HO_2 + M$$

In the liquid phase, in aqueous solutions, the radical HO_2 may be considered as a hydrated form of the superoxide anion-radical

$$O_2^- + H_2O \rightleftarrows HO_2 + OH^- \qquad p^K = 4.88$$

Evidently, the concentration of HO_2 is strongly dependent on the p^H of the solution.

In both phases, it is a strong oxidant agent. HO_2 radical is an important species in the gas-phase oxidation of hydrogen, hydrocarbons, and oxygenated organic compounds. In the majority of combustion reactions with air or molecular oxygen, its role relates mainly to the formation of hydrogen peroxide by self-recombination:

$$HO_2 + HO_2 \rightarrow H_2O_2 + O_2$$

The subsequent decomposition of H_2O_2 is a source of two more reactive OH radicals:

$$H_2O_2 \rightarrow 2OH$$

However, according to the available data, the recombination of HO_2 does not have a simple mechanism of bimolecular reaction. For a certain range of temperatures, the reaction has a negative temperature coefficient.[76] Different suggestions have been offered to explain the unusual kinetic peculiarities of this reaction. One hypothesis involved the formation of the H_2O_4 dimer complex, an intermediate, subsequently giving the vibrational excited dimer molecule.[77] Another suggestion was the formation of an H-bonded six-member transient complex, the decomposition products of which were oxygen and hydrogen peroxide.[76, 78] Note that hydrogen peroxide or intermediates formed from hydrogen peroxide may be decomposed in two ways: homogeneously and heterogeneously. Nalbandyan and co-authors showed not only the feasibility of the heterogeneous decomposition of hydrogen peroxide, but also the appearance of hydroperoxy radicals in the gas phase.[79] Carlier et al.[80] obtained evidence for the formation of halogen oxy-radicals in heterogeneous decomposition of gaseous hydrogen peroxide on NaCl and KCl related to the heterogeneous generation of radicals. Later, another interesting phenomenon was shown: the transfer (transportation) of solid materials (salts, oxides of metals) via the heterogeneous decomposition of hydrogen peroxide vapors to the gas phase.[81] Evidently, all of these observations relate to the formation of HO_2 radicals, which subsequently were transferred to the gas phase and detected by the matrix isolation EPR method.

The first attempts to reveal the heterogeneous reactions of HO_2 radicals also were related to investigations of gas-phase oxidation reactions of organic compounds. For instance, in the reaction of oxidation of tetramethyl butane with oxygen,[82] it was shown that HO_2 radicals were efficiently destroyed on the surface of the KCl-coated reaction vessel by the following reaction:

$$4HO_2 \xrightarrow{\text{KCl}} 2H_2O + 3O_2$$

Baldween et al. quantitatively confirmed the compliance of HO_2 decay to this reaction.[76, 82]

Heterogeneous reactions of HO_2 were also investigated in relation to the problems of atmospheric chemistry. The uptake coefficients of HO_2 were determined on the surfaces of the aerosol particles (for instance, $(NH_4)_2SO_4$ and NaCl), under ordinary conditions, using the LIF method.[83] A strong dependence of uptake coefficients on the humidity of samples was observed. For instance, on a sample of $(NH_4)_2SO_4$, it was 0.01 at 20% humidity and 0.19 at 75% humidity. The important role of HO_2 radicals in atmospheric processes will be discussed in Chapter 5.

In early investigations, the feasibility of forming adsorbed HO_2 radicals in oxidation of hydrogen with oxygen has been postulated and, then, evidenced by a number of kinetic and non *in-situ* spectroscopic data on different solid surfaces.[84] However, the direct spectroscopic observation of absorbed HO_2 radicals, formed in reaction conditions, appeared following the wide use of nanocatalysts in chemistry. The nanosized Au-clusters-catalyst, supported on silica or TiO_2, unlike traditional catalysts, exhibits activity in a number of oxidation reactions, including the reaction of hydrogen with dioxygen. The formation of HO_2 radicals (more exactly: the radical-like HO_2 species adsorbed on the catalyst), as well as H_2O_2 on the surface of this catalyst was shown using the vibrational spectroscopy by inelastic neutron scattering (INS), in interaction of $H_2 + O_2$ gaseous mixture with Au/TiO_2 catalyst, at 423 K.[85] It was suggested that HO_2 formed via a hydrogen-bonded complex with water, having frequencies 1525–1600 cm^{-1} in INS spectra.

3.4.2.5 Oxygen-Centered Peroxy Radicals

Among the oxygen atoms containing radicals, the organic peroxy radicals have great importance in oxidation reactions with molecular oxygen. The oxygen-centered organic peroxy radicals,[86–88] abbreviated as RO_2, where R is alkyl, aryl, or another hydrocarbon radical, contain an unpaired electron localized on the side oxygen atom. This class also involves the radicals of the type RC(=O)OO or RC(=O) $(CH_2)_n$OO or other functional groups. The formation of this type of radicals in the oxidation of hydrocarbons, alcohols, aldehydes, and other major classes of organic compounds with dioxygen, both in the gas and liquid phase, is well established experimentally.[89] The formation of peroxy radicals is predominant at the low-temperature heterogeneous decomposition for the majority of organic peroxide compounds.[100 in Ch. 2; 38 in Ch. 1] Moreover, in most cases, these radicals are responsible for the heterogeneous–homogeneous proceeding of the oxidation or decomposition reactions by transferring them to fluid phases (see also Sections 2.3.1 and 2.3.2).

Oxygen-centered organic peroxy radicals may be obtained in the thermal decomposition of almost any organic compound in the presence of dioxygen, at temperatures sufficient for formation of carbon-centered radicals R, as a decomposition product, by further combination with dioxygen. The methods of the generation, as well as detection, of the most abundant organic peroxy radicals such as CH_3O_2, $C_2H_5O_2$, $CH_3C(O)O_2$, and $CH_3C(O)CH_2O_2$, in both phases, are well known and described in the kinetic databases.[87–89, 100 in Ch. 2]

One of the general ways of the generation of RO_2 is decomposition (thermal or photochemical) of organic peroxide compounds ROOH or ROOR′ on the surfaces of the various solid substances[38 in Ch. 1, 89, 90] (Section 2.3.2). Heterogeneous-radical decomposition was established for a number of reaction systems, such as H_2O_2, ROOH (including peroxyacids), ROOR/noble or transition metals, their oxides, salts, zeolites, porphyrins, glass, quartz, and many other solid materials. This kind of decomposition reaction is usually accompanied by the formation of free radicals, part of which is able to "escape" from the surface of solid substances to fluid phases. In the presence of molecular oxygen in reaction medium, the radicals formed are mainly of the type RO_2. Radicals may be detected in the gas phase by EPR spectroscopy, using the matrix isolation method at cryogenic temperatures,[38 Ch. 1] and by *in-situ* measurements on the surfaces.[90–92]

An interesting case of the heterogeneous generation of peroxy radicals is decomposition of *tert*-butyl hydroperoxide on the NiO (black), investigated by EPR spectroscopy, mass spectrometry, and electroconductivity methods.[91] The decomposition of vapors of this hydroperoxide in the presence of small amounts (< 1%) of dioxygen begins at temperature 330 K and increases with the temperature, reaching about 70% of the initial amount, at 413 K. The main final products of the reaction were acetone, methanol, and H_2O. The radicals, which appeared in the gas phase, were detected by the matrix isolation EPR method and identified as typically RO_2 radicals. The chemical analysis showed that the quantities of the registered radicals in the gas phase were close to the amounts of the decomposed hydroperoxide and formed acetone in the reaction, at the temperature range of 388–413 K. The heterogeneous decomposition of hydroperoxide seemingly occurs by cleavage of O–O bonds of peroxide and formation of radicals on NiO

$$(CH_3)_3\,OOH \rightarrow (CH_3)_3\,O + OH$$

Then, the adsorbed radical RO undergoes further decomposition, giving acetone and methyl radicals

$$(CH_3)_3\,CO \rightarrow (CH_3)_2\,CO + CH_3$$

$$CH_3 + O_2 \rightarrow CH_3O_2$$

$$OH + (CH_3)_3\,OOH \rightarrow HO_2 + (CH_3)_3\,OH$$

The combination of CH_3 with dioxygen either on the surface or in the gas phase has been established experimentally in certain works.[38 in Ch. 2] Apparently, a selective formation of the RO_2 radicals (the yield was about 80%) may be a result of reaction of CH_3 on the surface (it is a reversible act in the gas phase; see Section 3.5.2.2).

As a source of acetyl peroxy radicals in the experimental investigations, either the thermal or the photochemical decomposition of biacetyl on the solid surface may serve. For instance, in the decomposition of biacetyl on the TiO_2 the detected radicals were fully characterized as peroxy species ($g_1 = 2.0345$, $g_2 = 2.0070$, $g_3 = 2.0010$).[90] Another example of the heterogeneous generation of peroxy radicals is decomposition of diacetyl $CH_3COCOCH_3$ on the walls of the quartz reactor at 773 K, permitting to obtain a peroxy radical concentration of about 10^{14} spins/cm^3.[92]

$$CH_3COCOCH_3 \rightarrow 2CH_3CO$$

$$CH_3CO + O_2 \rightarrow CH_3CO_3$$

In this case, the peroxy radicals were detected both in the gas phases and on the surface of SiO_2 (both at 77 K).[92] The adsorbed radicals were stable up to room temperature on crystalline KCl and KCl/SiO_2.

One of the first experimental proofs of the heterogeneous formation of the reaction products, via interaction between the adsorbed radical and reactant from the gas phase was obtained in studies of the decomposition of CH_3OOH (533–613 K), at atmospheric or low pressure, on the surface of Pt.[93] It was shown that the decomposition of CH_3OOH occurred by breaking the O–O bond on the surface forming adsorbed methoxy radicals and $OH_{(ads)}$. The assumed reaction of CO with $(OH)_{ads}$, was carried out in experimental conditions excluding homogeneous interactions. The residence time of the mixture in the capillary reactor was chosen 10–100 times lower than the average time of the homogeneous reaction of CO with OH radicals. In the presence of hydroperoxide, a significant amount of carbon dioxide was detected:

$$OH_{(ads)} + CO \rightarrow CO_2 + H_{(ads)}$$

On the other hand, the formation of CO_2 in the absence of CH_3OOH was not observed. A number of heterogeneous reactions of the adsorbed peroxy radicals are summarized in Table 3.2 (Section 3.5.4.3). The reactivity of the adsorbed alkylperoxy radicals was shown with CH_4 and acetaldehyde on TiO_2,[94] on the walls of the capillary reactor treated with boric acid,[95] as well as on the salts, such as KCl, NaCl, and NH_4NO_3.[96]

According to a great number of experimental results[38 Ch. 1; 100 in Ch. 2; 90–96] the appearance of radicals, particularly oxygen-centered peroxy radicals, in the thermal oxidation of hydrocarbons on the various surfaces starts with the activation of the C–H bond of the adsorbed organic compound. Note that in this case the mechanism suggestions of the formation and reactions of peroxy radicals are very different and often contradictory.[90, 96, 97] The main role was attributed to the active oxygen species in either the extrafacial or interfacial reactions of lattice oxygen. Apparently, in most cases, owing to the attack of weakly adsorbed O_2^- and O^- ion-radicals, the breaking of the C–H bond and the formation of alkyl radicals or surface alkyl species take place on the surface of the solid substance. The surface defects of metal oxides can play a significant role, being analogous to adsorbed oxygen. Once the alkyl radicals are generated, they can easily react with oxygen, forming alkylperoxy radicals, which can continue the reaction both on the surface and in the gas or liquid phases after desorption. In the case of alkenes, according to Haber's conception of oxidation reactions (3.5.2.1), the C=C bond reacts with the cation center of the solid substance, and other mechanisms of the further oxidation reaction takes place.

The following questions arise: What is the further fate of these radicals on the surface? Is a further chain-radical heterogeneous reaction feasible? All existing data indicate that one of the most widespread reactions of peroxy radicals on the solid surfaces with organic compounds is H-abstraction by the formation of the corresponding peroxide compound.

$$ROO + R'H \rightarrow ROOH + R'$$

According to data obtained on the TiO_2 surface,[96 and ref. within] the reaction may occur by multiplication of the peroxy radicals on the surface, part of which may "escape" to the gas phase. This implies a chain-radical mechanism of the heterogeneous reaction, as the homogeneous interaction may be excluded, choosing experimental conditions that exclude direct gas-phase interaction ROO with the organic molecule R'H. In the conditions excluding homogeneous reaction with R'H, a heterogeneous reaction may be initiated on the surface oxidation reaction by a radical pathway.[124 in Ch. 1]

The heterogeneous reactions by photostimulated generation of OH, HO_2, RO, RO_2, RCO_3, and other active radical intermediates in complex oxidation reactions on the surfaces of semiconductor materials, will be discussed in Chapter 4.

3.4.2.6 Anion and Cation Radicals

Carboxylate anion-radical $CO_2^{\bullet-}$ may be easily generated on the MgO surface in the presence of CO_2.[98] $CO_2^{\bullet-}$ can be characterized by the EPR method, including the so-called continuous wave EPR (CW-EPR) measurements.[99] Analysis of spectra and theoretical calculations show that $CO_2^{\bullet-}$ preferably is bounded at $(H^+)(e^-)$ centers formed near cationic corners.[100] The product of oxidation of $CO_2^{\bullet-}$ with oxygen mainly is carboxylate anion CO_3^{2-}.

Cation-radical $O_2^{\bullet+}$ (molecular term $^2\Pi_g$) is known as a very strong oxidant. Salts, such as $(O_2^+)MF_6^-$ (where M is As, Sb, Pt) containing $O_2^{\bullet+}$, exhibit high reactivity to aromatic amines, heterocyclic compounds containing azote or sulfur, and fluorochlorohydrocarbons (such as $CHClF_2$).[101] As an intermediate, O_2^+ does not have a significant role in "ordinary" heterogeneous catalytic oxidations with dioxygen.

$O_3^{\bullet-}$. For details on the adsorbed ozonide anion-radical on surfaces of the solid substances,[102] see Section 5.4.

3.4.2.7 Singlet Oxygen

The history of the heterogeneous reactions of singlet oxygen began from the famous experiments of Kautsky, in 1931. He was the first to prove the existence and reactivity of singlet oxygen.[103, 104] Kautsky observed photodynamic effects[refs. in Sec. 4.2] in an experimental system, where the photosensitizer (dye, trypaflavine) and substrate (leuco malachite green, which is colorless) were separated spatially and adsorbed on the different grains of silica gel. At the pressure of oxygen 10^{-3} mm Hg, the quenching of the fluorescence of photosensitizer had been observed. At the same time, in the presence of oxygen, under illumination, colorless leuco malachite became green. Experiments showed that oxidation occurred through the participation of the gas-phase-mediated molecules formed through the interaction of oxygen and a photosensitizer (dye, trypaflavine). According to Kautsky's explanation, the oxygenation of the substrate took place due to the intermediate, electronically excited dioxygen molecules, which were formed by energy transfer from photosensitizer to dioxygen. Passing through the gas phase and reaching to the surface of the silica gel grain containing adsorbed substrate (leuco malachite), the "metastable" molecules of dioxygen reacted with them, producing the oxygenated product.[103, 104]

Subsequently, a number of experimental results were obtained indicating the role of singlet oxygen as an active intermediate in oxidation reactions.[105]

The direct detection of singlet oxygen on the surface may be performed using one of the electronic transitions between singlet and triplet states of dioxygen, for instance, phosphorescence radiative transition $O_2(^1\Delta_g) \rightarrow O_2(^3\Sigma_g^-)$ at 1270 nm,[106] applying powerful lasers and sensitive photomultipliers at cryogenic temperatures. Although some researchers believe that presently there are no experimental methods for direct (*in situ*) detection of singlet dioxygen, adsorbed on the surfaces of the solid substances,[107] some works report the identification of singlet oxygen by spectral methods.[108, 109] For example, the generation of singlet oxygen was reported in adsorption of dioxygen on the surface of photosensitizer, microporous Si, composed from Si nanocrystals and having sizes of several nanometers. Evidence about the formation of singlet oxygen on the surface of this material was obtained by photoluminescence spectroscopy.[110] EPR may be used as an indirect method of singlet oxygen detection, when $O_2(^1\Delta_g)$, formed in the reactions, reacts with a number of organic scavengers and traps, becoming detectable (for instance, mainly with nitroxide radical adducts). However, spin-trapping agents, such as 2,2,6,6-tetramethylpiperidine (TEMP), which is more often used as a reagent for detecting $^1O_2$, do not totally satisfy the requirements for definitely identifying $^1O_2$, as the surface superoxide radical O_2^- also gives the same adduct–product nitroxide radical.[107]

Kazansky et al. clearly showed the formation of singlet oxygen on the surface of chromium oxide in their early EPR investigations.[111] Other evidences on the formation of singlet oxygen on the solid surfaces of metal oxide catalysts were represented in the literature beginning in the 1980s.[112] However, only a few examples evidence the role of singlet oxygen as a main active intermediate in the heterogeneous or heterogeneous–homogeneous oxidation of organic compounds. According to a number of works on heterogeneous catalysis,[113–116] singlet oxygen is an active intermediate of the oxidation of benzene to phenol,[113] toluene to maleic anhydride,[115] naphthalene to phthalic anhydride,[113] other aromatic hydrocarbons to products of the selective oxidation over metal oxide catalysts, such as activated quartz, vanadium, molybdenum, and tungsten oxides, as well as zeolites.[113–116] For example, it has been shown that the oxidation of toluene on the mixed vanadium and molybdenum oxides occurred by singlet oxygen, generated on the surface of catalysts,[114] which selectively reacted with toluene forming maleic anhydride and benzaldehyde. In oxidation of propylene over zeolites and SiO_2 supported V, Mo, Bi metal oxide catalysts, surface singlet oxygen, as the authors noted, participated in the formation of products in "mild" oxidation.[116] In all mentioned cases, however, the detailed mechanisms of the generation and further heterogeneous reactions of singlet oxygen were not clarified in these systems.

The reactions of singlet oxygen are apparently more important in heterogeneous photocatalytic oxidation, when it may be directly generated on the SiO_2[117] or TiO_2.[118] From the viewpoint of heterogeneous–homogeneous reactions, it is important that oxygen species adsorbed on the solid surface migrate to the gas phase. Therefore, the affiliated occurrence of both the heterogeneous and homogeneous reactions of singlet oxygen in the reaction media cannot be ruled out. Although there is an opinion[119] that there is no direct evidence for the act of transfer of singlet oxygen from the surface to the gas phase, the Kautsky experiment remains one of the first and best examples of the revelation of the homogeneous–heterogeneous mechanism in complex photocatalytic reaction. Moreover, the above-mentioned works indicate that singlet oxygen, generated in any way, immediately or intermediately, may participate in both heterogeneous and homogeneous reactions. Examples of the heterogeneous–homogeneous reaction with the participation of singlet oxygen were given in the review of Tachikawa and Majima,[120] showing the desorption of singlet oxygen from TiO_2 nanoparticle by the single-molecule, single-particle fluorescence imaging technique (Chapter 2). Another domain, where the heterogeneous reactions or deactivation of singlet oxygen play an important role, is atmospheric chemistry (see Chapter 5).

3.5 HETEROGENEOUS–HOMOGENEOUS REACTIONS OF OXIDATION WITH DIOXYGEN

As mentioned earlier, nearly all oxidation reactions with dioxygen of organic or inorganic matter are complex chemical processes involving many reaction intermediates. The rates of oxidation reactions are dependent on a large number of parameters and factors. The majority of the homogeneous reactions of dioxygen with the organic compounds are in many aspects well investigated and documented in the chemical literature. Most of the discussion in this section is devoted to the role of the heterogeneous reactions of oxygen and oxygenated intermediates in different oxidation reactions occurring as heterogeneous–homogeneous processes. Here, the bimodal reaction sequences are exemplified by the oxidation reactions of hydrogen and the main classes of organic compounds as alkanes, alkenes, alcohols, and aldehydes.

The first example, the oxidation reaction of hydrogen with dioxygen, is often considered a model reaction for more complex oxidation processes, in spite of the fact that it is itself a complex and "capricious" reaction.

3.5.1 CATALYTIC REACTION OF OXIDATION OF HYDROGEN WITH DIOXYGEN

Understanding the reaction mechanism of the oxidation of hydrogen with oxygen has fundamental importance. This reaction became a classical example showing how to reveal and explain many peculiarities of oxidation processes. The first bridges, indicating the relationships between homogeneous and heterogeneous chemical reactions, were established in investigations of this reaction (Section 1.3).

Although the intermediates of the heterogeneous and homogeneous oxidation of hydrogen with dioxygen often are the same, their functionalities are different in the gas phase and on the surfaces of the solid substances. Note that the problem is multifaceted, voluminous, and disputable in many aspects. Therefore, here we cannot give a complete presentation of the existing mechanism suggestions either in heterogeneous catalysis or in gas-phase reactions.

Please note that some important outlines of this classical example of the chain-branched reaction in gas-phase oxidation and combustion were presented in the introductory parts of Chapters 1 and 2.

3.5.1.1 Adsorption and Reactivity of Reactants on the Surfaces

Before making a brief presentation of existing relationships in heterogeneous and homogeneous interactions in the hydrogen + oxygen reaction, it is useful to introduce some data on the adsorption and reactivity of intermediates in this reaction on the solid substances.

Adsorption of molecular oxygen and formation of oxygen active species on the surfaces of solid substances (metals and metal oxides) were briefly presented in the previous sections of this chapter. As regards hydrogen, from a very great number of investigations, it is well known that, in general, on metal surfaces as well as on the majority of the metal oxides, the adsorption of hydrogen has a dissociative character in a wide range of temperatures.[112; 51 in Ch. 1; 121] Adsorption of hydrogen on the transition metals, but not on Au, Ag, and Cu,[112] is dissociative. On Pt(111) hydrogen dissociates nearly without an activation barrier,[122] mainly occupying hollow sites. The strongly adsorbed hydrogen atom, formed by the dissociation of molecular hydrogen, is less reactive on the surface than that in the gas phase. Moreover, all alkali (or like alkali) metals combine with atomic hydrogen, forming hydrides.[123] The adsorption heat of hydrogen is not very sensitive to the metal facet, but it is sensitive to the crystalline defects.

The nature and strength of the bond of the adsorbed hydrogen atoms with the solid-phase surfaces determine their mobility and reactivity. In some cases, even atomic hydrogen may be directly detected by the EPR method, as a radical-like species. For example, there are known the EPR spectra of the adsorbed hydrogen atoms on the surface of single crystals of different salts, such as NaCl and KCl (77 K),[124] CaF_2, K_2SiF_6,[125] phosphates, acidic glasses (solid acidic ice as sulfuric acid,[126] and silicon (100).[127] Their EPR spectra consist of two hyperfine lines, corresponding to atomic hydrogen.[126]

On metal oxide surfaces, atomic hydrogen is usually stabilized, being associated with oxygen of the surface hydroxyl groups. The $(H^+)(e^-)$ pair may be detected by the EPR method, as in the case of the polycrystalline samples of MgO surface.[128] In rare cases, a mobile surface and active adsorbed hydrogen atoms may be found, which can reduce cations (as in the case of Cu^{2+} ions to $Cu,^0$ for samples doped Nb_2O_5 and WO_3).[129]

It is considered that on the nonreducible metal oxides, such as ZnO, MgO, CaO, and SrO, at room temperature and above, hydrogen dissociates heterolyticaly.[130] It may be presented by the following simplified scheme:

$$H^{\delta-} \quad H^{\delta+}$$
$$| \qquad |$$
$$-Zn^+ - O^- -$$

Metal oxide surfaces are not perfect crystalline substances. Certain types of defects on the surface of oxide materials play an important role in adsorption and catalytic reactions on the surfaces.[131] The perfect, or nearly perfect, metal oxide crystals are almost inert toward the adsorption of hydrogen and further dissociation.[132] According to Züttel,[132] H_2 does not dissociate on the structurally perfect TiO_2, but H-atoms may form on the defect sites, and they are very mobile on the surface, having surface diffusion coefficient $D = 1.8 \times 10^{-3}\exp(-0.59ev/kT)$ $cm^{-2}s^{-1}$, which at 300 K is about 2×10^{-13} $cm^{-2}s^{-1}$.

The mobility of the adsorbed H-atoms on the Pt-catalyst was shown by different experimental methods, including direct observations by STM-images.[133] A simple and nice example, showing the mobility of adsorbed hydrogen on the surface, is the following observation by Henrici-Olive.[134] WO_3, at ambient temperatures, does not react with hydrogen, while small amounts of Pt-metal added to WO_3 "initiate" the adsorption of hydrogen on WO_3. The amount of 1% Pt on WO_3 is sufficient to obtain dark blue hydrogen tungsten bronze of the composition $H_{0.4}WO_3$, within 10 minutes.[134] It is clear that the hydrogen atoms formed on Pt in dissociative adsorption of hydrogen easily diffuse to the surface of WO_3, forming new complex compounds.

However, as mentioned earlier, being bonded to the surface, adsorbed hydrogen is less reactive than that in the gas phase. In this regard, the following question arises: Why is the heterogeneous catalytic reaction of hydrogen oxidation at low temperature faster, particularly for Pt-catalyst than in the gas phase? The adsorption of hydrogen molecules on the solid surfaces of noble metals, including on Pt, is dissociative. Evidently, the adsorbed hydrogen atoms, formed as a result of the dissociation of molecular hydrogen on the surface, are notably different from the hydrogen in the gas phase. The reaction in the gas phase between hydrogen and oxygen practically does not occur at the same low temperatures. It may be initiated in different ways, such as heating, irradiation, addition of free radicals, or application of other external energy sources. All of these means have one aim: to generate reactive intermediates, radicals, ions, or electronically excited species in the gas phase. The appearance of the radicals in the gaseous mixture initiates reactions, which can lead to an explosion, as the temperature also increases due to the thermal effects. In the presence of the solid surface, part of hydrogen molecules reached to the surface, generating adsorbed hydrogen atoms. Hence, in the same conditions, the amount of active intermediates increases on the surface. In spite of a less chemical activity, the created concentrations of active intermediates on the surface, are enough for the occurrence of the reaction with oxygen at a much higher rate, additionally being accelerated by the energy released in surface reaction.

One of the important reactions of the adsorbed H-atoms is their recombination. Recombination has an important role in the mechanisms of gas-phase oxidation and combustion processes, determining the main kinetic regularities of the system. The recombination reaction of hydrogen atoms is a very exergonic process, releasing about 432.2 kJ/mol energy.[123] This energy must be transferred to the third body; otherwise, the formed H_2 may not be stable and will again be dissociated. Thus, the recombination of H-atoms in the gas phase is possible by trimolecular collisions:

$$H + H + M \rightarrow H_2 + M^*$$

where M is a third body and M^* is its energetically excited state. The surfaces of solid substances in the reaction vessel (the walls of the reactor, catalyst) may serve as a third body in the gas-phase reactions. A number of substances, for example, metals, play the role of catalyst in recombination reaction of the hydrogen atoms. The catalytic activity of metals in recombination increases by the following order Sn, Cu, Ag, Cr, Fe, W, Pd, Pt.[123] There are also inhibitors of hydrogen atom recombination, such as Teflon, silicon, and phosphoric acid, which may often be used as materials for coating the walls of the reaction vessel in experimental studies by participation of atomic hydrogen.[123] Kazansky and Pariiskii[135, 136] were the first to study the recombination of H-radicals on silica gel (H-atoms were obtained by γ-irradiation of surface hydroxyl groups of silica gel) using the EPR method, freezing hydrogen-atoms at 77 K. It was observed that a complex signal was caused by silica gel defects, consisting of four "superimposed" lines. Also observed was the high reactivity of this radical toward ethylene and hydrogen molecules.

In early investigations of the combustion of hydrogen, Nalbandyan and Shubina[136] showed a noticeable change in the reaction rate in hydrogen oxidation using the rods, coated with different substances, or metallic wires, which were inserted into or removed from the reaction zone. Measuring the sticking coefficients (ε) of hydrogen atoms in the specially chosen conditions of reaction, they found that $\varepsilon = 1$ for graphite or glass, coated with $ZnO-Cr_2O_3$, at 496°C; $\varepsilon = 0.1$ for gold, at 440°C; $\varepsilon = 0.09$ for platinum and $\varepsilon = 6.10^{-3}$ for tungsten, at 540°C. Noting the analogy of the surface effects between the chemical reaction and nuclear processes, in controlling the rate of the processes, Semenov remarked: "These experiments are very reminiscent of controlling nuclear reactions with cadmium rods; this is to be expected, since in both cases we have branched chain reactions and the active particles (H atoms and neutrons) readily captured by rods."[27 in Ch. 1]

3.5.1.2 Suggestions of the Reaction Mechanism in Hydrogen Oxidation

Although the reaction mechanism of the gas-phase oxidation and combustion of hydrogen suggested by Semenov and Hinshelwood successfully described the role of solid surfaces[137] and the main kinetic peculiarities related to them, some new phenomena discovered after the creation of the fundamental theory of the branched-chain reactions, required to reconsider some entrenched viewpoints. Later, the mechanism of this reaction was completed and successfully developed by Nalbandyan and Voevodski in their work *Mechanism of Hydrogen Oxidation and Combustion* (1949), taking into consideration the details of the heterogeneous reactions of atoms and radicals in the oxidation of hydrogen.[30 in Ch. 1] As mentioned in Section 1.3, subsequently, investigating the heterogeneous factors in hydrogen oxidation and combustion, Azatyan and co-workers[41, Ch. 1; 43, Ch. 1; 138–140] showed that in certain reaction regimes, the walls of the reactor participated not only in the initiation and termination of the homogeneous chains, known previously, but in the branching and propagation stages of the chain reactions. These conceptual interpretations helped explain a number of observations regarding oxidation and the combustion reaction of hydrogen with oxygen, such as hysteresis in the first and second self-ignition limits; self-acceleration of reaction, when concentrations of reactants decrease during the reaction below the initial self-ignition limit of the combustion peninsula; anomalous high rates of combustion, unpredictable by theory; and various other critical phenomena. Experimentally, also confirmed were reversible and irreversible changes of the solid surface in its nonstationary state during oxidation and combustion. All of this means that the adsorbed atoms and radicals interact with the species from the gas phase, generating new radicals or radical-like species on the surface.[30, Ch. 1; 128; 140] They partially recombine, desorb to the gas phase, or again react with the gas phase or surface species. As a result, the surface of the reactor walls or other solid-phase substances, on the one hand, participate in propagation of the reaction chains and, on the other hand, supply radicals to the gas phase.

Thus, according to these investigations, the solid surface, principally, may participate in all four main stages of the branched-chain-radical oxidation of hydrogen (initiation, propagation, branching, and termination). Some reaction mechanism suggestions, related to the heterogeneous stages, as a supplement to the general schemes of the oxidation or combustion of hydrogen, were briefly represented in Section 1.3. Here, note again that, in any case, the overall process rests as a bimodal reaction sequence (heterogeneous–homogeneous and homogeneous–heterogeneous, subsequently).

Let us take one example of the experimental results that evidences the heterogeneous propagation of chains in the oxygen–hydrogen reaction.[140] It was shown that hydrogen, preadsorbed on the surface of the reactor walls, reacted with dioxygen from the gas phase. The products (H_2O_2 and H_2O) were detected by the near IR emission spectra of $D_2 + O_2$ mixture (D-deuterium). Here, the formation of products, in an initial stage, is caused by the reactions $H + O_2$ that provide the heterogeneous propagation of chains. As a result, the first limit of self-ignition of $D_2 + O_2$ or $H_2 + O_2$ mixtures decreases in this case. A part of the vibrational excited HO_2 radicals may be inhibited by the addition of small amounts of propylene.

These notions also have importance for understanding other heterogeneous catalytic oxidation processes, particularly catalytic combustion of hydrocarbons.[141]

Heterogeneous catalytic oxidation of hydrogen on the solid substances was a subject of two centuries of permanent investigations. The different aspects of the kinetic behaviors of this reaction were revealed, and the suggestions about the mechanism were periodically changed, depending on the level of the applied experimental technique. In recent years, the reaction has been investigated mainly using combination of the "traditional" methods of heterogeneous catalysis and a great number of methods, created in surface science.

In catalytic oxidation or combustion reactions, the temperature of transfer of the reaction to the gas phase or ignition of the gaseous mixture is related to many factors, including the nature of the solid-phase substances, rates of the heat accumulation and dispersion, composition and thermal conductivity of the mixture, and others.[141] In general, a low activity of the surface in oxidation of hydrogen may be favorable for the heterogeneous–homogeneous reactions in catalytic combustion.[84 in Ch. 1]

The hydrogen–oxygen reaction was investigated in more detail on the surfaces of noble metals. The most efficient catalysts are the VIII group elements. The catalytic activity decreases in the following order: Pt, Pd, Ir Os Rh, Ru, Ni.[34] The relatively more investigated reaction in this area may be considered catalytic oxidation of hydrogen on Pt-catalysts at a wide range of temperatures (40–1300 K).[142–156]

Depending on the reaction conditions, the oxidation of hydrogen over catalysts may occur in different regimes. The existence of kinetic and mass diffusion control regimes in catalytic combustion reactions (concerning also the catalytic combustion of hydrogen on Pt) can be seen in Figure 1.5 (in Section 1.5). If the conditions of heat accumulation over the catalyst are sufficient for the increase in temperature, the mixture may burn at the temperature that is lower than the combustion temperature. For instance, in the presence of catalysts, hydrogen reacts with oxygen, producing water at much lower temperatures than the temperature of the flammable combustion. According to data taken from various literature sources, the observed minimal temperature of the transfer of the heterogeneous reaction into a heterogeneous–homogeneous reaction, via the transfer of radicals to the gas phase, for hydrogen oxidation on Pt, is about 500°C and above.[141 and ref. 31 within] There is no experimental evidence on the bimodal reaction sequences over this catalyst at low (from cryogenic to room temperature) and middle temperatures (from the room temperature to 500°C) ranges. It has been suggested that, on Pt-catalyst, at high temperature and pressure (1000 K, pressure 1–1000 mTorr), the mechanism of hydrogen oxidation includes the formation of OH intermediates followed by the dissociative adsorption of H_2 and O_2 on the surface.[142] I this case, suggestions of the reaction mechanism indispensably include desorption of OH radicals from the surface and their further reactions in the gas phase. It has been suggested that the formation of water occurred in two different ways: $OH + H \rightarrow H_2O$ and $OH + OH \rightarrow H_2O + O$, in/on both phases. The kinetic models show that the overall rate of reaction strongly depends on the ratio of H_2 and O_2 as well as on the rate of decomposition of OH ($OH \rightarrow O + H$) that becomes an important stage of reaction at high temperatures.[142] Note that, at temperatures of 520–910 K and oxygen pressure 500 mBar, in-situ X-ray investigations show the formation of an α-PtO_2(8×8) layer on Pt(111). In vacuum conditions, OH desorbs from Pt(111) at 720 K.[143] At elevated temperatures that reach to about 1000 K, the desorption of OH radicals formed on the surface of Pt(111) is also established by the data from LIF spectroscopy.[144] Desorption of OH species formed from chemisorbed O and H atoms on the single-crystal Pt(111) and polycrystalline Pt foil surfaces also have been observed using the LIF method in UHV conditions at the temperature ranges 1227–1479 K for Pt(111) and over 1283–1475 K for Pt foil.[145]

The above-mentioned and many other data indicate that the OH radicals formed on Pt-catalysts may be desorbed from the surface to the gas phase only at elevated temperatures (beginning from about 480–500°C). The reactions of OH radicals play a key role in the hydrogen oxidation mechanism at both low and high temperatures and in both the gas phase and the solid surfaces. At elevated temperatures via the escape of radicals OH from the Pt surface, the homogeneous reaction of combustion of hydrogen may be initiated, producing and involving other radicals. The surface reconstruction and restructuring, and even formation of a new surface phase such as PtO_2 become more obvious at elevated temperatures.[142, 145]

In regard to the mechanism of reaction on Pt(111) at low temperatures (from cryogenic to room temperature), the main mechanism suggestions were related with the Langmuir–Hinshelwood and often the Eley–Rideal type of heterogeneous reaction as well.[146] Mechanism suggestions based on the Mars–van Krevelen type of schemes, seemingly have little importance in this case. Presently, it was well established that the main intermediates of the oxidation of hydrogen on the metal surfaces, particularly on Pt(111), were adsorbed as O (or $O^{\delta-}$), H, and OH species. At low temperatures ($130 \leqslant T \leqslant 165$ K), the formation of OH species on Pt(111) was clearly shown by time-resolved electron energy loss spectroscopy (TREELS) in the presence of preadsorbed oxygen atoms.[146] The surface reactions of OH may explain the autocatalytic character and other peculiarities of the reaction at low temperatures (T < 170 K). There is no experimental evidence about the role of HO_2, or H_2O_2 and about other species (for example, H_3O^+) at low temperature oxidation of hydrogen on metals.

The existence of HO_2 radicals and their role in hydrogen oxidation at low temperatures on Pt(111), are disputable.[147] The problem is that the calculated activation barrier for oxygen dissociation (0.3 eV) is close to experimental values only at low coverage of the surface with oxygen (about 0.15ML), but it is higher and is not in agreement with the experiment at higher coverages. To explain this disagreement, it has been proposed a new pathway of the heterogeneous formation of water via the initial formation of adsorbed hydroperoxy species on the surface.[147] In this case, the activation barrier of the whole reaction will be about 0,4 eV, in spite of 1 eV activation barrier in the case of direct reaction on the surface between H and O adsorbed species at above 170 K. Thus, the formation of hydroperoxy species may occur by the surface reaction (the Langmuir–Hinshelwood pathway between adsorbed O_2 and H). Then, reacting with H, HO_2 gives two OHs and, finally, water. Presently, this hypothesis is based only on calculations, and there is no experimental evidence proving this mechanism. Indeed, the formation of HO_2 radicals in the gas-phase reaction is beyond doubt after the appearance of the OH radicals in the reaction zone.

A pretty pattern of the surface catalytic reaction may be obtained by the application of the STM technique.[149, 150, 154] It allows a "full length "STM movie" of catalytic events on the surface, particularly for the oxidation of hydrogen with oxygen on Pt(111).[35, 149–151] At 110–300 K, on oxygen-covered Pt(111), the atoms form hexagonal and honeycomb structures that are reactive toward the adsorbed H_2O phase. $O(2 \times 2)$ islands grow in exposure on hydrogen (135 K), forming a layer. STM images permit the observation of the islands O, OH and H_2O.[150] The fast reaction O and H_2O give two OHs, in turn, forming water and H. Reaction O + H is a rate-determining and relatively slow process. Additionally, the reaction H_2O + O has the unexpected stoichiometry of 2:1. The first molecule of water reacts with O and gives OH, while the second molecule dissociates into H and OH. The reaction mechanism was suggested by Ertl et al.[150, 151] based on experimental evidence obtained by a number of methods of the surface science, including STM images. The following oxidation scheme for hydrogen on Pt(111) is considered more suitable for existing kinetic data, at T > 170 K:

1. $O_2 + {}^* \rightarrow 2O_{ads}$

2. $H_2 + {}^* \rightarrow 2H_{ads}$

3. $O_{ads} + H_{ads} \rightarrow OH_{ads}$

4. $OH_{ads} + H_{ads} \rightarrow H_2O_{ads}$

5. $H_2O_{ads} \rightarrow H_2O_g + {}^*$

6. $H_2O_{ads} + O_{ads} \rightarrow 2OH_{ads}$

where * is surface active center.

Although at 170 K, water partially desorbs, the autocatalytic behaviors of reaction is preserved due to the reaction $H_2O_{ads} + O_{ads}$. This reaction is faster than reaction 4. The front of OH propagates at velocity 15 nm/min in the experimental conditions of the work of Ertl et al.[151]

Another good example of the "full-length STM movie" was demonstrated by Matthiesen et al.[157] in their investigation of the H adatom reaction with O_2 on the $TiO_2(110)$ surface, at temperature about 100–200 K. The STM method permitted identification of the intermediates of reaction: OH, HO_2, H_2O_2, H_3O_2. The final product of reaction was the dimer of water molecule $(H_2O)_2$. Real-time STM images (about 560) allowed the survey of all stages of the reaction: adsorption, surface diffusion, heterogeneous reactions, and formation of the new products. The STM movie reveals the displacement distribution of intermediate species, and therefore, also the probability of jumping from one adsorption site to the other.

There are yet other reaction mechanism suggestions at the low-temperature catalytic oxidation of hydrogen on Pt. On the basis of the available data, obtained by applying a number of modern methods, including a number of methods of the surface science, Verheij[152, 153] and Somorjai[154] proposed mechanisms of low-temperature oxidation of hydrogen on Pt(111) and other noble metals, which are different than the above scheme, including reactions 1–6. Note that in these mechanism suggestions[152] a central role is attributed to the OH radicals and their reaction on the surfaces of Pt(111).

A very different mechanism of hydrogen oxidation on Pt-catalyst was proposed by L'vov and Galwey.[155] Their oxidation mechanism scheme, other than the Langmuir–Hinshelwood and Eley–Rideal types of reactions, was based on consideration of the reactions

$$\left(PtO_2\right)_s + 2H_2 \rightleftarrows \left(Pt\right)_g + 2H_2O$$

$$\left(Pt\right)_g + O_2 \rightleftarrows \left(PtO_2\right)_g \rightleftarrows \left(PtO_2\right)_s$$

The first step is oxidation of hydrogen, whereas the second step is regeneration of PtO_2 on the catalyst. This mechanism suggestion explains the surface restructuring effect, the autocatalytic character of the reaction, the depressive influence of water to the rate of the process, and the phenomenon of the threefold variation of Arrhenius parameters with temperature in the 300–600 K range. In the same work, L'vov and Galwey describe the hydrogen oxidation on Pt in the framework of the so-called congruent dissociative vaporization (CDV) phenomenon, using thermochemical data and calculations. Here, however, there is no comment about the bimodal reaction sequences. In fact, according to the authors' descriptions, the oxidation reaction of hydrogen on Pt is a biphasic, multistep process (heterogeneous reaction, transfer to the gas phase of Pt atoms, homogeneous reaction, condensation of PtO_2 on the surface of Pt metal catalyst). Therefore, the overall process is a sequence of heterogeneous and homogeneous reactions, accompanied by mass transfer steps from one phase to other.

Although at low and middle temperatures the oxidation of hydrogen on the Pt-catalyst was considered to be a "purely" heterogeneous reaction, the diameter of the vessel and wall temperature have well-manifested effects, which, at first glance, is reminiscent of the "wall effect" in homogeneous reactions.[156] However, all these observations may be explained considering that desorbed water has a depressing effect on the reaction.

About half a century ago, in one of the works[156] on oxidation of hydrogen over the platinum wire, it had been observed the effect of the diameter of vessel on the reaction rate, strong retardation of the reaction rate by water, effect of the wall temperature on the rate of reaction, and, in addition, the non-reproductibility of results. Thin platinum 15 cm wire (diameter 0.2 mm) was used as a catalyst in a cylindrical vessel. The "wall temperature" had the following effect. Initially, it increased with the diameter of the vessel, and then it decreased (when the diameter was changed about 1–5 cm). On the other hand, at wire temperature 375 K, the rate decreased with the increase in wall temperature. Also observed was an initial increase and further decrease of the rate with the increase in the diameter (wire temperature was 250 K and unchanged). The mechanism suggestions were related to the "depressing effect" of desorbed water.

In 1817, Sir Humphry Davy's discovery of the oxidation of hydrogen on Pt opened up an area which he described as a "new and curious series of phenomena." Fast forward two hundred years to 2010 when scientists compared the results of the measurements of the Arrhenius parameters in this reaction in the last 50 years. At least 16 different values of the activation barrier in 15 works (12–80 kJ/mol) in oxidation on Pt(111) have been found.[155] There is no unique mechanism suggestions not only for the "purely" heterogeneous reaction at low-temperatures, but also for the heterogeneous–homogeneous reaction at elevated temperatures. Yet, there are many inconsistent angles in this reaction.

3.5.2 Oxidation of Hydrocarbons with Dioxygen

As shown earlier, the bimodal reaction sequence was observed in a great number of investigations of the oxidation of hydrocarbons with dioxygen in the presence of solid substances. Gas-phase transformations of methane to methanol,[158] singas, ethylene, formaldehyde; ethane to ethylene;[159] propane to propanol, acetaldehyde; toluene to benzadehyde; xylene to phthalic anhydride; propene to acrolein; and many other oxidation processes with dioxygen, over the metal oxide catalysts, in certain conditions, are heterogeneous–homogeneous processes.[160–166]

Among the hydrocarbons, methane has special importance[159] as the main component of natural gas and raw material for the chemical and energy industry. On the other hand, the oxidation reactions by the participation of methane are the subject of investigations in atmospheric chemistry, photochemistry, biochemistry, and different areas of industry and technic.

These brief notes on the gas-phase oxidation of hydrocarbons concern mainly the light alkanes, as examples of heterogeneous–homogeneous reactions with the participation of solid phases. However, their oxidation also involves the reactions of unsaturated hydrocarbons. Note that, in general, the main peculiarities of the bimodal oxidation of hydrocarbons were observed not only for gas/solid, but also for liquid/solid reaction systems.[166]

3.5.2.1 Notes about Oxidation of Hydrocarbons

The long-time investigations of the oxidation of light alkanes (methane ethane, propane, butane) with dioxygen in the gas phase, as well as in the liquid phase, have confirmed the chain-radical character of these reactions.[164, 165] As a rule, they are chain reactions with the degenerate branching of chains in both phases. Among many factors strongly affecting the rate and distribution of products, surface effects, such as dependence of the oxidation rate on the nature of the walls, S/V, or diameter of the reaction vessel, even the degree of "aging" of the interior surface of the reactor, were periodically revealed in investigations in different periods of times, during about a century.[165] Many examples, showing surface effects in the oxidation of hydrocarbons before the 1960s, were represented and discussed in Stern's excellent monograph.[164] At that time, most explanations of the surface effects were related to the heterogeneous reactions of the initiation and termination of chains in the autocatalytic oxidation of hydrocarbons. Some manifestations of the surface effects were assigned the role of the heterogeneous reactions of peroxides, particularly the different rates of their heterogeneous decomposition on the solid surfaces of a different nature.[165]

The following example, taken from the chemical literature of the 1960s, demonstrates the heterogeneous factors in the gas-phase oxidation of hydrocarbons, particularly propane.[166] The products of the partial oxidation of propane in the gas phase, in the glass, borosilicate, or silica reactors, at 300–500°C and atmospheric pressure in fuel-rich systems, were composed of a mixture containing CO, CO_2, C_3H_6, C_2H_6, C_2H_4, H_2 in the gas phase and CH_2O, CH_3CHO, CH_3OH, H_2O and H_2O_2 in the liquid phase, respectively.[166] It was found that the heterogeneous factors (surface/volume, S/V, nature of the walls of the reaction vessel) strongly affected the rate of reaction. Coating the walls of the reactor with KCl or NaCl reduced the rate of reaction but did not affect the composition of products, except for the yield of H_2O_2. On the other hand, the kinetic data showed a dependence of the induction period on the nature of the walls. The surface effects of the walls were assigned not only to the heterogeneous initiation and termination of chains, but also to the reaction of the decomposition of H_2O_2. As propylene was one of the main products of the reaction, it was suggested that the formation of H_2O_2 was a result of the following reactions

$$C_3H_7 + O_2 \rightarrow C_3H_6 + HO_2$$
$$HO_2 + C_3H_8 \rightarrow C_3H_7 + H_2O_2$$

Although the latter reaction was not the only way of the formation of H_2O_2 and minor products (C_2H_5OH, CH_3CHO) may also give H_2O_2, their role was not decisive.

Homogeneous decomposition of H_2O_2 was not known at the given conditions. Some kinetic peculiarities in this reaction were explained by hypothesizing the heterogeneous decomposition of H_2O_2, and further transfer of radicals to the gas phase. Later, Grigoryan and Nalbandyan investigated the heterogeneous decomposition of

H_2O_2 in greater depth, applying the EPR method to detect radicals in the decomposition of H_2O_2.[78 in Ch. 2] It was found that the hydroperoxy radicals appearing in the gas phase, as a result of the heterogeneous decomposition of H_2O_2, exceeded their homogeneous formation by a factor 10^4.[89 and within] Unfortunately, at that time, the contribution of heterogeneous stages to homogeneous oxidation in this reaction was not fully known.

Later, in the 1980s, it was found that the oxidation of hydrocarbons on a great number of heterogeneous catalysts, including metals and metal oxides, in certain conditions, had a well-manifested heterogeneous–homogeneous nature.[68 in Ch. 1; 84 in Ch. 1; 128 in Ch. 1; 167; 168; 169] Previously, the majority of these reactions were considered to be "purely" heterogeneous-catalytic processes. Examples include partial and total oxidation of hydrocarbons, mentioned above (see the references in 3.5.2), which occurred below the combustion temperature of hydrocarbons. In these reactions, the heterogeneous and homogeneous processes are coupled by both intermediates and energy transfers between two phases. A number of them were reviewed by Lunsford[129 in Ch. 1] and co-workers,[167, 168] as well as by Garibyan and Margolis.[84 in Ch. 1] The main subject of these investigations was oxidative coupling of methane, although the heterogeneous–homogeneous oxidation on the oxide catalysts was observed for many other oxidation reactions of saturated, unsaturated, and aromatic hydrocarbons. Direct evidence of the heterogeneous–homogeneous occurrence of catalytic reactions was obtained by the finding of free radicals in the gas phase, generated on the surface of catalysts, by the EPR method combined with the matrix isolation technique,[84 in Ch. 1; 129 in Ch. 1] often accompanied by mass-spectroscopic and chemical analysis of products and kinetic measurements. In these pioneering works of the heterogeneous–homogeneous catalytic oxidations, however, it was not clear which were essential parameters of the system favoring the transformation of a heterogeneous catalytic reaction into a heterogeneous–homogeneous reaction. Not only the nature of the catalyst, its surface characteristics, size of particles, but also the temperature, ratio of the initial reactants, type of reactor, kind of inert gases, and combination of many other reaction parameters were found to be important in these transformations. The transfer of the heterogeneous reaction into the heterogeneous–homogeneous process, for instance, was observed in the gas-phase oxidative creaking of ethane on a Pt-catalyst.[170] Previously, the role of heterogeneous factors in the gas-phase oxidative creaking of ethane (600–750°C) was investigated in the quartz reaction vessels with different diameter for a dilute mixture of ethane/$O_2 = 2$, and the results were compared with the reactions over a more active catalyst as Pt. In the gas phase, the reaction was known as a chain degenerate-branched process. The overall conclusion was that the increase of S/V through variation of the diameter or by packing the reactor with the quartz pieces most likely inhibited the reaction, indicating the heterogeneous termination of the homogeneous chains in the gas-phase reaction. On a Pt-catalyst, below 400°C, in similar conditions, the reaction selectively produced CO_2 and water. In 650° > T > 400°C, the products were CO, CO_2, H_2, while for T > 650 the formation of ethylene increased and CO_x decreased. At the same time, the preliminary heterogeneous process gradually transformed into heterogeneous–homogeneous process with the increase in the temperature. The surface effects in this work were explained primarily by the heterogeneous nature of the initiating stage. The authors of this work[170] remarked that the role of the surface was limited by the initiation of the exothermic reaction of oxidation. It was also found that with the increase in temperature, the further homogeneous gas-phase process gradually became "independent," little influenced by the surface reactions, in some exception of the chain termination reactions, as in many "ordinary" chain-radical reactions.

Further development in this area was related to achievements in heterogeneous catalysis and surface science. As dioxygen and hydrocarbon are two principal reactants, their adsorption and activation are primary, indispensable conditions of the reaction on the solid surface. Dioxygen may be activated in several ways (thermal, catalytic, mechanical, radiative, including light, or their combinations), forming either singlet oxygen or other oxygen species, as oxygen atoms, radicals, or ion-radicals, including their different forms in the adsorbed state (Section 3.4.2; Table 3.1). All these species are candidates to oxidize organic compounds partially or completely, by consecutive reactions on the surface.

The activation of the adsorbed hydrocarbon is realizable in similar ways. In the presence of the oxide catalysts, hydrocarbons may be activated thermally in their adsorbed forms, becoming a precursor for the formation of new intermediates of the oxidation. The forms of adsorbed dioxygen mentioned earlier, such as oxygen atom, superoxide-ion, and radical-ion O^- in the adsorbed state are able to abstract the hydrogen atom from the hydrocarbon, yielding free radicals or radical-like alkyl and alkoxy species. As a result, a surface reaction of a radical nature may take place, often by the partial transfer of part of the radicals into fluid phases. If the temperature rises, catalytic combustion may also take place. Later in this chapter, it will be shown that the reactivity of alkanes is essentially different from that of alkenes on the metal oxides. Usually, alkanes, unlike alkenes, are weakly adsorbed on the surface. In the presence of strong oxidants, such as O_2^- and O^-, which also are weakly adsorbed, the alkanes form alkyl radicals or alkyl species by homolytic breaking of the C-H bond. Strongly adsorbed alkenes, owing to the C=C interaction with the metal oxide surface, may form an allylic complex through a partial charge transfer, resulting in the heterolytic activation of allylic C-H.[135 in Ch. 1]

It is obvious that the activation and functionalization mechanisms of the C-H bond of hydrocarbons on the heterogeneous catalyst is one of the cornerstones for understanding heterogeneous–homogeneous oxidation in the systems of hydrocarbon–oxygen on metal oxides. Historically, the problem of C-H bond activation was investigated over a very long period of time, in homogeneous catalysis and in organometallic chemistry.[171] From the point of view of heterogeneous catalysis, in the 1970–1980s, Polish chemist J. Haber (1930–2010) developed the concept of electrophilic and nucleophilic reactions in the oxidation of hydrocarbons on transition metal oxides.[172] According to this concept, the heterogeneous oxidation of hydrocarbons with oxygen may be divided into two types of reactions: electrophilic and nucleophilic. Elucidating the catalytic role of the surface of metal oxides in oxidation of hydrocarbons, Haber presented the following conceptual description. Adsorbed molecular oxygen, in dependence on the nature of metal oxides, forms active oxygen species, including singlet oxygen, superoxide ion-radical, or oxygen atom. They are all electrophilic species. Different spectroscopic evidences of their existence, as well as reactivity toward adsorbed organic compounds, including hydrocarbons, were given earlier in Section 3.4. Oxidation through this species may be named extrafacial, unlike interfacial interaction with lattice O^{2-} oxygen ions, which are nucleophilic species. If nucleophilic oxygen reacts with the adsorbed organic molecule forming oxygenated product and the latter is desorbed from the surface, the formed surface vacancy may be filled with oxygen from the gas phase. The suggested redox reaction (reduction of metal-ion by organic molecules and its reoxidation by gas-phase oxygen) corresponds to the Mars–van Krevelen type of mechanism (see Chapter 1, Section 1.9). Haber considered that the redox reaction might be not only consecutive, but also concerted, although the catalytic oxidation reactions were usually known as stepwise processes.

In both cases, the oxidation reaction of alkane must include the rupture of one of the C–H bonds of hydrocarbons. On the solid surface of the metal oxide, all adsorption parameters of alkane are dependent on the coordination state of the metal cation in interaction either with the hydrogen atom or the C–atom. If coordination is through one hydrogen atom of alkane, a complex of the σ (M...H – C...) type may be formed. Unsaturated hydrocarbons adsorb forming π-complexes with the coordination atom of metal oxide. Carbon atom becomes prone for nucleophilic addition of O^{2-} or O_2^{2-}, which are not oxidizing agents (lattice oxygen). They can insert into activated hydrocarbon, forming oxygenated species, as well as a vacancy, in the surface reaction.[173] The resulting oxygenates can desorb or react with other species on the surface.

According to Haber, the hydrocarbons may be coordinated with the transition metal ions by two or three hydrogen atoms, forming σ-complexes. For instance, the C-H bond of hydrocarbons may be coordinated by the surface ion forming a triangular structure (coordinated through electrophilic metal ion, forming two electrons for three center bonds), analogous to the gas-phase trimer H_2 ... H^+. For example, carbocation $C_3H_7^+$ forms this kind of structure, exhibiting unusual bridging. Oxidative addition of alkane to metal ion may take place, when the metal ion is capable of interacting with the hydrocarbon by back donation of electrons to the antibonding orbital of the C-H bond (see Section 3.4.1.3).

In activation of the adsorbed hydrocarbon by abstraction of the hydrogen atom on the metal oxides, principally there exist two possibilities of cleavage of C-H bonds, homolytic and heterolytic. Heterolytic cleavage of C-H takes place when the hydrocarbon is adsorbed through the C-H bond forming the following complexes on the acid–base pair of surface sites: (a) formation of carbanion with the abstraction of proton, and (b) formation of carbocation with the abstraction of hydride anion

$$
\begin{array}{ccc}
\underset{/}{\overset{\diagdown}{C}} \overset{-\sigma}{-} \overset{+\sigma}{H} & \qquad & \underset{/}{\overset{\diagdown}{C}} \overset{+\sigma}{-} \overset{-\sigma}{H} \\
\overset{+n}{M} \quad O^{2-} & & O^{2-} \quad \overset{+n}{M} \\
a & & b
\end{array}
$$

Accordingly, further reactions will take place through the participation of ionic forms of the fragments of hydrocarbons. Haber also described three cases (here, denoted as c, d, and e) of the possible homolytic rupture of the C-H bond, two of which (d and e) lead to the formation of free radicals or radical-like species. They may take place:

(c) when the transition metal cation is electrophilic and there is a vacant coordination site; the C-H bond may interact with the metal cation forming a complex either by two-electron donation with three center bonding or by back donation of the electron pair of the antibonding d-orbital of the metal ion

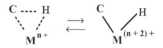

(d) when the cation of the surface is easily reducible, possessing the basic properties; the formation of alkyl radical and proton therefore becomes feasible:

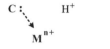

(e) when the atomic or molecular ion-radical forms of oxygen are capable of producing alkyl radicals by the reactions of hydrogen atom abstraction from the hydrocarbon.

Haber noted that, in general, the transition metal oxides are nonstoichiometric compounds, and their composition, particularly their surface structure, depends on the composition of the gas- or liquid-phase environment. Therefore, an equilibrium exists between the gas-phase oxygen and lattice O^{2-}, on the one hand, and between the vacant sites and adsorbed forms of oxygen, on the other hand. The rise of temperature shifts the equilibrium to the dissociated form of oxides, and the adsorption of electrophilic oxygen species rises. Haber believed that "the surface adapts itself to the requirements of the reaction, generating a more facile pathway for a concerted rearrangement of electrons and nuclei." Haber's conceptual statements were related to the restructuration and reconstruction of the oxide catalysts during reaction, as well as to the sensitivity of the reaction to the surface geometry and structure.[172, 174]

In certain cases, the size of the catalyst particles changes the reaction rate. The stepwise catalytic reaction may proceed on different faces of the crystal with different rates, as the quantity and nature of active sites, as well as various crystalline defects, are different on diverse plans of the crystalline structure. Moreover, the concentrations of terraces, steps, and kinks are different depending on particle sizes. Boudart named this phenomenon "structure sensitivity of the heterogeneous reaction."[175] For example, the oxidation of ethylene to ethylene oxide on the silver catalyst is considered a structure- or surface-insensitive reaction, while the hydrogenation of ethylene on the Ni-catalyst is structure sensitive.[175] The sensitivity of the reaction to the surface geometry and structure was hypothesized in 1929 by Balandin.[176] He predicted that successful catalytic action required that the accommodation of molecule on the surface geometrically corresponds to the surface atom groups, named by him multiplets. The best example was hydrogenation of benzene on the metallic Ni-catalyst. It appeared that the adsorption of aromatic ring on the metal surface occurred through the so-called "sextet complex" (ensemble); i.e., adsorbed benzene was laid on the flat surface of the catalyst. Many years later, it was found that in hydrogenation of benzene on the metallic Ni-catalyst, the favorable configuration on Ni(111) and Ni(001) was in exact agreement with the Balandin prediction: 2.48 A spacing between the Ni atoms *on the surface.[177] The Balandin theory is now of only historical importance, serving as an archetype of the concept of structure-sensitive reactions. The importance of the geometric—in modern terminology, structure and steric factors in heterogeneous catalysis, especially in the catalytic oxidation of hydrocarbons—was confirmed by many other examples.[178] However, according to some opinions, this phenomenon is not general in heterogeneous catalysis.[96 in Ch. 2] Obviously, the heterogeneous constituent of the heterogeneous–homogeneous reaction may also be structure sensitive or insensitive. Examples of the effect of the particle size on the rate of bimodal reaction (structure-sensitive reaction) are given by Sinev,[179] for instance, for iso-propane conversion over γ-Al_2O_3 and alumina-supported V-containing catalyst at 550°C.*

General regularities of the activation and functionalization of the C-H bond of hydrocarbons, particularly of alkanes on the metal oxides, may be interpreted on the basis of both thermodynamic and kinetic data. Investigating several heterogeneous–homogeneous reactions of oxidation of alkanes on the metal oxides, Sinev[179] proposed the following relationships between thermodynamic and kinetic data. It was considered that the thermal effect ΔH (enthalpy) of reaction in activation of a C–H bond of alkanes was the sum of energy of the expenditure E_{ex} and energy of stabilization E_{st} of the species on the surface of the solid substance, particularly on the catalyst

$$\Delta H = E_{ex} + E_{st}$$

In the first approximation, E_{ex} is the energy required to pass an activation barrier for alkane molecules. From known thermochemical data, the approximate values of E_{ex} were estimated, and it was shown that the reaction of homolytic dissociation of C-H bonds had minimal energy E_{ex} in comparison with cases of heterolytic abstraction of the H atom on a strong basic or Lewis acidic center. If one considers an ionization of alkane, the estimated values of E_{ex} will be much higher than in the case of the homolytic dissociation of C-H. For instance, in the reaction

$$[O] + RH \rightarrow [OH] + R \tag{i}$$

$$E_{ex} = D_{R-H} \tag{i'}$$

where D_{R-H} is dissociation energy of the C-H bond in alkane, for instance, in CH_4 equal to $E_{ex}(i') = 431$ kJ/mol (reaction i). In the case of the heterolytic breaking of the C-H bond, occurring by the charge transfer, as in the reactions

$$[O^{2-}] + RH \rightarrow [O^{2-} \cdots H^+] + R^- \tag{ii}$$

and

$$[M^{n+}] + RH \rightarrow [M^{n+} \cdots H^-] + R^+ \tag{iii}$$

Correspondingly, E_{ex} will be equal to

$$E_{ex} = D_{R-H} + I_{H^+} - I_{R^-} \tag{ii'}$$

and

$$E_{ex} = D_{R-H} - I_{H^-} - I_{R^+} \tag{iii'}$$

where $[M^{n+}]$ is the metal center and I_{H^+}; I_{R^-}; I_{H^-}; and I_{R^+} are ionization potentials and electron affinities of R and H species. According to the data presented in the work of Sinev, for methane E_{ex} (ii') = 1630 kJ/mol, and E_{ex} (iii') = 1308 kJ/mol (calculated on the basis of the thermodynamic data given in Refs.[179, 180]). It is obvious that homolytic activation of the C-H bond is an energetically more favorable pathway than the heterolytic one. In other words, E_{st} is minimal to compensate E_{ex} in the case of the homolytic rupture of the C-H bond. It is noteworthy that the energy of binding hydrogen on the surface of the oxide catalyst mainly compensated E_{ex}, having 250–470 kJ/mol for the strength of O-H in the reactions of methane (see Ref-s[within 179]) for the majority of oxide catalysts (such as Li_2O/MgO, K_2O/Al_2O_3, and PbO/Al_2O_3 catalysts for oxidative coupling of methane-OCM). Therefore, when E_{ex} is comparable with E_{O-H}, the formation of R, seemingly is facilitated.

The thermodynamic data concerning other possible reactions, such as ionization of alkanes with the participation of the hole $[h^+]$ centers forming the $[h^+ \ldots e^-]$ pair, also were examined

$$[h^+] + RH \rightarrow [h^+ \cdots e^-] + RH^+$$

for which E_{ex} is equal to the ionization potential $RH - e^- \rightarrow RH^+$. In another case, for alkanes other than methane, when the synchronous abstraction of two hydrogen atoms by the reaction

$$[M^{n+} \cdots O^{2-}] + C_nH_{2n+2} \rightarrow [M^{n+} \cdots H^{\delta-} \cdots C_nH_{2n} \cdots H^{\delta+} \cdots O^{2-}]$$

occurs, giving either an unsaturated hydrocarbon and hydrogen, or hydrocarbon and water, the values of lowest limits of ΔH_{ol}-ΔH_{al} (the differences of ΔH of olefin and initial alkane formation) were estimated. In two recent cases, the comparison of thermodynamic data for E_{ex} in homolytic and heterolytic dissociation of C-H bonds leads to the same conclusion: that the radical pathway is thermodynamically more favorable than other ways by the charge transfer. Analogous analysis of thermodynamic data for light alkanes, other than methane (ethane, propane, butane and iso-butane), gave similar results.[179]

As a general conclusion from this analysis of thermodynamic data, it may be mentioned that nearly all important kinetic parameters of the complete or partial oxidation of hydrocarbons, including the oxidative coupling of light alkanes, as well as oxidative dehydrogenation reactions, are dependent on two important factors: (1) the affinity of the surface active site to the hydrogen atom $E_{[O-H]}$ and (2) the oxygen-binding energy $E_{[O]}$ on the surface of solid substances, catalysts or inhibitors. Here, it must be explored the feasibility of the expected observation of the heterogeneous–homogeneous proceeding of reaction. In other words, the nature of the solid substance—more precisely, the nature of the active sites—determines the possibility of forming the radical-like or other species in favor of the heterogeneous–homogeneous occurrence of the process. However, it is noted that steric factors, consideration of the isomerization reactions, energies of the co-adsorption of other compounds, and a number of other factors are able to change the basic thermochemistry of the C-H bond breaking in oxidation reactions of the organic compounds. Apparently, a number of factors make possible the heterolytic rupture of C-H bonds in heterogeneous reaction on some oxide catalysts, for instance, as in the case of the activation of butane on Cr(III)-O surface sites.[181]

Summarizing a great number of experimental data, Kung[182] gave the following qualitative description of the desired surface properties in activation of C-H bonds, producing oxygenated products in the partial oxidation of light alkanes. Weakly adsorbed surface oxygen must be a strong Bronsted base and the cation of oxide a soft acid. Subsequently, for the desorption of products, oxygenated or nonoxygenated, including radicals, oxygen must be in deficit or in limited available amounts. On the other hand, the soft acidic properties of cation sites are desirable to prevent the further combustion of hydrocarbon on the surface.

Another important problem that arose beginning in the 1980s involving oxidation of hydrocarbons by mixed heterogeneous and homogeneous mechanisms, was the revelation of the complex interactions between the components of two phases, including the active intermediates. In this regard, beginning with the early investigations of heterogeneous–homogeneous catalytic reactions, it was clear that "one reaction could not occur without the other, so the system can properly be described as involving an interaction between homogeneous and heterogeneous processes."[183] However, in early investigations, in heterogeneous–homogeneous oxidation of hydrocarbons on the oxide catalysts, without excluding the possibility of the heterogeneous branching and propagation of homogeneous chains, the existing experimental results were explained in terms of the generation and termination of homogeneous chains on the surface. Presently, there are no sufficiently proven examples of the oxidation of hydrocarbons, where the homogeneous chains additionally propagate the reaction by heterogeneous pathways. However, there is sufficient evidence showing that the interaction of homogenous chain reactions with the solid surface is not limited only to the initiation and termination stages. For instance, an interesting case of the surface reaction, indirectly indicating the feasibility of the surface radical reactions, other than the generation and termination of chains, was found in the oxidation of some saturated and unsaturated hydrocarbons, such us methane, ethane, butane, ethylene, and propylene with another oxidant, H_2O_2, at relatively low temperatures (420–730 K).[184] The reaction vessel (cylindrical tube, the length of which was about 20 cm) was packed by the broken pieces of silica gel. The oxidation of the given hydrocarbons with H_2O_2 began at different temperatures. It was shown that already in the initial (entry) part of the reactor (about 1 cm away from the entrance along axis of reactor) hydrogen peroxide entirely decomposed in the absence of hydrocarbons. At the same time, it was observed that for a mixture of hydrocarbon + H_2O_2, the progression of the reaction through the entire length of the reactor was different for various hydrocarbons. In the case of ethane, propane, and propylene, the reaction expanded, with the increasing rate in the reactor mainly forming acetaldehyde from ethane, acetone from propane, and propylene oxide from propylene, while in the case of methane and ethylene the rates of the formation of products (methanol and acetaldehyde, correspondingly) were unchanged in different parts of the reactor. Obviously, the hydrogen peroxide decomposes through the heterogeneous pathway on a SiO_2, by the formation of radicals, part of which diffuse to the volume. They were detected by the EPR method. In other experiments in analogous conditions, carried out by the same method, it was also known that the oxidation of the mentioned hydrocarbons with H_2O_2 produced radicals RO_2 and HO_2 in the gas phase. Surprisingly, in the experimental conditions described, the radicals, detected in the gas phase from the olefin + O_2 mixture, were only HO_2 radicals, and their amount was less than that in the decomposition of H_2O_2 in the absence of hydrocarbon. It was also known that at the chosen temperatures, for each mixture, the reactions between the hydrocarbon and H_2O_2 in the gas phase, between the grains of SiO_2, were absent or were at least very slow. According to the authors of this work, these inconsistent results may be explained by the hypothesis of the existence of heterogeneous reactions between the hydrocarbons and radicals, reaching to the surface of the neighboring SiO_2 grains from the gas phase. In other words, appearing in the gas phase, the radicals did not develop a chain reaction in the volume, but passing through the gas phase, finally, reacted on the surface. In this regard, the authors' conclusion reminds the sequence of reactions and diffusion processes in the famous experiment of Kautsky (Section 3.4.2.7). The authors suggested the following scheme for the heterogeneous reactions on the SiO_2 grains:

$$H_2O_2 + 2Z \rightarrow 2(ZOH)$$

$$ZOH + H_2O_2 \rightarrow Z(H_2O) + HO_2$$

$$ZOH + H_2O_2 \rightarrow ZHO_2 + H_2O$$

$$Z + RH \rightarrow ZRH$$

$$ZR + O_2 \rightarrow ZRO_2 \rightarrow Z + RO_2$$

$$ZA + ZB \rightarrow products$$

$$ZRO_2 + ZRH \rightarrow ZR + other\ products$$

where A, B = OH, HO_2, R, RO_2, and Z are the active centers of the surface of SiO_2.

Presently, there are no unified mechanism suggestions for the oxidation of hydrocarbons on the heterogeneous catalysts, and therefore, also for the oxidation in the regime of the bimodal reaction sequence. In a first approximation, all metal oxide catalysts in oxidation of the light hydrocarbons may be divided into two groups: reducible and nonreducible oxides. Usually, it has been considered that the activation of the C–H bond in oxidative dehydrogenation on the reducible metal oxides, as transition metal oxides, occurs by the redox mechanism, and on the nonreducible metal oxides, as alkali or earth-alkaline metal oxides, by the mechanisms involving the adsorbed oxygen species.[185] However, this is not a general rule (see Section 3.4.2.2). In this context, without précising the exact nature of the surface site, the suggestions of the reaction mechanism on the reducible metal oxides, corresponding to the Mars–van Krevelen concept (see Section 3.4.1.2), may be presented by the following scheme:[185]

$$1/2O_2 + Z \rightarrow [Z\text{-}O]$$

$$C_nH_{2n+2} + [Z\text{-}O] \rightarrow C_nH_{2n+1} + OH\text{-}[Z]$$

$$C_nH_{2n+1} + [Z\text{-}O] \rightarrow C_nH_{2n} + OH\text{-}[Z]$$

$$2OH\text{-}[Z] \rightarrow H_2O + [Z\text{-}O] + [Z]$$

This scheme assumes the formation of OH and carbanion by redox reactions.

In the oxidation of light hydrocarbons on the metal oxides with the cations of variable oxidation state, the following sequence of the formation of intermediates and products may be expected

$$C_nH_{2n+2} \xrightarrow{[o]} [H_{2n+1}]_{ads} \xrightarrow{[o]} [C_nH_{2n}]_{ads} \xrightarrow{[o]} [C_nH_{2n}O]_{ads} \xrightarrow{[o]} [C_nH_{2n}O_2]_{ads} \xrightarrow{[o]} CO_2, H_2O$$

$$\downarrow \qquad\qquad \downarrow \qquad\qquad \downarrow \qquad\qquad \nearrow$$

$$olefin \rightarrow aldehyde \rightarrow acid\ or\ anhydride$$

where [O] is an active oxygen species.[178]

The schemes of the proposed mechanisms for the heterogeneous oxidation of light alkanes, for instance, for oxidative dehydrogenation reactions, suitable mainly for nonreducible metal oxides, are usually presented as the following (for simplicity, the indexes *ads* were not mentioned, but the scheme completely corresponds to the presentation of the heterogeneous reactions.)[185]

$$C_nH_{2n+2} + [O] \rightarrow C_nH_{2n+1} + OH$$
$$C_nH_{2n+1} + O_2 \rightarrow C_nH_{2n} + HO_2$$
$$C_nH_{2n} + [O] \rightarrow [C_nH_{2n}O]$$

as well as

$$C_nH_{2n+2} + 2O_2 \rightarrow C_nH_{2n+1}O_2 + HO_2$$
$$C_nH_{2n+1} + O_2 \rightarrow C_nH_{2n}O_2H$$
$$C_nH_{2n+1}O_2 + C_nH_{2n+2} \rightarrow C_nH_{2n+1}O_2H + C_nH_{2n+1}$$
$$C_nH_{2n}O \rightarrow CO_x + H_2O$$
$$2HO_2 \rightarrow 2OH + O_2$$
$$C_nH_{2n+2} + OH \rightarrow C_nH_{2n+1} + H_2O$$
$$2C_nH_{2n+1} \rightarrow C_{2n}H_{2(2n+1)}$$

Not all of the reactions in this scheme are proven by kinetic, spectroscopic, or other experimental data. Some of them, however, are supported by experimental evidence. The schemes were discussed on the example of a number of catalysts, such as V-Mg-oxides.[185]

In the above scheme, an important role was assigned to $C_nH_{2n+1}O_2$ radical-like species, as well as $C_nH_{2n+1}O_2H$. In oxidation of ethane on rare earth and alkaline metal oxides, for instance, above 653 K, it has been suggested that alkyl peroxy radicals and alkyl peroxide may be formed directly in interaction with O_2 by the abstraction of H-atom from hydrocarbon or hydrocarbon radicals.[170, 185] Escaping from the surface, they usually react near the surface zone of the catalyst.

3.5.2.2   Oxidative Coupling of Methane

Chemistry of the oxidation reactions of methane with dioxygen or air is very rich in its diversity.[158, 179] Besides the complete oxidation or combustion, applied mainly as a source of energy, by the formation of CO_2 and H_2O, the partial oxidation reactions of methane, such as transformations to methanol, formaldehyde, singas, or other useful products, were always priority areas of chemical research. In general, the partial oxidation of methane has a large spectra of products: C_2H_4; CH_3OH; HCHO; HC(O)OH; C_2H_6; H_2; higher hydrocarbons and their oxygenates, CO; CO_2, and H_2O. Different types of reactions, heterogeneous, heterogeneous–homogeneous, and homogeneous may occur in a chemical system involving a methane–air/oxygen mixture, in the presence of solid-phase substances, including catalysts or inhibitors. Depending on the nature of the solid phases, the reaction type also depends on the temperature and ratio of the reactants. The combustion of methane (non-catalytic) is considered a high-temperature homogeneous oxidation (T > 873 K), while partial oxygenation, oxidative coupling, and oxidative dehydrogenation may also be heterogeneous–homogeneous or heterogeneous processes.[158] Among partial oxidation reactions of methane, the oxidative coupling and oxidative dehydrogenation processes have special importance from the practical point of view, promising useful chemical synthesis.

The reaction of the oxidative coupling of methane with molecular oxygen, occurring mainly on the metal oxides at 500–900°C, is as follows[179]:

$$2CH_4 + 1/2O_2 \rightarrow C_2H_6 + H_2O \qquad \Delta H = -177 \text{ kJ/mol}$$

It is usually accompanied by the reaction of dehydrogenation of ethane, producing ethylene

$$C_2H_6 + 1/2O_2 \rightarrow C_2H_4 + H_2O \qquad \Delta H = -105 \text{ kJ/mol}$$

The overall reaction of the conversion of methane to ethylene by oxidative coupling is highly exothermic.

$$2CH_4 + O_2 \rightarrow C_2H_4 + 2H_2O \qquad \Delta H = -281.8 \text{ kJ/m}$$

Although this reaction has not yet had large-scale industrial application, it opens an attractive opportunity to resolve one of the main problems of synthetic chemistry: to furnish the chemical industry with basic raw materials such as ethylene.

Note that not only ethane and subsequently ethylene, but other important chemical products of the oxidative co-condensation, may be obtained through the participation of methane. For example, using the heterogeneous catalysts, such as alkali metal salts on the supports (group I and group II metal halides, for instance, 1–1.5% KBr/SiO_2), acetonitrile with methane and oxygen transforms to propionitrile and acrylonitrile at 700–800°C,[186, 187]

$$CH_3CN + CH_4 + O_2 \rightarrow CH_3CH_2CN + CH_2 = CHCN + HCN + H_2O$$

toluene with methane and oxygen, to styrene, ethylbenzene, and some other products.[187]

$$C_6H_5CH_3 + CH_4 + O_2 \rightarrow C_6C_5CH = CH_2 + C_6H_5C_2H_5 + C_6H_6 + C_6H_5OH + H_2O$$

Although there are many other good examples of the successful synthesis of valuable products by the participation of methane and oxygen,[158, 187] the oxidative coupling of methane is the most investigated among them and may be considered a well-proven example of the heterogeneous–homogeneous reaction on a great number of the metal oxide catalysts.

Oxidative coupling of methane (OCM) on the metal oxides or on the mixture of metal oxides was intensively investigated beginning in the 1980s. The formation of C_2 hydrocarbons in the oxidation of methane was first observed by Nersissyan, during his doctoral research (before 1979) work[188] that is not well known by English readers.[189] Previously it was reported that in the oxidation of methane over alumina and silica, the solid substances generated free radicals escaping to the gas phase.[190] In English-language research, Mitchell and Waghorne in their USA Patent reported the formation of the coupling products by oxidation of methane in 1980.[191] Further intensive investigations in this field were directed primarily toward exploration of selective catalysts for the oxidative coupling of methane giving ethane and, by further oxidative dehydrogenation, giving ethylene as unique process. An enormous number of solid substances—about several hundred—were tested as catalysts in order to design industrially suitable process in these investigations.[192] By 2011, the number of scientific publications in this field had reached about 2840 (2700 articles and 140 patents).[192]

Note that the aim of this section is not to review all or even a part of this enormous data source, but to demonstrate the importance of considering the heterogeneous–homogeneous mechanisms involved in investigating heterogeneous catalysis.

Catalysts that are active in the oxidative coupling of methane may be divided into several groups:[193] reducible metal oxides ($NaMnO_4/MgO \cdot SiO_2$); nonreducible metal oxides (Li/MgO, SrO/La_2O_3); halogen-containing solid oxides (BaF_2/Y_2O_3); and solid electrolytes ($SrCe_{0.9}Yb_{0.1}O_{3-x}$). Among them, MgO- and La_2O_3-based catalysts containing alkali (Cs, Na) and alkaline-earth (Sr, Ba) metals as dopants are known to be more selective in OCM.[192]

In the review of Zavyalova et al.,[192] the maximal conversion of methane to C_2 hydrocarbons was 72–82%, and the respective yields were 16 to 26%. Presently, however, these limits have been found to be remarkably enhanced.

Oxidative coupling of methane on the metal oxides may be divided into high (600–750°C, often –900°C) and low temperature (<600°C) processes. In early investigations of OCM, the type of reaction (heterogeneous or heterogeneous–homogeneous) was a subject of lively discussions. Contradictory facts proving either one or another opinion, Bhasin and Campbell[194] explained by the differences between the reaction parameters and specific behaviors of catalysts in different conditions, first of all, by the differences between the temperature and partial pressure of reactants. They found that the heterogeneous–homogeneous mechanism operates at relatively high temperatures and that only the heterogeneous mechanism may be predominant at low temperatures.

The most important fact uncovered in the majority of investigations devoted to the oxidative coupling of methane on a great number of metal oxides, their mixtures, and other more complex catalysts on the base of the metal oxides, rests the heterogeneous–homogeneous character of the overall process.[190, 195] The appearance of CH_3 radicals (precursor of ethane), generated on the surface, provided the main evidence for a process having a typically bimodal character. Furthermore, some interesting kinetic features were observed in the investigations of the oxidative coupling of methane on a variety of the oxide catalysts. For instance, on the one hand, in certain reaction conditions, the selectivity of the formation of C_2 hydrocarbons decreases with the increase in the concentration of oxygen. On the other hand, the selectivity increases with the increase in temperature in certain intervals. In general, however, the selectivity of C_2 products decreases with the increase in the conversion of methane.

The primary reaction of OCM is activation and functionalization of the C–H bond of methane.[196] The energy of the C–H chemical bond in methane is very high, 439 kJ/mol for H_3C–H (an average value is 414 kJ/mol), indicating that methane is a very stable compound in ordinary conditions. The activation of methane by rupture of the C–H bond is a rate-determining step for OCM processes. In this regard, investigation of the mechanism of activation of the C–H bond in methane molecules has a special theoretical interest.[197]

There are no general or unique mechanism suggestions of the activation of the C–H bond of methane on the metal oxides completely explaining all experimental observations of oxidative coupling. In general, the transformation of methane may proceed by the following two types of catalytic reactions, beginning with either homolytic cleavage of the C–H bond

$$\left[Cat - O^-\right] + CH_4 \rightarrow \left[Cat\right] - OH^- + CH_3$$

or heterolytic cleavage of C–H bond (Section 1.9)

$$\left[M^{2+}O^{2-}\right]+CH_4 \rightarrow \left[M^{2+}\right]-CH_3^- + OH^-$$
$$\left[M^{2+}\right]-CH_3^- \rightarrow \left[M^{2+}+e^-\right]+CH_3$$

In both cases, the surface of oxide catalyst generates methyl radicals.

One attempt to explain the mechanism of the heterolytic cleavage of the C–H bond is related to the acid–base properties of metal oxides.[182] Taking into account the fact that methane is weakly acidic, we may suggest that the catalyst or site on the surface must be strongly basic.[198] Although there are some basic metal oxides that are relatively active in the oxidative coupling of methane (for instance, SnO_2[199], MgO[200]), it is not the only important factor in this reaction. Moro-oka,[201] investigating the properties of acid–base sites on metal oxides, found that heterolytic cleavage was less preferable than homolytic cleavage. CH_3 and proton may be adsorbed without charge changes of metal ions

$$CH_4 + M^{n+}O^{2-} \rightarrow M^{n+}CH_3^- + O^{2-}H^+$$

Earlier, in Section 3.5.2.1 it has been mentioned that in many cases, thermodynamic data indicate a preferentially homolytic cleavage of the C-H bond.

In any case, independent of the precise nature of the active sites on the surface, all existing experimental data are evidence of the formation of CH_3 radicals or radical-like species on the majority of oxide surfaces. The "escape" of methyl radicals to the reaction volume makes possible the formation of ethane by recombination (coupling) of CH_3 radicals. The recombination of two methyl radicals in the gas phase is nearly a barrierless reaction, apparently, occurring through the participation of the "third body" (any M), accepting the excess of the released energy. Consequently, ethane undergoes dehydrogenation-forming ethylene. This pathway is essential in the majority of the heterogeneous–homogeneous reactions of the oxidative coupling and the dehydrogenation of methane in variety of metal oxides by oxygen, at least, at 600–750°C.[198, page 463]

One of the best catalysts of the oxidative coupling of methane is alkali metal doped metal oxides.[129 in Ch. 1] The oxidative coupling of methane over Li/MgO (Li-doped) occurs through the formation of CH_3 radicals, transferred to the gas phase. In the presence of oxygen, part of methyl radicals transform to CH_3O_2 radicals. These radicals were detected by the method of matrix-isolation EPR and also their presence was confirmed by beam mass-spectrometric data, reported by Lunsford et al.[129 in Ch. 1; 202] Other evidence about the role of CH_3 radicals was obtained in experiments with isotopically labeled methane CD_4 in the presence of the same catalysts. These experiments confirmed the formation of C_2H_6, CD_3CH_3, C_2D_6 as coupling products. However, direct experimental proof of the existence of methyl radicals in the gas phase in reaction conditions of the oxidative coupling of methane, generated on the catalyst Li/MgO, was not first obtained until 2013, using synchrotron VUV(vacuum-UV) photoionization mass spectrometry.[203] The reaction, yielding mainly C_2 hydrocarbons, was carried out at 750°C, using a mixture of 8% CH_4 and 4% O_2 in argon.

Lunsford[129 in Ch. 1] gave the following interpretations to experimental kinetic observation in the reaction of the oxidative coupling of methane on alkali metal modified (doped) catalysts. As the reaction $CH_3 + O_2 \rightarrow CH_3O_2$ is reversible, the rise of the temperature leads to shifting the equilibrium to the left, thereby increasing the selectivity of the formation of ethane.[129 in Ch. 1; 202] On the other hand, increase in the concentration of O_2, in certain limits (named "oxygen window"), will also help increase ethane selectivity.[204]

According to the authors of the review of Horn et al.,[205] the nature of the surface active site remains open, while, without any proof, Lunsford et al.[129 in Ch. 1, 202] postulated that the active sites of the surface are Li^+O^- on MgO. It has been hypothesized that the oxygen-centered radicals on the metal oxides (MO) may be sites of the hydrogen abstraction from methane by the homolytic cleavage of the C–H bond. In other words, the following surface radical reaction was suggested:[183]

$$MO/-O^{\bullet} + CH_4 \rightarrow MO/-OH + CH_3^{\bullet}$$

Here, the desorption from the surface of the catalyst of CH_3 requires 227 kJ/mol energy.[205] The adsorption of dioxygen results in the formation of adsorbed superoxide radicals on the surface, releasing 191 kJ/mol energy. It has been suggested that the coupling of O_2 adsorption with this reaction makes the overall reaction nearly thermally neutral:

$$\left[Mg^{2+}O^{2-} \right] + CH_4 + O_2 \rightarrow (O_2^{\bullet-}) \left[HO^-Mg^{2+} \right] + CH_3^{\bullet}$$

In this and other mechanism suggestions, the further dehydrogenation of ethane to ethylene, as well as formation of the secondary oxygenated products, through reactions of CH_3O_2 radicals in the gas phase, were expected.

Lunsford and co-authors[202] also showed that in a number of metal oxides, such as ZnO, CeO_2, La_2O_3, and Li/MgO in the presence of gas-phase oxygen, the sticking coefficient of methyl radicals increased. It may happen as a result of the following reaction of methyl radicals via the electron transfer

$$M^{(n+1)+}O^{-2} + CH_3^{\bullet} \rightarrow M^{n+}(OCH_3)^-$$

or

$$CH_3^{\bullet} + O_2 + O_{(s)}^{2-} \rightarrow CH_3O_{(s)}^- + (O_2^{\bullet-})_{(s)}$$

The further reactions of methoxy species usually produce carbonate species at room temperature, which decompose only at high temperatures by the formation CO, CO_2. Thus, the presence of oxygen in the gas phase in different amounts drastically changes the reaction pathway.

The above suggestions, regarding the mechanism, may not be considered commonly accepted, as many authors, accepting the involvement of $CH_3^{\bullet}$ radicals in the process, suspect the role of these radicals in the gas-phase formation of the main reaction products, such as ethane in OCM. In their review, Horn and Schlogl[205] remark that the coupling in the gas phase is an "inefficient step" and may take place in the presence of a third collision partner, while direct formation of HCHO, then H_2 and CO by consecutive reactions, may be much more efficient. Therefore, they consider that "whether this coupling step occurs in the gas-phase or at the catalyst surface cannot be concluded from the product pattern (e.g., the formation of the symmetrically substituted coupling products in labelling experiments with CH_4/CD_4 mixtures, as C_2H_6, CH_3CD_3 C_2D_6 and C_2H_4, CH_2CD_2, C_2D_4), also not from the mere detection of gas-phase $CH_3^{\bullet}$ radicals."[205] However, these arguments are not completely convincing. The recombination of methyl radicals on the surface, nearly with the same probability, may be "inefficient" owing to "the high activity of methyl radical attack by active species, which leaves less time for methyl radical dimerization," as mentioned by Cavani and Trifiro.[206]

Surface oxygen directly bonds with methyl, for instance, on Rh(111) p(2x1)O, beginning from 100 K. Further reactions lead to the formation of CO (400 K), CO_2 (500 K) and small amounts of HCHO (290 K).[207] Theoretical calculations show that the presence of surface oxygen dramatically changes the energetic state of the adsorbed methyl radicals.[208]

From our point of view, in this case, a somewhat exhaustive elucidation may be given by the investigation of this reaction in the condition, excluding homogeneous recombination of methyl radicals, if these conditions are available experimentally. In any case, understanding the detailed reaction mechanism requires a much more detailed, simultaneous investigation of both heterogeneous and homogeneous reactions and their interactions.

From the practical point of view, the main problem with OCM processes rests on the low and limited C_2 selectivity[209] in formation of ethane. As mentioned earlier, selectivity decreases with the increase in conversion of methane, well demonstrated for a large number of catalysts in OCM.[192] None of the efforts over the last three decades to obtain a yield of ethane or ethylene and a conversion of methane, appropriate to the industrial application, has given successful results. The essential reason for this failure, we believe, lies in the indispensable existence of the homogeneous constituent of the heterogeneous–homogeneous catalytic reaction. Apparently, the highly exothermic gas-phase radical reaction in the presence of dioxygen is one cause for the low yield of ethane. On the other hand, catalysts activating methane, nearly by equal efficacy, are able to activate products of OCM in their further transformations.[192] In this context, note again that only the revelation of the detailed mechanism of the heterogeneous–homogeneous processes may open a desired way to successful industrial application.

OCM processes also open alternative pathways in solving the problem of the conversion of methane to aromatic compounds.[210] As aromatics are liquid-phase compounds, it is also an appropriate means to obtain liquid-phase hydrocarbons from natural gas. For example, interesting results were obtained by applying the catalyst, composed from Mo-oxides, supported on ZSM-5, in dehydroaromatization of methane to liquid aromatics, primarily benzene and hydrogen.[211, 212] In primary reaction, methane transforms into ethylene, and further reactions on the catalyst produce a mixture of compounds containing 70–80% aromatic hydrocarbons. Apparently, the reaction occurs through the participation of O-atoms of the anchored Mo-oxides. Catalyst deactivates during the reaction by the formation of coke; however, it may be completely regenerated by dioxygen treatment after the reaction.

In addition, among the partial oxidation reactions of light alkanes, the oxidative coupling of methane is not only a reaction of oxydehydrogenation, occurring by the heterogeneous–homogeneous advancement of the process. Reactions of oxydehydrogenation of ethane, propane, iso-butane, and others with dioxygen on the certain metal oxide catalysts also are complex heterogeneous–homogeneous processes, the gas-phase constituent of which is a radical reaction.[206, 213] For example, in oxydehydrogenation of ethane giving ethylene

$$C_2H_6 + O_2 \rightarrow C_2H_4 + H_2O$$

on the metal oxides of IA and IIB groups, including the Li^+/MgO catalysts for oxidative coupling of methane, the generated C_2H_5 radical transfers to the gas phase (T > 600°C). Transition metal oxides of f-elements (rare earth metals), as well as transition metals of d-elements (e.g., vanadium), often containing Mo and Nb, also are selective catalysts in oxydehydrogenation of light alkanes.[206]

3.5.2.3 Catalytic Oxidation of Propylene

One of the industrially important reactions, exhibiting heterogeneous–homogeneous characteristics, is selective oxidation of propene to acrolein on the bismuth-molybdate catalysts (different compositions of Bi_2O_3 and MoO_3) at about 300–450°C[214–216]

$$CH_3CH = CH_2 + O_2 \rightarrow CH_2 = CH\text{-}CHO + H_2O$$

High selectivity of the formation of acrolein (90–95%) over bismuth-molybdate catalysts, with air or oxygen, was usually observed in this reaction.[214] Minor products were carbon oxides, acrylic acid, acetaldehyde, and formaldehyde.

One of the interesting peculiarities of this oxidation is the occurrence of the reaction on the catalyst in the seemingly absence of oxygen. Indeed, it may happen when lattice oxygen O^{2-} participates in the reaction.[217] The vacancies formed may be replenished by dioxygen from the gas phase, according to the Mars–van Krevelen type of mechanism. The role of lattice oxygen in this reaction was established by a number of experiments, including isotopically labeled oxygen ^{18}O.[217, 218]

It was assumed that the primary reaction was the formation of symmetrical allylic complex on the surface by the abstraction of hydrogen atom of propylene

$$CH_3\text{-}CH = CH_2 + M^{n+} + O^{2-} \rightarrow (H_2C \cdots CH \cdots CH_2)_{ads} + M^{(n-1)+} + OH^-$$

Here, the adsorbed π-allylic complex $(H_2C \cdots CH \cdots CH_2)_{ads}$ is a nearly electroneutral radical-like specie. For instance, the first evidence of its formation was obtained by kinetic isotope effect[217, 218] and isotopic carbon ^{13}C atoms experiments.[219]

Apparently, the second stage of the stepwise oxidation of propylene is abstraction of the second hydrogen atom with the incorporation of lattice oxygen in hydrocarbon structures forming acrolein.[220]

Investigating the kinetic peculiarities of this reaction, Keulks[221] reported evidence of the heterogeneous–homogeneous oxidation of propylene on the oxide catalysts. It was found that the formation of products in oxidation was dependent on postcatalyst volumes, which were variable in experimental conditions. Keulks suggested that following the formation of the allylic complex on the surface, the second stage might be formation of allyl peroxy radicals or radical-like species with molecular oxygen adsorbed from the gas phase. Consequently, the formation of allyl hydroperoxide and its further decomposition gives acrolein.

$$CH_2 = CH\text{-}CH_2OOH \rightarrow CH_2 = CH\text{-}CHO + H_2O$$

Mass-spectrometric evidence of the surface-generated allyl radical, appearing in the gas phase, was obtained by Dolejsek and Novakova.[222] Soon, the Lunsford group of investigators[223] obtained strong evidence of the heterogeneous–homogeneous nature of the oxidation of propylene by molecular oxygen on Bi_2O_3 and γ-bismuth molybdate catalysts (at 713 K). Heterogeneously generated allyl and allyl peroxy radicals were detected by the EPR method, applying the matrix isolation technique. Apparently, the formation of 1,5-hexadiene by dehydrodimerization propylene on Bi_2O_3 in this experiments was the result of the recombination of two allyl radicals, generated on the surface. It was also shown that allyl and allyl peroxy radicals were in equilibrium and that small amounts of oxygen favor the formation of allyl and relatively more amounts of allyl peroxy radicals. The transfer of π-allyl radicals into allyl peroxy radicals was well demonstrated through comparison of the characteristics of their EPR spectra. One conclusion, in this regard, was that MoO_3 played the role of a sink for radicals produced elsewhere in the system.

Relatively recently, on the basis of some theoretical investigations (by the so-called DFT + U, variant of the density functional theory), it was reported[224] that in the primary reaction of hydrogen abstraction by lattice oxygen of the bismuth molybdate catalyst, the most active site would be bismuth perturbed molybdenyl (Mo = O) oxygen. According to this work, the allyl radical may freely diffuse across the catalyst surface and form allyl or alkoxy molybdenates, becoming the precursor of acrolein, in the abstraction of the second hydrogen. Apparently, only Mo ions will be reduced in this case, and the presence of Bi ions may favor the required electronic environment. On the other hand, the formation of acrolein only on Bi_2O_3 was negligible,[223, 225] but it is remarkable at phase composition: 1 : 01 : 0.008 corresponding to the amounts of γ-bismuth molybdate, α-bismuth molybdate and MoO_3.[223] However, the exact nature of the active sites of the formation of allyl radicals, as well as the mechanism of their desorption from the surface, remains unclear.

Although, due to its industrial importance, this reaction has been investigated for more than 60 years, the mechanistic details regarding neither the heterogeneous constituent nor the homogeneous constituent of the overall heterogeneous-homogeneous process have yet to be fully established.

3.5.2.4 Conclusions about Heterogeneous–Homogeneous Oxidation of Hydrocarbons

The above examples of the heterogeneous–homogeneous reactions of oxidation concern a large class of organic compounds: hydrocarbons. The peculiarities of the bimodal character of these reactions, in certain conditions, are caused by the generation of alkyl radicals on the solid surfaces. The primary stage of the activation of hydrocarbon in oxidation, in the presence of solid phases begins by their adsorption or co-adsorption with oxygen, forming surface complexes. Cleavage of the C-H bond of hydrocarbon, occurring in either an homolytic or heterolytic pathway, is in dependence on the nature of the solid phases, results in the formation of the alkyl and other radical-like species on the surface. At least three general pathways may be predicted in further transformations of these species: heterogeneous oxygenation; dehydrogenation (which may be accompanied by isomerization); and desorption from the surface and diffusion to the fluid phases. All of these transformations occur in the presence and under the influence of dioxygen and reactive oxygen species on the surface, in defined quantities, determining the fate of surface species. Often, the probability of their desorption increases with the temperature related to the entropic factor. Although the details of mechanisms of the generation and desorption of radicals and other species from the surface are not well known, in certain reaction conditions, their appearance in the gas phase is a well-established experimental fact. The radicals, in the gas phase, either recombine or initiate a homogeneous chain reaction, depending on the reaction conditions. In this regard, one of the most important stages is the formation of alkyl peroxy radicals. Apparently, the following important reversible reaction exists in the fluid phases:

$$R + O_2 \rightleftarrows RO_2$$

On the other hand, the radicals RO_2 are precursors of hydroperoxide, which usually are responsible for the degenerate branching of chains in the general schemes of the mechanism of the oxidation of hydrocarbons.

Thus, in general, in partial oxidation and oxidative dehydrogenation of alkanes on the oxide materials, the concurring reactions of primary active species determine the type of oxidation (oxygenation, oxidative dehydrogenation, or other). As the surface intermediates have very different reactivities, the desorbed chemical entities may be radicals or stable oxygenated products.

The described general scenario of the progression of the oxidation of hydrocarbons in the presence of the solid substances is presently the base of the accepted concepts of the heterogeneous–homogeneous reaction. Unfortunately, we have limited information on the heterogeneous reactions by the participation of active surface intermediates. Almost nothing is known about the interactions of the gas-phase and solid-phase reactions. However, the existing experimental data and reaction modeling results indicate that the radical reactions, including chain-radical reactions, are nearly analogous to the gas-phase reactions and may be advanced on the surfaces in interaction with the gas-phase reaction. On the other hand, presently, there is no well-proven example of the oxidation of hydrocarbon on the catalysts occurring by the "purely" radical mechanism analogous to the mechanisms of the gas-phase processes, repeating classical schemes, in exception of some polymerization (polycondensation) reactions on the catalyst (also by the participation of dioxygen), mentioned in Section 2.4.2.

It should be considered a serious omission when, in the mechanism suggestions of the oxidation reactions of alkanes over catalysts, the homogeneous constituent of the heterogeneous process is not taken into consideration. For instance, Védrine and Fechete[178] state that in oxidative dehydrogenation, "at very high temperatures, the ethane molecule reacts on a catalyst to produce an alkyl radical, which desorbs from the surface to undergo homogeneous gas-phase reactions. At lower temperatures, the alkyl species remain adsorbed on the surface." This generalized notion about the probability of the heterogeneous–homogeneous pathway of the oxidation reactions is disputable. Note that Lin and Bent[226] investigated the desorption of alkyl, particularly methyl radicals, generated on the surface of Cu(111) at low temperatures (T < 400 K) and discussed different aspects of this feasibility. The probability of the desorption of radicals, generated on the solid surfaces, depends not only on the temperature, but also on a number of other parameters of the reaction system, such as chemical nature of the surface or nature of the sites on the surface, ratio of the initial reactants, and total pressure. Therefore, the kinetic models of the partial oxidation of hydrocarbons must consider the possibility of the existence of the homogeneous constituent not only at very high temperatures of heterogeneous catalytic reaction, but also at relatively low temperatures, in certain other reaction conditions, and its inverse influence (feedback) to the heterogeneous reaction.

Finally, note one of the important peculiarities of the oxidation of hydrocarbons in a great number of metal oxides catalysts. Nearly always, the partial oxidation of hydrocarbons is accompanied by the formation of products, such as alcohols, ketones, aldehydes or carboxylic acids, and organic peroxides or H_2O_2, at least, in small amounts. As will be shown in the next Section 3.5.3, their heterogeneous reactions usually play a significant role in establishing the heterogeneous–homogeneous mechanisms of hydrocarbon oxidation. For instance, under comparable conditions, the probability of the "escape" of radicals from the surface of metal oxide catalysts to the gas phase in the oxidation of alcohols, aldehydes, and carboxylic acids is 5 to 10 times higher than for hydrocarbons.[227]

3.5.3 OXIDATION OF ALCOHOLS WITH DIOXYGEN

In certain conditions, the oxidation of alcohols with dioxygen over the heterogeneous catalysts, noble metals Pt and Ag, mainly on the support materials, as well as transition metal oxides (Fe_2O_3, TiO_2, V_2O_5, Cr_2O_3, MnO_2, NiO, CuO, ZnO and others) are heterogeneous–homogeneous reactions. [84 in Ch. 1; 228] Like oxidation of hydrocarbons, the reaction intermediates, formed on the surface, are able partially to desorb and diffuse to the fluid phases, continuing the reaction in the both phases. In this section, we will attempt to demonstrate the existence of some interrelationships between the heterogeneous and homogeneous reactions in the complex mechanisms of the oxidation of alcohols.

Usually, the main products of the partial oxidation of normal alcohols are aldehydes and, those of secondary alcohols are ketones.[229, 230] Depending on the temperature and other parameters of the reaction, as well as the activity of the solid-phase catalyst, the oxidation of alcohol may be partial or complete. The further oxidation of alcohols to aldehydes or ketones, in the presence of dioxygen, may continue to the corresponding carboxylic acids, which, in their turn, may be oxidatively decomposed to the products of the complete oxidation, CO_2 and H_2O. Note that there are remarkable differences between the gas- and liquid-phase oxidations of alcohols in the presence of the heterogeneous catalysts. In the case of the liquid-phase reaction, alcohols may be oxidized by dioxygen beginning from the ambient temperatures, while in the gas phase, at relatively elevated temperatures. The differences were well demonstrated in the example of the oxidation of ethanol with dioxygen on the nanostructured Pt-catalyst in the both phases ($60°C$).[231] For instance, it was shown that the turnover frequency was two orders of magnitude higher in the liquid phase than in the gas-phase oxidation of ethanol to acetaldehyde and CO_2. Correspondingly, the activation energy was five times higher in the gas–solid interface than that in the case of the liquid–solid interface. Additionally, the observed quantities of CO_2 were two times higher in the gas phase than in the liquid medium. Analogous differences were also observed in the oxidation of isopropanol to acetone over Pt-catalyst in two phases.[232] In this regard, we may expect important differences in the mechanisms of the oxidation of other alcohols over the Pt-catalyst in the cases of the gas and liquid phases. Apparently, these differences are caused primarily by the so-called molecular orientation deviation of alcohol molecules on the platinum surface, observed and investigated by the sum frequency vibrational spectroscopy.[231, 232] Here, attention will be paid mainly on the gas-phase oxidation of alcohols with dioxygen over the heterogeneous catalysts, as the mentioned effects in the case of the liquid phase do not exist in the gas phase.

3.5.3.1   Heterogeneous Generation of Intermediates in Oxidation of Alcohols

The first gas-phase partial catalytic oxidation of alcohol, methanol, producing formaldehyde, on a Pt-wire was discovered by Hofmann in 1867.[233] Another noble metal, Ag, was for a long time an industrially important catalyst in the production of formaldehyde, at 600–725°C, by partial oxidation of methyl alcohol (80–90% selectivity).[234] Experimental investigations of the oxidation of methanol with oxygen on the noble metal surfaces demonstrate that the reaction intermediates are alkoxy,[235] often also dioxymethylene O-CH$_2$-O species.[234] Even early investigations (by the temperature-programmed desorption spectroscopy, using deuterated reactants) of methanol conversion over Ag(110) monocrystals showed that adsorbed methanol formed methoxy CH_3O and H surface species.[235] In turn, methoxy species on the solid surface may be decomposed into CH_2O (formaldehyde) and H-atoms. Further reactions depend on the surface coverage of oxygen, as well as temperature and many other parameters of the surface reaction. Formation of alkoxy species on the surfaces of solid-phase catalysts may be considered a more common pathway in adsorption and further oxidation of small alcohol molecules, even on the surfaces of different metals, metal oxides, and many supported catalysts. The data obtained by a number of extensive studies, applying the methods of the surface science—X-ray photoelectron diffraction (XPD), reflection absorption infrared spectroscopy (RAIRS), near-edge X-ray absorption fine structure (NEXAFS), temperature-programmed desorption (TPD), low-energy electron diffraction (LEED), scanning tunneling spectroscopy (STM), high-resolution electron energy loss spectroscopy (HREELS), energy scanned photoelectron diffraction (PED), surface extended X-ray absorption fine structure (SEXAFS), etc.—and theoretical calculations showed the formation of metoxy species, particularly on Cu(110), Cu(111), Cu(100), Ag(111), Ni(111), Ni(110), Pt(111), and many other metallic surfaces in the oxidation of alcohols (references for these works are given in the work N'dollo et al.[236]).

On Pt(111) methanol adsorbs through the O-atom, forming methoxy species by dissociative chemisorption

$$CH_3OH + 2Pt \rightarrow CH_3O\text{-}Pt + H\text{-}Pt$$

in a relatively small coverage (0.36 ML).[237] The oxygen atom of methanol is bonded with the Pt-surface in the atop configuration (the O-C axis is tilted 63° from the surface normal), and the H-atom moves to the adjacent atop position).[236, 237] Experiments in UHV conditions, in adsorption of methanol on Pd(111),[238] showed that at low temperature (100–200 K), dissociative adsorption was the only way of reaction, while with the increase of temperature (T > 300 K), the probability of breaking CO in alcohols increased. Also observed was the formation of CO, carbon deposit, and formyl (CHO), at millibar pressure on the single crystal surfaces. The final products of oxidation were formaldehyde, CO_2, and H_2O, as the dehydrogenation is accompanied by the oxidation by surface oxygen species. Surprisingly, the formation of ethanol in the oxidation of methanol was also observed. One pathway for the formation of ethanol may be the breaking of C-O in alcohol by the formation of CH_x (x = 1 – 3). Also to be considered is that the gas-phase formation of ethanol takes place after the desorption of CH_3 radicals to the gas phase, in this case.[236, 238] However, this is a minor channel in the overall reaction, and the main pathway of methanol oxidation, according to many investigations, especially at low temperatures, remains the formation of alkoxy (particularly methoxy) species on the surface, as the main intermediate. In this regard, nearly all suggestions concerning the mechanism of the oxidation of alcohols on pure metals also are based on the reactions of alkoxy species by "purely" heterogeneous but not heterogeneous–homogeneous mechanisms, especially at low and medium temperatures (300–650°C). McCabe[239] suggested that, in the presence of oxygen, the following sequence of the formation of surface intermediates might explain the heterogeneous reactions of the oxidation of alcohols on the surface of Pt

$$CH_3O \rightarrow H_2CO \rightarrow HCO \rightarrow CO \rightarrow CO_2$$

Analogous sequences of reaction intermediates have also been proposed for the oxidation of alcohols on the more "modern" catalysts, as Au or AuPd nanocatalysts. Formerly, "ordinary" Au was considered to be inactive in the oxidation of alcohols.[236]

3.5.3.2 Bimodal Reaction Sequences in Oxidation of Alcohols on Metals and Metal Oxides

Nearly all the suggestions regarding the reaction mechanisms in oxidation of alcohols on the noble metals are based solely on "purely" heterogeneous reactions. In this regard, the noble metal catalysts, supported by the metal oxides, exhibit somewhat other peculiarities, depending on reaction conditions. For example, the oxidation of propanol was investigated on Pt/γ-Al$_2$O$_3$ (0.64% Pt).[123 in Ch. 1] Conversion of the gaseous mixture alcohol/air was started at 473 K, while, in the absence of a catalyst, the detectable quantities of products (CO, CH_4, ethylene) was observed only at 770 K. The main products of the catalytic reaction for 15 vol.% of O_2 and 20 vol.% of alcohol mixtures were propionic aldehyde and CO_2. It was shown that the overall process was a heterogeneous–homogeneous reaction, initiated by the catalyst. Evidence of the radical and chain character of the gas-phase constituent of reaction was obtained by the detection of radicals, applying the matrix isolation EPR method. The recorded spectra were assigned to the peroxy radicals (RO_2). It was also shown that only in a certain ratio of initial reactants in the gaseous mixture the reaction occur by the "purely" heterogeneous pathway. In the case of the increased concentration of oxygen in gaseous mixtures, in a certain range, the homogeneous reaction became observable. Kinetic data showed bell-shaped curves of the temperature dependence on the concentration of radicals in the gas phase. Analysis of products showed the existence of the negative temperature coefficient (NTC) of the reaction rate at a certain interval of the temperature variation. Analyzing the EPR spectra, their kinetic behaviors and kinetic data on the accumulation of products and especially, the dependence of the rate of reaction on the pressure, Ismagilov et al.[123 in Ch. 1] concluded that changes in the leading active centers in the gas phase took place with the increase in temperature. They proposed the following sequence of the formation of radicals: initially, RCO and RCO_3 radicals were formed, and then, in secondary reactions, they were transformed into RO_2 radicals:

$$RCO + O_2 \rightarrow RCO_3$$
$$RCO_3 + RH \rightarrow RCO_3H + R$$
$$R + O_2 \rightarrow RO_2$$
$$RCO_3H \rightarrow RCO_2 + OH$$
$$RCO_2 \rightarrow R + CO_2$$
$$RCO \rightarrow R + CO \qquad\qquad\qquad (m)$$

Apparently, the appearance of the RCO radicals on the heterogeneous catalyst is a result of the dissociation of alcohol with the subsequent oxidative dehydrogenation.

It was also assumed that the further transformation of R produced olefin:

$$R + O_2 \rightarrow R'(olefin) + HO_2$$

In this scheme, reaction (m) may be mainly responsible for the decrease in the reaction rate with the increase in temperature (the NTC region), notably limiting the branching of chains.

Another important observation in this work[123 in Ch. 1] was the revelation of temperature oscillations on the catalyst in a certain ratio of the initial reactants, arising together with the periodic self-ignition of the reaction mixture, which may be assigned to the interaction of homogeneous and heterogeneous processes. According to the same authors,[123 Ch 1] the main role of the catalyst at low-temperature oxidation was initiation of the homogeneous reaction by the radicals, transferred to the gas phase from the surface. In general, the catalytic combustion of the reaction mixture over the same catalyst was found to take place at only T > 1000 K, while in the given case, the heterogeneous–homogeneous reaction, was observed at 600–700 K. Nothing was known about the mechanism of the formation of radicals on the surface and their transfer to the gas phase, or about interactions of the radicals formed in the gas phase with the heterogeneous reaction. The EPR spectra, recorded by freezing the gas-phase mixture,[240, 241] were a complex mixture, at least, of RO, RO_2, RCO_3, and HO_2 radicals, the initial origins of which were unclear in this case.

The next example of the overall heterogeneous–homogeneous reaction was oxidation of propanol on the $CuCr_2O_4/\gamma$-Al_2O_3, (673–773 K) catalyst, exhibiting analogous kinetic behaviors.[123 in Ch. 1] Ismagilov et al.[123 in Ch. 1] concluded that at low pressure the reaction was a heterogeneous process, and at atmospheric pressure it became heterogeneous–homogeneous. The contribution of the homogeneous component increased with the increase in the ratio O_2/ROH. Here, the homogeneous process was a typically chain–branched reaction.

One important peculiarity of the oxidation of alcohols on noble metal supported catalysts, such as Pt/γ-Al_2O_3, was the fact that the generation of radicals and, consequently, the heterogeneous initiation of homogeneous chains may be experimentally observed only at high alcohol concentration in the mixture.[123 in Ch. 1] In contrast, at low concentration of initial reactants and relatively low temperatures, the reaction occurs as a "purely" heterogeneous process. Therefore, the following question arises: Why, in some conditions, the process is "purely" heterogeneous but in other conditions a heterogeneous–homogeneous reaction on the same catalyst? Moreover, how does the "purely" heterogeneous reaction on a Pt-supported catalyst transform to the heterogeneous–homogeneous reaction in oxidation of alcohol? In this regard, important information was obtained through investigation of methanol oxidation by the EPR matrix isolation and infrared spectroscopic methods, on the surface of Pt catalyst, supported on Al_2O_3 and SiO_2, at 770–920°C.[240–242] The work[240] was one of the rare examples of the investigation of heterogeneous–homogeneous reactions, where the homogeneous and heterogeneous constituents and their interactions were studied in comparable conditions. Here, the reactions of methanol and oxygen were investigated separately on the support materials (in the absence of Pt), on a pure Pt, on Pt supported by Al_2O_3 or SiO_2, as well as on their mechanical mixture (Pt + support). The role of the composition of the initial mixture and the temperature dependencies of the rates of reactions were investigated in the mentioned systems applying both methods. The formation of different types of surface methoxy species, as well as formate groups, were observed by the FTIR (Fourier-transform infrared) method. Three of them (denoted as I–III) are terminal and bridging methoxy species. They had the following spectral characteristics on Al_2O_3 at 523 K.

(I) CH_3O–Al_{oct} (v_s(C–H) = 2806 cm^{-1}), terminal

(II) CH_3O–Al_{tetr} (v_s(C–H) = 2825 cm^{-1}), terminal

(III) $CH_3O < (Al)_2$ (v_s(C–H) = 2845 cm^{-1}), bridging

Correspondingly,

$$\delta_{as}(C\text{–}H) = 1460 \text{ cm}^{-1}, \; r_{||}(CH_3) = 1185 \text{cm}^{-1},$$

$$\delta_s(C\text{–}H) = 1440 \text{ cm}^{-1}, \; v(C\text{–}O) = 1095 \text{ cm}^{-1}.$$

At the same time, the isolated methoxy groups with v_{as}(C–H) = 2997 cm^{-1}, v_{as}(C–H) = 2959 cm^{-1}, v_s(C–H) = 2857 cm^{-1}, and δ(CH$_3$) = 1450 cm^{-1} were detected. Additionally, the hydrogen-bonded groups on Al_2O_3 were identified as v(O–H) = 3400–3550 cm^{-1}. These data on spectral characteristics evidence the presence of a CH_3O-Al fragment on the surface, differently bonded with the surface. The nearly analogous overall image was obtained for SiO_2 support.[240] The FTIR spectra were recorded in both the absence and presence of oxygen. If the temperature increases (573–773 K), the intensities of spectral lines of methoxy of types II and III decrease and the intensity of type I increases. At the same time, some spectral evidence was obtained about the formation of formate groups, which in turn may be decomposed to CO_2 and H_2O. On this basis, the following sequence of the formation of intermediates and products (considering also the participation of lattice oxygen) was suggested:

$$CH_3O < (Al)_2 \; (bridge) \rightleftarrows CH_3O{-}Al \; (terminal)$$

$$CH_3O{-}Al \; (400 \; K) \rightarrow HCOO{-}Al(500 \; K) \rightarrow CO_2 + H_2O$$

$$CH_3O{-}Al \; (400 \; K) \rightarrow CH_2O_{ads} \, (500 \; K)$$

In the presence of oxygen, it was suggested that the formation of formate-methoxylate anion complexes stabilized on the surface of Al^{3+} ions:

$$[HCOO{-} \, ...CH_3O{-}]$$

The proton formed on the surface may be stabilized on the neighboring O^{2-} center, giving the surface hydroxyl group.

On the silica gel surface, alcohol is chemisorbed on the siloxane bridges or bonded by hydrogen.[240] The methoxy formed is stabilized on the isolated silicon cations. It is important that radicals in the gas phase were detected only from the surfaces of the support samples (SiO_2 and Al_2O_3, in the absence of Pt on their surfaces) and their mechanical mixture with Pt, but not on Pt supported by SiO_2 and Al_2O_3. This observation was explained by the following assumptions. The methoxy radicals were formed chiefly on Pt, but consequently, they readily diffused to the surface of the support.

The contribution of CH_3O and CH_3O_2 radicals in the detected EPR spectra has been estimated, and the following data were obtained.[123 in Ch.I; 240, 242] The radicals were detected for a methanol mixture with oxygen 10:1 over a mechanical mixture Pt + SiO_2 and in a ratio of 5:1 on Pt + Al_2O_3. The spectra were assigned to CH_3O radicals (98% and 100% for Al_2O_3 and SiO_2, respectively). In all other compositions of the gaseous mixture (2:1, 1:1 and 1:2), the detected radicals were assigned mainly to CH_3O radicals, with small amounts of CH_3O_2 on both supports (7–24% for SiO_2 and until 12% for Al_2O_3). Radicals were not detected in the mechanical mixture over both support materials for methanol/O_2 composition 1:10. The activation barrier of radical generation was about 40 kJ/mol. Additional evidence about the origin of the radicals detected in the gas phase was obtained by a comparison of FTIR and EPR spectroscopic data.[240,241]

The formation of radicals on SiO_2 in the oxidation of methanol was represented as

$$\equiv Si - O - Si \equiv + \, CH_3OH \rightarrow \, \equiv Si - OCH_3 + \, \equiv Si - OH$$

$$2(\equiv Si - OCH_3) + O^* \rightarrow 2 \, CH_3O^{\bullet}_{(gas)} + \, \equiv Si - O - Si \equiv$$

where O^* is the surface oxygen species. It is not surprising that in comparable conditions on the bare Pt surface the radicals do not desorb from the surface, as Pt is a very active catalyst for complete oxidation of alcohols. As coverage of the surface is a determining factor in the overall rate of the heterogeneous process, the strong dependence of the quantities of intermediates on the ratio of initial reactants is not surprising. According to the authors' opinion,[240] on the "bare" platinum surface, the reaction is directed toward the complete oxidation of methanol rather than to the generation of intermediates, producing radicals transferred to the gas phase

$$CH_3OH + 1/2O_2 \rightarrow CO + H_2O + H_2$$

$$CO + 1/2O_2 \rightarrow CO_2$$

also considering that

$$\equiv Si - OCH_3 \rightarrow CO + H_2 + \equiv Si - H$$

Based on the spectroscopic data, it has been suggested[240] that the generation of radicals to the gas phase, in the case of a mechanical mixture Pt + support, was more than that in the case of support, and apparently occurred as a result of the following reactions

$$O_2 + Pt \rightarrow 2O_{(ads)}(Pt)$$

$$CH_3O(Si) + O(Si) \rightarrow CH_3O_2(Si) \rightarrow CH_3O_2 + (Si)$$

According to this scheme, the main role of the Pt-surface was attributed to the activation of dioxygen. The authors[240, 241] mentioned an increase in the concentration of radicals in the gas phase, explaining it by the existence of a super-equilibrium concentration of intermediate species in the interface of a Pt/support region for mixed catalysts. It has also been suggested that in the case of the mixed catalyst (Pt + support), the boundary layer between the Pt^0 aggregates and support had specific properties. The methoxy may be generated on the surfaces of both the support and catalyst. They tend to be stabilized by a spillover mechanism, diffusing from the support to the metal-support boundary and from the metal to boundary layer. If the majority of active sites of the support are occupied with methanol or other species, formed in the decomposition of methanol, the concentration of radicals in this layer will be increased. This may open an additional pathway for the formation of radicals, which can be transferred to the gas phase. In this case, accumulation of the super-equilibrium concentrations of radicals in the boundary layer may be expected.

Formation of radicals on a γ-Al_2O_3 surface was concluded in the investigation by *in situ* IR spectroscopy, in adsorption of ethanol in a methyl formate-methanol-dimethyl ether system.[242] Formaldehyde was formed via the oxidation of a methoxy group by oxygen activated on the surface defects of alumina (100–300°C).

Thus, at least, three types of radicals were observed in these heterogeneous–homogeneous reactions, in a certain composition of the reaction mixture: methoxy, methyl peroxy, and formyl radicals. In all cases, the primary radical formed on the surface was methoxy radical. Methyl peroxy and formyl radicals are the results of the further reaction of methoxy radicals with the oxygen atoms, formed mainly in the presence of adsorbed oxygen on Pt-atoms:

$$CH_3O + O^* \rightarrow CH_3OO^{\bullet *} \rightarrow CH_3OO^{\bullet}_{(gas)}$$

where O^* is the electronically excited oxygen atom. Here, it was hypothesized[242] that the formation of the electronically excited radical $CH_3OO^{\bullet *}$ adsorbed on the surface takes place

$$CH_3O^{\bullet *} \rightarrow HCO + H_2O$$

Apparently, in the case of alumina, the active centers will be Al^{3+} in the coordination sphere.

The appearance of the methoxy radical in the gas phase may be considered to be a result of the decomposition of the methoxy complexes formed on the surface of alumina in the adsorption of alcohols.[242]

Additionally, the formation of the adsorbed O atoms in the absence of Pt may occur by the dehydration mechanism.

Thus, the radicals that appeared in the gas phase may be generated by the support material, and, from this point of view, they are products of a secondary reaction of surface methoxy species with activated oxygen. On the other hand, methoxy species on Pt readily transform to the final products. In any case, phenomenologically, the overall reaction may be classified as a heterogeneous–homogeneous process. Here, it is obvious that the main precursor for the formation of both the gas-phase radicals and surface intermediate species is the same methoxy species. It was shown that RO and RO_2 were generated only on the surface of SiO_2, while their total concentration depended on the quantity of Pt loaded on the surface.[242]

We have seen two different and inconsistent behaviors in alcohol oxidation on the Pt-catalist, supported by Al_2O_3, containing 0.64 weight % Pt· (in the work of Ismagilov et al.[123 in Ch.1]) and 3.1-2.6 weight % Pt (in the work of Pak[243]).[244] One of these catalysts, in certain conditions, is able to oxidize alcohols by a typically heterogeneous–homogeneous reaction, the gas phase constituent of which was the chain-radical reaction, evidenced by the detection of radicals by the matrix isolation EPR method in the gas phase. On the second catalyst, containing a greater amount of Pt-atoms on the support, nearly in the same conditions, the oxidation of alcohol is a "purely" heterogeneous process. How may these contradictory observations be explained? According to Lunina et al.,[244] it was considered that the number of coordinatively unsaturated centers on alumina decreases in commensurable quantities with the support of Pt-atoms. Therefore, the possibility was not excluded that platinum-atoms block or also occupy the active centers of alumina, responsible for the generation and transfer of peroxy radicals to the gas phase. According to estimations given by Lunina et al., the decrease in the rate of the homogeneous constituent by factor 2.3, corresponds to the decrease in the number of active sites by the same factor, when the amount of Pt-atoms supported by alumina is higher. Thus, the quantitative changes of the active component on the support may lead to the qualitative changes in complex reactions.

In general, the same radical CH_3O on the gas phase (accompanied with the small amounts of radicals RO_2) and on the surfaces of Pt + support (Al_2O_3 and SiO_2) systems was detected as the main intermediate in the bimodal oxidation of methanol.[242] Adsorbed methoxy radicals, which are active and mobile on the surfaces of metal oxides, are precursors of the radicals desorbing from the surface.[241] For the majority of metal oxides, the cationic sites of acidic character of different strength play a stabilizing role in the formation of radicals from the chemisorbed alcohol.[242]

It should also be mentioned another important precondition in catalytic oxidation of alcohols to the type of reaction (heterogeneous or heterogeneous–homogeneous). The adsorption and reactions of organic molecules, particularly of alcohols on metal oxides, strongly depend on the concentration of surface hydroxyl groups. In ordinary conditions, the majority of the metal oxides contain adsorbed or chemisorbed water. Depending on the temperature of the pretreatment of samples, used in the heterogeneous catalysis, the surfaces of metal oxides contain different quantities of hydroxyl groups. Moreover, hydroxyl groups differently bonded with the surface also exhibit different reactivity (see Section 1.9). On the other hand, water formed in oxidation, even in small amounts, drastically changes the properties of the surface. Apparently, some discrepancy between the conclusions of the results of UHV experiments and high-pressure ("ordinary") conditions relates to the blocking of active centers by intermediates and products. The reaction of methyl radicals on oxygen layer on the surfaces of NiO(111) and NiO(100) were investigated by X-ray photoelectron spectroscopy and temperature programmed desorption methods at low temperatures (120–170 K).[245] It was found that carbon coverage on the surface indicated the formation of hydrocarbon chains. Apparently, they were formed by polymerization of methylene species.

All of these notes on the peculiarities of the bimodal oxidation of alcohols, in general, also refer to other catalytic systems. For instance, oxidation of n-alcohols[123 in Ch. 1; 240; 242; 246] at temperatures ranging from 653 to 823 K (P = 14 Pa) occurs by formation of radicals on the surfaces of TiO_2, V_2O_5, Cr_2O_3, MnO_2, Fe_2O_3, Co_3O_4, NiO, CuO, ZnO[123 in Ch. 1] and many others, a part of which desorb from the surface as RCO and RO_2 radicals.

By generalizing the above observations, it may be stated that the main intermediates in the oxidation of alcohols were adsorbed radical-like species or coordinatively bonded surface species on the surface of metals and metal oxides. Two main pathways of reaction may be observed, resulting either heterogeneous formation of aldehydes or desorption of oxygen-centered radicals. As mentioned earlier, aldehydes were the main products of the partial oxidation of alcohols on metals and metal oxides, and they were one of the links in successive oxidation of the hydrocarbons. However, their transformation to oxygenates has its own peculiarities, some of which will be presented in the next section.

3.5.4 OXIDATION OF ALDEHYDES WITH DIOXYGEN

The first investigations of the oxidation of aldehydes with air or oxygen were those by Justus von Liebig (1803–1873), dating back to 1835, notifying the formation of acids in this reaction.[247] He was also the discoverer of the silver-mirror reaction (the reaction of silver nitrate with an aldehyde, resulting in the formation of carboxylic acid and reduced metallic silver on the glass of the vessel). In the liquid phase, including in solvents, the majority of aldehydes may be oxidized by oxygen at room temperature and above. In oxidation of nearly all aldehydes, the main intermediate products are peroxyacids, finally, giving the corresponding acids. Like oxidation of alcohols, the phase state of reactants has a significant role in their oxidation. In general, the temperature of the gas-phase oxidation of aldehydes is higher than that in the liquid-phase oxidation in homogeneous conditions. In the gas phase, formaldehyde may be notably oxidized with molecular oxygen at T > 250°C, while other aldehydes (C_2, C_3) will be oxidized at relatively low temperatures (T > 100°C).[248] The main products, like liquid-phase oxidation, are peroxyacids and acids. Apparently, the differences of oxidation in the liquid and gas phases relate primarily to the mechanisms of the stabilization of the products in the reaction media.

Particularly, in a liquid-phase solutions, the stabilization of the reaction products of the oxidation of aldehydes occurs by the participation of solvent molecules. The significance of the solvent effects in the energetics of the reaction system was well demonstrated by an example given by van Santen and Neurock.[126 in Ch.1] The homolytic dissociation of the OH bond in acetic acid, in the gas phase, may occur at elevated temperatures, and the energy required for the formation of $CH_3C(O)O$ and H radicals, as an endothermic reaction, is about 440 kJ/mol. In the same phase, the heterolytic dissociation by the formation of $CH_3C(O)O^-$ and H^+ ions requires 1532 kJ/mol energy. In water solution, acetic acid dissociates by the interaction with water molecules. Due to the hydrogen bonds, stabilizing the products, the dissociation becomes less endothermic, as the solvation is an exothermic process. It has been estimated that the energy of interactions of 12 molecules of water in stabilization of the products of the dissociation of CH_3CO_2H is enough to obtain nearly a thermoneutral process in aqueous media. The corresponding experimental value for the dissociation of acetic acid in water is 1.6 kJ/mol.

The thermal and photochemical oxidation in both the gas and liquid phases are chain-radical reactions, with the degenerate branching of chains, known even from the early investigations.[248, 249] The first investigations of the oxidation of low aldehydes (formaldehyde, acetaldehyde, propionaldehyde) revealed the catalytic or inhibiting role of the solid substances in the reaction media. It also became evident that, in general, the oxidation of aldehydes, under certain conditions, particularly at low temperatures, had well-manifested bimodal (heterogeneous–homogeneous or homogeneous–heterogeneous) character, although each aldehyde had its own kinetic peculiarities in oxidation with molecular oxygen in both phases.[247]

From the point of view of kinetic peculiarities and reaction mechanism considerations, in dependence on the temperature, the gas-phase oxidation of acetaldehyde and propionic aldehydes may be divided into low (T < 200°C), intermediate (200° < T < 400°C), and high-(T > 400°C) temperature regions. These temperature ranges are relative, as the rates of oxidation of aldehydes are in strong dependence not only on the temperature, composition of the reaction mixture, and pressure, but also on the diameter and material of the reaction vessel, essentially changing the temperature of the initiation of oxidation. The cause of the very slow oxidation of formaldehyde at temperatures of 220–250°C in the gas phase, according to Dixon and Skirrow,[248] may at least partially be explained by the formation of HO_2 radicals in the initiation stage.

$$HCHO + O_2 \rightarrow HO_2 + CHO$$

$$HCO + O_2 \rightarrow HO_2 + CO$$

At low temperatures, the HO_2 radicals are not able to propagate homogeneous chains by the reaction

$$HO_2 + HCHO \rightarrow HCO + H_2O$$

having an activation barrier $E_a = 46.7$ kJ/mol.

The gas-phase oxidation reactions of aldehydes were investigated in two types of reactors coated with either inorganic acids or alkali metal salts.[248] These two types of surfaces exhibit different behaviors in heterogeneous initiation of the reaction, decomposition of peroxides, and toward the termination reactions of peroxy radicals (see Section 1.3). In general, the rates of these reactions are more in the case of the salts than in the case of the inorganic acids.[248]

The main focus is on the low temperature gas-phase oxidation of formaldehyde, acetaldehyde, and propionaldehyde, as a number of interesting examples of the heterogeneous–homogeneous reactions were found in their oxidation with dioxygen.

3.5.4.1 Heterogeneous and Heterogeneous– Homogeneous Oxidation of Formaldehyde

In general, the gas-phase oxidation of formaldehyde with air or oxygen, like other aldehydes, is a sensitive reaction toward the surface effects, namely, the area and nature of the surface of the reactor walls.[30 in Ch. 1, 248, 250–253] There are essential differences between the low-temperature and high-temperature oxidation of formaldehyde. In both cases, the reaction in the gas phase is a chain-radical process, starting with the initiation of HCO radicals. According to the mechanism suggestions[252] at 300–370°C, the reaction occurs by the formation of performyl radicals HCOOO (by the reaction $CHO + O_2 \rightarrow HCOOO$) and, consequently, performic acid HCOOOH, but at high temperatures, by HO_2 radicals ($HCO + O_2 \rightarrow CO + HO_2$) and, consequently, H_2O_2.

In the gas phase, at around 320°C, an increase of S/V by packing the glass or quartz vessel with pieces of the same materials leads to a multiple decrease in the initial rate of oxidation, in comparison with the empty reaction vessel.[253–255] Bone and Gardner made an interesting observation.[256] Sudden cooling of the reaction mixture of formaldehyde and dioxygen (275°) in ice water after the 4.5, 15, and 30 minutes of mixing showed that amounts of performic acid and dioxidimethylperoxide ($CH_2(OH)OOCH_2OH$) decreased up to undetectable quantities, but formic acid, H_2. H_2O, CO, CO_2 increased in passing from a short to a long duration experiment.

As at low temperatures the gas-phase interaction of formaldehyde with oxygen was excluded,[253, p. 128] it may be assumed that the formation of performic acid or other peroxide compound was a surface reaction. In this regard, Axford and Norrish,[257] based on kinetic data at 337°C, suggested that the decomposition of performic acid was also a heterogeneous reaction. However, surprisingly, they did not observe a dependence of the reaction rate on the diameters of the Pyrex vessels used.

The above-described investigations were periodically documented and analyzed by Lewis and von Elbe,[253] and mention was made of the existence of a number of contradictory opinions in the detailed mechanism suggestions of the low-temperature oxidation of formaldehyde. However, it was conclude that the low-temperature oxidation of formaldehyde in the gas phase, seemingly was a surface-initiated reaction.

At relatively high temperatures (623–823 K), according to the kinetic data obtained by Nalbandyan and Vardanyan,[30 in Ch. 1] comparison of reaction rates in the reactors of the first and second types (Section 1.3) showed that in the case of a CsCl-coated surface, the maximal rate of reaction was 7 to 8 times more than that in boric acid-coated vessel, and the products were only CO_2 and H_2O. In this reaction, the EPR method (freezing the radicals from the reaction zone) showed the formation of HO_2 radicals (but not HCO_3 or HCO, postulated previously by other investigators). It was shown that HO_2 radicals were leading active centers in the gas phase. Existing kinetic data confirmed that H_2O_2 was responsible for the branching of chains. It had also been concluded that, unlike acetaldehyde and propionaldehyde, in the case of formaldehyde, peracid, HCO_3H (performic acid) was not formed. Therefore, the branching of chains by heterogeneous decomposition reaction did not take place. The branching, apparently, occurs by the radical decomposition of H_2O_2. The heterogeneous–homogeneous nature of reaction, in the given case, relates primarily to the ratio of the rates of the heterogeneous initiation and termination of the leading active centers, radical HO_2. Conversely, in the vessels precoated by salts, the formation of H_2O_2 and HO_2 radicals was not observed, and the major products were CO_2 and H_2O, in the mentioned conditions. Moreover, in this reaction vessel, the reaction lost its autocatalytic nature.

Even today, all suggestions of the mechanism of the surface reactions in the oxidation of formaldehyde remains largely hypothetic, as, in our knowledge, there is no complete and unified mechanism suggestion proven by experimental data in the quartz, silica, or other types of reactors.

The heterogeneous catalytic oxidation of formaldehyde[258, 259] over a great number of catalysts, noble metals (Pt, Pd, Au, Ag, usually, supported on SiO_2, TiO_2, Al_2O_3, molecular sieves), and transition metal oxides (MnO_2, Co_2O_3 V_2O_5) were described as "purely" heterogeneous but not as heterogeneous–homogeneous processes. However, like oxidation of alcohols, in catalytic oxidation of formaldehyde, the surface may generate radicals or radical-like species.

Like other aldehydes, formaldehyde adsorbs either as an $\eta^1(O)$-structure or an $\eta^2(C, O)$-structure on the majority of metals and metal oxides[260] (Figure 3.15).

At least two types of reactions may be seen in the oxidation of formaldehyde on all variety of metal and metal oxide catalysts: partial (formic acid, often also carbonate) and complete (CO_2, H_2O) oxidation. Immediately note too that the adsorption and further transformation of formaldehyde, as well as those of other aldehydes on the surfaces of solid substances are in essential dependence on the presence or absence of oxygen or reactive oxygen species on the surfaces of the solid substances.

The mechanism suggestions of the heterogeneous oxidation of formaldehyde on the surfaces of a number of metals and metal oxides, such as Ag, Cu, TiO_2, NiO, ZrO_2, and ThO_2, are based on the formation and further reactions of adsorbed oxy- and

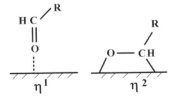

FIGURE 3.15 η^1 and η^2-aldehyde configurations in adsorption on the solid surfaces.

dioxy-methylene species in the presence of pre- or co-adsorbed oxygen (oxygen atoms, or active oxygen species). Schematically, it may be presented as follows:[259, 261]

$$H_2CO_{gas} \rightarrow H_2CO_{ads} \xrightarrow{+O} H_2COO_{ads} \rightarrow HCOO_{ads} + H_{ads}$$

Although there is no direct experimental proof of the existence of H_2COO_{ads}, its two frequencies in the IR region are known for several metals.[262] Furthermore, the interaction of Schwartz's reagent, $Cp_2Zr(H)Cl$, has been investigated, with labeled $^{13}CO_2$, performed by the *in situ* variable-temperature ^{13}C NMR (nuclear magnetic resonance) method, showing the formation of H_2COO_{ads}.[263] Obviously, the formation of the adsorbed formate species are in competition with the decomposition, dimerization, or polymerization reactions of formaldehyde, in the presence of oxygen species.

In the absence of oxygen, the adsorption of formaldehyde on metals and metal oxides occurs either by a dissociative pathway into H and HCO species or by transformation to a CH_2O- (oxymethylene) species. Images obtained by the STM method,[264] in adsorption of formaldehyde at the cryogenic temperatures, show that on the partially reduced $TiO_2(110)$, formaldehyde preferentially adsorbs on the bridge-bonded oxygen (O_b), or on the vacancy defect site (usually, denoted as V_o), forming oxymethylene species CH_2O-. Additionally, the STM-images confirm the diffusion of this specie along the Ti row of nanocrystal.[364]

Apparently, the decomposition of radicals and the formation of formic acid on the surface relate to concurring reactions

$$HCOO_{ads} \rightarrow CO_2 + H_{ads}$$
$$HCOO_{ads} + H_{ads} \rightarrow HCOOH$$

A theoretical investigation[266] of the co-adsorption of formaldehyde and atomic oxygen, as well as their reactions on the Cu(111), showed that, if the oxygen atom was adsorbed on the single atom of Cu on fcc position and formaldehyde was stabilized on a long-bridge site, the formation of dioxymethylene was thermodynamically favorable. All below-surface reactions are rewritten without an index denoting surface. It was estimated that

$$H_2CO + O \rightarrow H_2CO_2 \qquad E_a = 36 \text{ kJ/mol}$$

and the product was 33 kJ/mol more stable than the pair of co-absorbed $H_2CO + O$. This permits prediction that the desorbed products from Cu(111) will be formate species.

$$H_2CO_2 \rightarrow H + HCO_2$$

It is interesting to compare these results with the results of the adsorption and further reactions of formaldehyde on partially reduced surfaces such as $TiO_2(001)$.[264] Like other metal oxides, formaldehyde adsorbs on TiO_2 by the donation of σ lone pair of electrons of $C = O$ to the cation. The mechanism suggestions are related to the Cannizzaro-type reaction, giving methanol and formate. In order to explain the experimental data, it has been suggested that at around 550°C, on the partially reduced surface of TiO_2, part of the adsorbed formaldehyde decomposes[262, 264]

$$HCHO \rightarrow H_{ads} + C_{ads} + O_{(ads \rightarrow lattice)}$$

Therefore, in the presence of oxygen, the formation of CO, CO_2, as well as H_2O, are expected products.[267]

Presently, nearly all mechanism suggestions in catalytic oxidation of formaldehyde are based on the Mars–van Krevelen type of reaction schemes, assigning an essential role to the lattice oxygen reactions, forming dioxymethylene or formate species on the surface. The decomposition of formate produces adsorbed CO, and, subsequently, CO_2

$$HCHO \rightarrow CHOO^- \rightarrow CO \rightarrow CO_2$$

According to the review of Bai et al.,[259] the Pt-catalyst, containing alkali metal K or Na (1–2%) on the support TiO_2 (T < 80°C), is more active in formaldehyde oxidation with oxygen than the Pt/TiO_2 catalyst. Discussing the mechanism of the oxidation of formaldehyde on this catalyst led to the suggestion that the increase in the activity related with the existence of free hydroxyls was

$$HCHO \xrightarrow{\;O_2\;} (HCOO)_{ads} + (^\bullet OH)_{ads} \rightarrow CO_2 + H_2O$$

An analogous reaction pathway was suggested for other catalysts in the oxidation of formaldehyde, as $K\text{-}Ag/Co_3O_4$.[259] It was suggested that on the Au/CeO_2 catalyst, the formation of water was a result of the combination of free OH and H radical-like species on the surface, formed by decomposition of formate species. The latter oxidation also was accompanied by the formation of carbonate and hydrocarbonate species, deactivating catalytic surface. Although the term *radical* or *radical-like species* was not used by the authors, the participation of free OH (or H), in terminology adopted in Section 2.1, it is a heterogeneous reaction of the radical-like species (about free OH radical species on the surfaces; see also Section 3.4.2.3). In this regard, dioxymethylene and formate intermediates may, in certain cases, also be considered as adsorbed radical-like species.

Returning to the problem of the possibility of the heterogeneous–homogeneous pathway in oxidation at low temperatures, we should note that, although there are no examples of this pathway in heterogeneous catalytic oxidation of formaldehyde, it does not exclude the probability of the transfer of radicals or radical-like species to the postcatalyst volume, in favorable conditions. To our knowledge, this possibility has not been investigated, at least for the majority of catalysts of recent generations.[258, 259] One reason is that most selective heterogeneous catalysts for oxidation of formaldehyde operate at very low temperatures (T < 80–90°C), while in the above given examples (previous section) of the heterogeneous–homogeneous reactions of formaldehyde occur only in relatively high temperatures (T > 270°C). On the other hand, the above-mentioned intermediates usually formed from formaldehyde, may be formed also in the oxidation of other compounds, for instance, in catalytic oxidation of alcohols (see the previous section), and are able to initiate homogeneous gas phase reactions.

3.5.4.2 Heterogeneous and Heterogeneous–Homogeneous Oxidation of Acetaldehyde and Propionaldehyde

According to Dixon and Skirrow's review,[248] the low temperature (below 150°C) oxidation of acetaldehyde and propionaldehyde, in the gas phase, quantitatively corresponds to the stoichiometry of the reaction

$$RCHO + O_2 \rightarrow RCO_3H$$

In the acid-coated vessel, at 70°C, this stoichiometry is well pronounced (99.5% for a half reaction). However, the strong dependence of the reaction rate on the nature of the walls, indicating the existence of the heterogeneous constituent of reaction,[248, 268] considerably changes this stoichiometry. In early investigations conducted in the 1930s, nearly all authors mentioned the multiple acceleration of the reaction rate in the reaction vessel precoated by KCl, which was known to be an effective surface in the termination of radicals. For example, Pease[269] showed that at low-temperature oxidation of acetaldehyde and propionaldehyde in the reaction vessel with a relatively developed surface (about a tenfold increase of S/V by packing the vessel with the glass pieces) and precoated with KCl, the maximal rate of reaction increased about 3–5 times, and the proportions of the obtained products changed notably. Pease attempted to explain these observations, hypothesizing the existence of the heterogeneous stages of the initiation and decomposition of the intermediate product, peroxyacides, by surface reactions. It was also found that the oxidation of acetaldehyde and propionic aldehyde had relatively low activation barriers of overall reaction, about 15 kcal/mol, indirectly indicating the heterogeneous contribution in reaction.[269, 270]

Low-temperature gas-phase oxidation of aldehydes was investigated during several decades by many scientific groups. The results obtained until the 1950s were reviewed by Emanuel,[271] followed in the mid-1970s by Dixon and Skirrow,[248] and a decade ago by Nalbadyan and Vardanyan,[272] in light of the new experimental results of that time in this area of knowledge. These works demonstrated the chain-radical nature of these reactions with the degenerate branching of chains, as well as the proposition of the overall mechanism of their oxidation at different temperature ranges. Another important highlight of that time was the statement regarding the overall heterogeneous–homogeneous nature of these reactions.

The main evidence on the chain-radical character of these reactions was obtained by detecting radicals in the gas phase, applying the matrix isolation EPR method, and freezing radicals from the reaction zone.[273] The investigations of the spectra and kinetic behaviors of radicals, chemical analysis of reactants and products in different conditions, and extensive studies of the influence of heterogeneous factors led to the conclusion that the leading active centers at the low-temperature gas-phase reaction were CH_3CO_3 or $C_2H_5CO_3$ radicals.[7, 38 in Ch. 1, 272] These findings were in favor of the previously proposed mechanism of low-temperature oxidation of acetaldehyde and proionaldehyde,[248] a brief version of which may be represented as

$$RCHO + O_2 \rightarrow RCO + HO_2 \qquad \text{initiation}$$
$$RCO + O_2 \rightarrow RCO_3$$
$$RCO_3 + RCHO \rightarrow RCO_3H + RCO \qquad \text{propagation}$$
$$RCO_3H \rightarrow RCO_2 + OH \qquad \text{degenerate branching}$$
$$RCO_2 + RCHO \rightarrow RCO_2H + RCO$$
$$OH + RCHO \rightarrow RCO + H_2O$$
$$RCO_3 \rightarrow pr. \qquad \text{termination}$$

Previously, it was considered that the initiation and termination of chains had heterogeneous in nature.[248] Soon, it became clear that the surface effects in the complex process cannot be explained by the contribution only of these surface reactions.[272] A more important finding in this regard was the discovery of the heterogeneous radical decomposition of peroxyacid, the main intermediate of this reaction, on the surface of the wall of the reaction vessel with the transfer of part of the radicals to the volume of reactor. Extensive studies of this phenomenon on the surfaces of different nature, combined with the great number of kinetic data on the oxidation, concerning the radicals and final products, permitted the conclusion that at low-temperature oxidation of acetaldehyde and propionaldehyde, the stage of the chain branching was heterogeneous[272]

$$RCO_3H \xrightarrow{\text{wall}} RCO_2 + OH \quad \textit{degenerate branching}$$

The important role of the mentioned phenomenon for understanding other interesting observations regarding the oxidation of aldehydes and other oxygenated compounds was described in Section 2.3.2.2.

Further investigations of heterogeneous factors at low-temperature oxidation of aldehydes led to the conclusion that more complex processes existed on the surface, in interaction with the gaseous chain reaction.[274] Back in the 1930s, Steacie et al.[270] (in 1934) investigated the low-temperature oxidation of propionic aldehyde in the gas phase in a Pyrex reactor, increasing S/V by packing the vessel with the glass pieces (403–430 K). They observed the increase of the reaction rate with the increase of S/V, as well as essential changes in the compositions of the products and the overall kinetic behaviors of the reaction. It was concluded that the homogeneous chain reaction of the oxidation of propionaldeyde may be gradually transferred to the surface with the increase of S/V, without any additional explanations. Nearly a half century ago, the same reaction of the oxidation of propionaldehyde was again studied, now using the EPR matrix isolation method, combined with the freezing of radicals from the reaction zone for detection of radicals in the reaction vessels packed with the glass and precoated by KCl.[52 Ch. 1] An increase in the maximal rate of reaction was observed about 7–8 times with the increase of S/V (about 20 times). The overall reaction again was the autocatalytic process in the reactors with different S/V, but the main final product was propionic acid with a high yield (up to 90%). The peroxy radicals were detected in all reaction vessels with developed S/V, and on the basis of overall analysis of kinetic data they were assigned to the RCO_3 radicals. Being leading active centers of the gas reaction, they exhibited interesting kinetic behaviors with the increase of S/V. The data showed that the maximum rate of the process increased with the increase of S/V and the induction period and the time of reaching the maximum rate of reaction were reduced. The concentrations of both $C_2H_5CO_3H$ and $C_2H_5CO_3$ radicals reached their maximum values at the moment of the maximum rate of the aldehyde consumption. Surprisingly, in the region of relatively high values of S/V, the fast rise of the consumption rates of C_2H_5CHO was not accompanied by a remarkable decrease in the concentration of peroxyacid in the gas phase. This observation was inconsistent with the previously well-established fact of the heterogeneous decomposition of peroxyacid on the surface of walls. On the other hand, KCl was known as an effective surface in the termination reaction of peroxy radicals; therefore, in conditions of increasing S/V, a decrease in the concentration of radicals will be

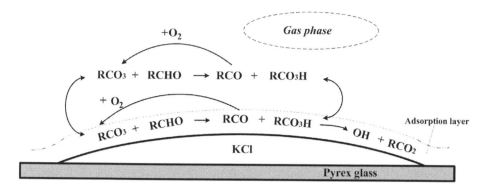

FIGURE 3.16 Proposed stages of the heterogeneous and homogeneous propagation of chains in the reaction of low-temperature (423–503 K) thermal oxidation of aldehydes in the Pyrex vessel precoated with KCl.[85 in Ch. 1]

expected, whereas they practically were unchanged in the gas phase. In the same conditions, the rate of reaction increased about 8 times.[52 in Ch. 1] These controversial results may be explained by assuming the existence of another way leading to peroxyacid formation than the homogeneous reaction. It may be the heterogeneous reaction

$$RCO_3 + RCHO \xrightarrow{\text{wall}} RCO_3H + RCO \quad \textit{heterogeneous propagation}$$

This hypothesis was equivalent to accepting the concept of the heterogeneous propagation of chains (Figure 3.16) or at least to the heterogeneous formation of the main intermediate, peroxypropionic acid, in this reaction.[52 in Ch. 1; 85 in Ch. 1] Comparisons of the rates of the homogeneous constituent of reaction with the heterogeneous one showed that with the increase of S/V, the homogeneous reaction gradually transferred to the surface, and at high values of S/V, the rate of heterogeneous reaction prevailed over the homogeneous one. For instance, in the packed reaction vessel with the increased S/V (about 20 times, compared with the empty vessel), the ratio of the rates of the heterogeneous and homogeneous constituents became about 3:1.[275]

Sections 1.8 and 3.5.2.1 elucidated the differences between surface-sensitive and surface-nonsensitive reactions. According to the data obtained,[275] the dependence of the specific rate on the surface area, or S/V, in this reaction, corresponds to the case of the surface-sensitive reactions.

The observed transfer of the homogeneous reaction to the surface and the progression of the heterogeneous reaction, together with homogeneous reactions, is a rare phenomenon. In this case, the occurring chain reaction itself opened a new, heterogeneous pathway to continue the transformation of the initial reactants. Obviously, it is essentially different from the case of the heterogeneous catalytic reaction, where some intermediates, but not the occurring reaction, transfers to the other phase.

Here, the heterogeneous–homogeneous nature of the overall process is obvious, but the mechanism of heterogeneous constituent of the reaction remains unclear in some aspects.

Relatively more convincing experimental data indicating the possibility of the heterogeneous chain propagation in oxidation reaction of CH_3CHO and C_2H_5CHO

were obtained in conditions, excluding the homogeneous propagation of chains, in the given temperature. In this aim, the oxidation was carried out in a capillary vessel precoated with KCl at 423 K.[124 in Ch. 1] The main final products were acetic and propionic acids, respectively. The amount of intermediate RCO_3H in the gas phase became considerable only when the temperature increased to 523 K, when the homogeneous propagation of the chains became feasible. The peroxy radicals were detected in the volume by the EPR matrix isolation method, even at low temperatures, but they rapidly decreased with the rise in temperature and were not detected above 573 K. The presence of radicals in the volume was the sign of its formation on the surface of the capillary tube, as their formation in the gas phase was excluded in this case. At the same time, the amount of the final product increased, indicating an increase in the reaction rate on the surface. The initiation of this heterogeneous process by peroxy radicals (obtained by the radical decomposition of *tert*-butyl hydroperoxide on the NiO surface,[75 in Ch. 2, 276] showed the increase in the rate of reaction in the conditions excluding homogeneous propagation of the chains. The calculation showed that at least several dozen molecules of the final products (acid) formed per one radical of the initiator.[277] All these facts favored the conclusion about the existence of chain propagation on the surface, together with the homogeneous ones.[124 in Ch. 1, 277]

To obtain information about the direct interaction between the peroxy radicals and aldehyde on the solid surface (SiO_2, 5% KCl/SiO_2 in oxidation of aldehydes), the experiments were carried out using the EPR and IR spectroscopy and mass-spectrometry methods.[276] A model reaction of C_2H_5CHO absorbed on a silica surface with CH_3O_2 radicals was studied at 373 K.[276] The experimental data indicated that in the absence of O_2 the reaction proceeds by breaking the C-H bond in the aldehydic group and forming newly adsorbed product, probably having ketonic structure. The interaction between the adsorbed CH_3CO_3 radicals on the SiO_2 surface and acetaldehyde in a range of the temperature 423–503 K was investigated in the work.[278] The radicals were obtained by heterogeneous radical decomposition of diacetyl on the SiO_2 and KCl/SiO_2. The EPR signal detected on the surface of the SiO_2 was attributed to the adsorbed peroxy radicals.

The decomposition pathway of diacetyl may be the following

$$CH_3CO\text{-}COCH_3 \rightarrow 2CH_3CO$$
$$CH_3CO + O_2 \rightarrow CH_3CO_3$$
$$CH_3CO \rightarrow CH_3 + CO$$
$$CH_3 + O_2 \rightarrow CH_3O_2$$

Thus, the mixture of radicals RCO, RCO_3, RO_2 may be detected on the surface. The radicals, detected by the EPR method in the gas phase, also will be mixture of radicals.

A gradual decrease in the intensity of the EPR signal was observed when the vapors of CH_3CHO were submitted to the ampule, containing a sample of SiO_2. During this process, the formation of new oxygenated compounds was detected by mass-spectrometry (CH_3OH) and IR spectroscopy ("free OHs" and unknown products) on the surface.[278] The interaction proceeds faster on the surface of KCl/SiO_2 than on an aerosil surface. Thus, in all the mentioned model reactions it is obvious the heterogeneous interaction between the radicals and aldehyde on the surface[85 in Ch. 1, 272, 274–178]

Later, analogous evidences were obtained in the system of peroxy radicals (RO_2) and aldehyde on TiO_2.[279]

Summarizing all these results in the reactions of the oxidation of aldehydes,[280] the mean values of the length of heterogeneous chains were estimated based on the experimental data in which the rate of overall reaction was compared with the rate of the model reactions between the adsorbed form of peroxy radicals and aldehyde. The calculations showed that the adsorbed forms of peroxy radicals are able to develop a chain, the length of which is equal to about 50. This result was in good agreement with the earlier results obtained in the heterogeneously initiated oxidation of aldehyde by peroxide compound, under conditions excluding the homogeneous propagation of chains.[124 in Ch. 1] As follows from the results obtained in this work, the mean length of a heterogeneous chain practically does not depend on the temperature in the 400–425 K range. Meanwhile, the rate of the overall heterogeneous reaction increases with the temperature. It may be assumed that the number of heterogeneous chains increases and, hypothetically, that a partial branching (like degenerate branching in the gas phase, in terminology of the chain reactions) of chains on the surface may take place in this case.

In the region of the negative temperature coefficient (NTC) of the rates, the influence of the heterogeneous factors in the oxidation of propionaldehyde was studied using the EPR method at 623–833 K, under flow conditions and at atmospheric pressure.[281] It was shown the effect of the surface nature on the position of the rates in the NTC region. In addition, the data indicated the heterogeneous consumption of aldehyde with the participation of peroxy radicals.

An interesting result was obtained in the study of the kinetic behavior of C_2H_5CHO oxidation in the gas phase, at 803 K, depending on the ratio S/V of the reaction vessel.[282] The increase in S/V led to the increase in the concentration of H_2O_2 and the decrease in the concentration of HO_2 radicals in the volume. Simultaneously, the rise of the consumption rate of C_2H_5CHO was observed. The data were explained by the existence of the interaction of HO_2 radicals with aldehyde on the surface of the reaction vessel.

Almost one and a half decades later, Jenkins and Murphy[283] investigated the thermal and photoreactivity of TiO_2 at the gas–solid interface with aliphatic and aromatic aldehydes. They clearly confirmed the existence of the heterogeneous-radical propagation of chains in the oxidation (thermal or photochemical) of aldehydes by the reaction of adsorbed RCO_3 radicals and aldehydes. Reaction was carried out on a partially reduced TiO_2 samples, in both thermal and photochemical conditions (273 K). Based on the EPR data, it was concluded that the surface could participate in both the initiation and propagation stages of aldehyde oxidation. The radicals RCO_3 were obtained by photochemical pathway in the oxidative decomposition of aldehydes. In the absence of molecular oxygen, even under UV irradiation, radicals were not detected by the EPR method on the surface. Only in co-adsorption of aldehyde and oxygen on a partially reduced TiO_2 (rutile), even at 100 K the peroxyacetyl radicals were detected and characterized. The heterogeneous formation of RCO_3 radicals was presented as

$$RCHO + TiO_{surf.}^- \rightarrow RCO_{ads.}^\bullet + TiOH^-$$

$$RCO_{ads.}^\bullet + O_2 \rightarrow RCO_{3\ ads}^\bullet$$

Thus, the kinetic data on the low-temperature oxidation of aldehydes indicate the essential contribution of the reaction on the solid surface in the complex heterogeneous–homogeneous reaction, demonstrating that, in certain conditions, the contribution of the heterogeneous reaction may prevail over the homogeneous one. In this regard, it may be expected the transformation of the heterogeneous–homogeneous reaction to the heterogeneous catalytic reaction in favorable conditions.

3.5.4.3   Intermediates in Heterogeneous Oxidation of Acetaldehydes and Propionaldehyde

This section presents some heterogeneous reactions of the adsorbed aldehydes with dioxygen and active oxygen species, involving oxygen-centered organic radicals, formed in the oxidation of aldehydes on the surface or diffused from the gas phase to the surface. They are examples of model reactions, simulating the heterogeneous stages of the heterogeneous–homogeneous reaction, involving also chain propagation presented in Section 3.5.4.2. They provide knowledge about the formation of possible intermediates in the surface reactions of aldehydes higher than formaldehyde.

Adsorption of aldehydes. The adsorption structures for formaldehyde presented earlier in Figure 3.15, are also characteristic for acetaldehyde and propionaldehyde on the majority of the metals and metal oxide surfaces.[284] For C_2 and higher aldehydes, the η^1-structure may be either $\eta^1(O)$ or $\eta^1(H)$. In the second case, aldehyde is bonded with the surface by either aldehydic or nonaldehydic H-atom (CH_3 or CH_2).

The metallic catalysts in the heterogeneous oxidation of aldehydes are elements of IB and VIII groups. In general, both structures may coexist on the metals; however, in the case of IB group elements, for instance, Cu and Ag, aldehydes adsorb mainly as the η^1-structure.

The differences in adsorption of aldehydes on the metals of different groups (IB and VIII) may be explained by the "reduced availability" of d electrons for back donation into the C=O(π*) in the case of IB group metals.[284]

The adsorption structures of aldehydes on metals, as well as metal oxides, depend not only on the parameters, as temperature and pressure, but also on coverage of the surface, presence of oxygen or oxygen active species, surface hydroxyl groups, and many other preconditions. Therefore, the reactivity of aldehydes on the surface depend on all of these parameters and preconditions. Depending on the presence of oxygen or other species on the surface, one adsorption structure of aldehydes may even transform into another.

On Pt(111) acetaldehyde adsorbs by both structures η^1 and η^2 (at low coverage, 1/9 ML, as η^2, and as η^1 through the hydrogen bond at high coverage, 1/3 or ¼ ML).[237] On Ag(111) aldehyde adsorbs as $\eta^1(O)$ structure (80 K), according to the data obtained by temperature programmed desorption (TPD) and reflection absorption infrared (RAIRS) spectroscopy.[285] On the oxygen-modified surface of Ag, the formation of the bridging acetate species, the precursor of acid, were observed.[285] On Ru(001) and Ni(100), the $\eta^2(O)$-structure is the predominant form of adsorbed species,[284] while in the presence of pre- or co-adsorbed oxygen, the η^1-structure becomes predominant for transition metals. The $\eta^2(O)$-structure reacts by the breaking of C-H (of CHO group) and by further decomposition (decarbonylation), oxidation, or condensation reactions. In the case of 1B group metals (mainly η^1-structure), as a rule, the decarbonylation does not take place.[284]

On the Pd-catalyst, Barteau et al.[286] observed changes of the reaction pathways of adsorbed C_1-C_3 aldehydes, depending on the presence or absence of oxygen atoms on the surface: decarbonylation on a clean surface and carboxylate formation on an O(2 × 2)-covered surface. It was shown that oxygen played multiple roles, transforming the adsorption structure $\eta^2 \rightarrow \eta^1$, carrying out nucleophilic oxidation to carboxylate, stabilizing the carboxylate by lateral interactions, etc. Interestingly, the O-atom initiates polymerization of formaldehyde but not that of acetaldehyde or propionaldehyde. Using the TPD and LEED methods, the adsorption of CH_3CHO was investigated on Pd(110)[287] in a molecular beam reactor. At low temperatures (T < 200 K), acetaldehyde was weakly adsorbed on the surface of Pd(110) in the form $\eta^1(O)$ and more strongly in the form η^2-(C,O), at T > 200 K. Below 270 K, no products were desorbed to the gas phase. It was suggested that acetaldehyde, adsorbed as the η^2-(C,O) type of structure, decomposed beginning at T > 200 K. At T = 270–400 K, the products were methane and H_2, which might be formed as a consequence of the primary surface reactions

$$CH_3CHO \rightarrow CH_3 + CO + H$$

$$H + H \rightarrow H_2$$

$$CH_3 + H \rightarrow CH_4$$

At around 400 K, however, the rate of reaction passed through the maximum, the mechanism of reaction changed, and the products became CO and H_2. Carbon deposits increased on the surface[287]

$$CH_3CHO \rightarrow CO + 2H_2 + C$$

On the metal oxides such as TiO_2,[288] CeO_2, and Al_2O_3, according to data from FTIR spectroscopy and mass spectrometry[289] acetaldehyde is bonded mainly by an H-bridge form (by aldehydic hydrogen) and more strongly with lone pair of O-atom electrons on the Lewis acid centers (cation) at 300–673 K. In heating, the different types of reactions occur through formation of condensation (crotonaldehyde by β-aldolization), dehydrogenation, and oxidation products. Investigations showed that the oxidation of acetaldehyde resulted in the formation of acetate species on the surface.

On the amorphous silica gel, according to micro calorimetric, IR, TPD data, and DFT calculations,[290, 291] acetaldehyde adsorbs by two hydrogen bonds (27 kJ/mol), formed by the donation of lone pair electronic density of carbonyl oxygen to hydrogen atoms of the hydroxyls of the surface. IR-spectroscopic data show resultant shifting to the lower wavenumbers. Adsorption of aldehydes on silica surfaces and further transformation (T < 120°C) occur by the hydrogen bonds. Theoretical investigations have produced the same results.

In general, on the hydroxylated surfaces of metal oxides, acetaldehydes and propionic aldehydes are adsorbed through hydrogen bonds, via either the H-atom of aldehyde

$$CH_3C(O)\text{-H}...OH\text{-Si}...$$

or the O-atom of C=O[237]

$$CH_3(H)C\text{-O}...HO\text{-Si}...$$

In the absence of pre- or co-adsorbed oxygen on the samples of stoichiometric oxides of the transition metals,[237, 292] aldehydes adsorb through lone pair of electron of oxygen using d or f empty orbitals of metal ions (in η^1 mode), forming a surface complex

$$RCHO + M^{n+}\text{-}O... \rightarrow \left[RCHO\text{-}M^{n+}\text{-}O\text{-}...\right]$$

In this case, the dissociative adsorption of aldehyde takes place on the metal oxide

$$[RCHO\text{-}M\text{-}O] \rightarrow RCOO\text{-}M + H_{ads}$$

H_{ads} may react with another intermediate, producing alkoxide on the surface, especially for formaldehyde on the TiO_2 and UO_2; the predominant reaction is[292]

$$RCHO\text{-}M\text{-}O + H_{ads} \rightarrow RCH_2O\text{-}M$$

In the absence of oxygen, for acetaldehyde and propionaldehyde, the main pathway of the reaction on a number of metal oxides is the surface analog of the crotonic condensation, giving unsaturated aldehydes.

The crotonic condensation of acetaldehyde on a SiO_2 occurs by the following reaction

$$2CH_3CHO \rightarrow CH_3CH=CHCHO + H_2O$$

Analogous reaction of propionaldehyde on a SiO_2 produces 2-methyl-2-penten-1-al

$$2C_2H_5CHO \rightarrow CH_3CH_2CH=C(CH_3)\text{-}CHO + H_2O$$

These reactions may be accelerated in the presence of KCl (5%) on the surface (with about 100% selectivity) at room temperature and above.[293] There are good examples of the formation of symmetric unsaturated hydrocarbons R-CH=CH-R from aldehydes on the surfaces of a partially reduced TiO_2 (ethylene from formaldehyde on TiO_2(110), butane from acetaldehyde TiO_2(001), hexatriene (CH_2=CH-CH=CH-CH_2) from acrolein on Ti(001), stilbene from benzaldehyde on Ti(001), etc.).[292] Apparently, the surface reactions occur through the formation of diolates with the rupture of bond in the $\eta^2(C=O)$– structure and oxidation of the partially reduced surface

where M is a metal ion.

The situation described above completely changes in the presence of adsorbed oxygen species on the surfaces of the metal oxides, in certain quantities. The presence of oxygen and active oxygen species in the reaction media or on the surface, first of all, favors the generation of acyl- and alkyl-oxy, acetate and other oxygenated intermediates. In this case, the above-mentioned condensation reactions have no essential importance. As a rule, the main final products of the oxidation on metal oxide catalysts become corresponding acids and/or acetates. For example, the main products of the partial oxidation of acetaldehyde and propionaldehyde on a VO_x/TiO_2 (at 120–280°C, in the presence of water vapors) were acetic and propionic acids. Lower acids and CO also were detected.[294]

In order to reveal the effect of molecular oxygen in the reactions of preadsorbed acetaldehyde on a TiO_2 (anatase) at 243 K (P_{O2} = 14 Torr, during 85 minutes), FTIR method was used.[295] Several adsorbed configurations of acetaldehyde, as well as the products of its surface transformations, were observed in these conditions. Among them, it may be seen that two forms of adsorbed species (acetaldehyde and crotonaldehyde) are present on the surface. The exposure of the sample under oxygen produces important evidence about the formation of carboxyl groups on the surface. Decrease of $v(C\text{-}C)$ at 1720 cm^{-1}, simultaneously with the decrease of intensities of $v(C\text{-}CH_3)$ at 1122 and 1064 cm^{-1}, indicates the formation of carboxylate species possessing a strong peak $v_a(OCO)$ at 1580 cm^{-1} and a medium intensity peak $v_s(OCO)$ at 1424 cm^{-1}. These changes were accompanied by the decrease and slight shift of $v(CH)$ at 2980 cm^{-1}, probably owing to the formation of acetate structures. In the same conditions, the photooxidation of preadsorbed aldehyde on TiO_2 also produces carboxylate species. The desorbed product may be either CO_2 or corresponding acid or acetate. Nearly the same picture may be observed in the case of other oxide surfaces such as SiO_2 and ZnO.[295] A theoretical investigation of the thermal transformations of acetaldehyde on $TiO_2(110)$ (rutile) demonstrates that adsorbed acetaldehyde may undergo dehydration, giving an unsaturated hydrocarbon (propene) in a reduced surface, but on an "oxidized"(apparently, nonreduced), surface, may give highly stabilized acetate.[296]

Radicals or radical-like species in the reactions of adsorbed aldehyde with oxygen. One of the ways of the formation of organic radicals in adsorption of aldehydes on the metal oxides is their reaction with surface oxygen active species, including radical-like surface OH species. Theoretical study[297] of the interactions of C_2-C_5 aldehydes on silicate clusters predicts that the main reaction pathway may be abstraction of aldehydic H-atom of the adsorbed aldehydes by the following reaction:

$$\text{Cluster-SiO}_2\ldots O(H)CR + OH \rightarrow RCO + H_2O + \text{Cluster-SiO}_2$$

In fact, here, RCO is an adsorbed radical-like species. At the same time, beginning with propionaldehyde, abstraction of alkyl hydrogen of aldehydes also becomes possible.

Radicals, stabilized on the reduced TiO_2, were detected by the EPR method in a model reaction of co-adsorption of dioxygen and aldehyde at 100–250 K. The EPR spectra obtained were assigned to RCO_3 radicals[298] (unstable T > 250 K). It was identified as a radical with the characteristic g-factor values: $g_1 = 2.012$, $g_2 = 2.008$, $g_3 = 2.003$.[298, 299] The RCO radicals were not detected in these experiments. At the present state of our knowledge, RCO can be detected only by the matrix isolation method at cryogenic temperatures.[300–301] In the gas phase, CH_3CO is detected by application of visible adsorption spectroscopy (490–660 nm, at 298 K).[302] On a TiO_2 surface with preadsorbed O_2, the formation O_2^- (Ti^{3+} centers), reacting with adsorbed RCO was shown. Zhao et al.[303] showed that the formation of RCO radicals was possible in adsorption of acetaldehyde on Pt(111). Here, the $\eta^1(O)$–structure of aldehyde is predominant on the surface in heating up to 150 K, and it undergoes dehydrogenation by α-C-H, producing a radical, structured as acetyl $\eta^1(O)$ species. They identified these species by the HREELS spectra method, assigning $v = 1647$ cm^{-1} to the $v(CO)$-acetyl and comparing it with the $v(CO)$-acetyl bond in $[PtBr(COCH_3(PEt_3)_2]CCl_4$, at 1630 cm^{-1}. The further heating until 350 K led to the decarboxylation reaction. Then, using the results of the TPD experiments, the formation of CH_4, CO, with gradual dehydration of the CH_3 group until carbon formation, was established. Thus, RCO is a very reactive species and may not be easily detected in the adsorbed state. However, it is obvious that only RCO may be a precursor for the formation of RCO_2 and RCO_3

radicals (radical-like species), depending on their surface reactions of oxidation. The carboxylate and peroxycarboxyl species may be formed both in thermal reactions and under irradiation conditions in photochemical reaction. Carter[97, 300] showed that O_2^- was primarily responsible for the formation of RCO_3 radicals stabilized on Ti^{3+} sites.

A nearly analogous picture was observed in the case of SiO_2 (aerosil). However, in this case, the adsorbed aldehyde was relatively more stable and did not undergo chemical transformations until 373 K in the absence of oxygen. In heating, it transformed into crotoaldehyde, but in the presence of oxygen-active species or oxygen-centered radicals it transformed into oxygenated products.[276, 278]

The decomposition reactions of RCO_2 also may be a source of radicals in the reaction media. Acetate radical stabilized on MgO(111) was reported by Xu and Koel.[304] In the absence of oxygen, RCO may be decomposed, while in the presence of oxygen it may be oxidized to acetate or peroxyacetyl radical with oxygen or other oxygen species.

$$RCO_2 \rightarrow R + CO_2$$
$$R + O_2 \rightarrow RO_2$$

In the case of RCO_3, the radical nature of the heterogeneous constituent of reaction, as shown by the above examples, may be considered well established by the experimental investigations.

Heterogeneous reactions of peroxy radicals (RO_2 and RCO_3) with aldehydes. A number of heterogeneous reactions of peroxy radicals (RO_2 and RCO_3) with aldehydes on the surfaces of different solid substances are summarized in Table 3.2.

As Table 3.2 shows, the reactions of peroxy radicals with C_2-C_3 aldehydes at temperatures of 100–500 K were investigated on the different solid surfaces, including surfaces of acidic or basic oxides and salts, as well as oxides of transition metal (TiO_2) or semimetal (SiO_2) oxides. In the works cited in the table, the main investigatory technique used was the EPR method applied either by *in situ* measurements on the solid substances or by the freezing of radicals from the gas phase (matrix isolation of radicals). The most important result of all of these works was the fact that the oxygen-centered peroxy radicals, being generated on the surface or diffused from the gas phase to the surface, are reactive with acetaldehyde or propionaldehyde by the heterogeneous pathway. Moreover, in some cases—for instance, in the reaction of acetaldehyde with peroxy radicals on the TiO_2 surface—there was an increase in the number of radicals in the gas phase, as a result of the heterogeneous reaction. This observation, named "multiplication of radicals," is additional evidence for the heterogeneous propagation of chains at low-temperature oxidation of aldehydes (Ref. 5 in Table 3.2). Kinetic data gathered through chemical analysis confirmed this conclusion. This fact also shows the feasibility of the "degenerate-branching of chains" in heterogeneous reactions of aldehydes. As mentioned in Section 3.5.4.2, the branching of heterogeneous chains was suggested in the thermal, low-temperature oxidation of propionic aldehyde in the reactors with a developed surface area (increased S/V) by analyses of the kinetic data (Ref. 6 in Table 3.2).

Thus, we have substantial direct and indirect evidence for the feasibility of heterogeneous propagation and degenerate (partial) chain branching in the oxidation of

Table 3.2 Heterogeneous reactions of the oxygen-centered organic peroxy radicals ($R = CH_3$, C_2H_5) with acetaldehyde and propionaldehyde on the surfaces of solid substances.

Aldehyde	Radical	Solid Substance	Temperature	Reference
Acetaldehyde	RO_2, RCO_2	SiO_2	178–216 K	1
Propionaldehyde	RO_2	KCl/SiO_2	373 K	2
Acetaldehyde, propionaldehyde	RCO_3	SiO_2, KCl/SiO_2	423–503 K	3
Acetaldehyde	RO_2 RCO_3	capillary reactor, treated with TiO_2	room temperature	4
Acetaldehyde	RO_2, RCO_3	capillary reactor, treated with H_3BO_3		5
Acetaldehyde, propionaldehyde	RO_2	capillary reactor, treated with KCl	423–523 K	6
Acetaldehyde, other aldehydes	RCO_3	TiO_2	100–250 K	7
Acetaldehyde	RO_2	capillary reactor, treated with KCl	297–353 K	8
Acetaldehyde	RO_2	NH_4NO_3	room temperature	9

References in Table 3.2

1. Tavadyan L. A. Study by the EPR-method of the ability of molecular mobility of peroxy radicals on the surface of silica gel. *Kinetica i cataliz* (Kinetics and Catalysis, in Russian), 1983, v. 24, 2, p. 396–402.
2. *Bakhchadzhyan R. H., Vardanyan I. A., Nalbandyan A. B. Studies of the reaction between the adsorbed propionaldehyde and peroxy radicals by the method of IR-spectroscopy. *Khimicheskaya Fizika* (Chemical Physics, in Russian), 1986, v. 5, 3, p. 393–396.
3. *Bakhchadzhyan R. H., Gazaryan K. G., Vardanyan I. A. Formation and reactions of acetylperoxy radicals on the solid surface. *Khimicheskaya Fizika* (Chemical Physics, in Russian), 1991. v. 8, 5, p. 659–663.
4. Vardanyan I. A., Manucharova L. A., Jalali H. A., Tsarukyan S. V. Interaction of CH_3O_2 radicals with CH_3CHO and CH_4 on TiO_2 surface. Chemical Journal of Armenia, 2012, v. 6, 1, p. 132–135.
5. Vardanyan I. A., Arustamyan A. M., Martirosyan A. S., Tsarukyan S. V. Interaction between peroxy radicals and acetaldehyde on solid Surfaces and its role in the oxidation of aldehydes. Russian Journal of Physical Chemistry A. 2016, v. 90, 4, p. 744–747.
6. Bakhchadjyan R. H., Vardanyan I. A, Nalbandyan A. B. Low temperature oxidation of aldehydes in conditions excluding homogeneous propagation of chains. *Doklai Academii Naul Armianskoy SSR* (Reports of the Academy of Sciences of the Armenian SSR, in Russian), 1988, v. 56, 4, p. 170–173.
7. Charles J. A., Murphy D. M. Thermal and photoreactivity of TiO_2 at the gas-solid interface with aliphatic and aromatic aldehydes. Journal of Physical Chemistry B, 1999, v. 103, 6, p. 1019–1026. 10.1021/jp982690n
8. Manucharova L. A., Tsarukyan S. V., Vardanyan I. A. Reactions of CH_3O_2 radicals on solid surface. Int. J. Chem. Kin., 2004, v. 36, 11, p. 591–595.
9. Jalaly H. J. An EPR study of heterogeneous reaction of CH_3O_2 radicals with organic compounds: Effect of organic compound and surface nature. Physical Chemistry and Electrochemistry, 2013, v. 2, 1, p. 31–37.

*Author's note: This is an incorrect form of my family name that appeared in English translations some of my articles from the Russian language literature.

acetaldehyde or propoionaldehyde in certain experimental conditions, in the presence of the solid-phase substances via the following reactions:

$$RCHO + RO_2(RCO_3) \rightarrow RCO + RO_2H(RCO_3H)$$

Multiplication of radicals may also take place through further heterogeneous decomposition reactions of RO_2H or RCO_3H:

$$RO_2H \rightarrow RO + OH$$

$$RCO_3H \rightarrow RCO_2 + OH$$

3.6 CONCLUDING NOTES ABOUT THE BIMODAL REACTION SEQUENCES IN THERMAL OXIDATION

The examples given above demonstrate that thermal oxidation of organic compounds, such as hydrocarbons, alcohols, and aldehydes, and their structural analogs (derivatives) with molecular oxygen in the presence of solid substances, in certain conditions, may occur as bimodal (heterogeneous–homogeneous or homogeneous–heterogeneous) reactions. The homogeneous constituents of these complex reactions, as a rule, are free radical, primarily chain-radical processes, certain stages of which may take place on the solid phases. The totality of a great number of reaction parameters (temperature, composition of the initial mixture, nature of the solid surfaces, S/V ratio of the reaction vessel, and others) determine the heterogeneous–homogeneous occurrence of the complex oxidation process with dioxygen. The heterogeneous constituents of reactions, depending on conditions, may have different functionality in the overall process. Principally, the solid phase may participate in all four main stages of the chain-radical reactions of oxidation: 1—initiation; 2—propagation; 3—branching or degenerate branching; and 4—termination of chains. In particular cases, the overall heterogeneous–homogeneous or homogeneous–heterogeneous reactions may include one or more, or even all four of the heterogeneous stages. (Each of these cases was illustrated in the above sections.) "Purely" heterogeneous and "purely" homogeneous reactions, in the oxidation of organic compounds, may be considered as two extreme cases of complex reactions, occurring in special conditions, excluding the contributions of either heterogeneous or homogeneous reactions. We have already mentioned some examples where reaction conditions were chosen in the aims of excluding either homogeneous (see examples, excluding the homogeneous propagation of chains in Ref. 6 in Table 3.2) or heterogeneous constituents (see Section 1.4, "wall-less like reactor"). In some cases, the formation of the main products may occur in both phases as parallel processes ($H_2 + O_2^{\text{ref. in 3.5.1}}$ and $RCHO + O_2^{\text{ref. in 3.5.4}}$). Control over the rate ratio of the heterogeneous and homogeneous constituents has vast importance in chemical engineering (for examples, see Section 6.8).

With regard to the mechanism suggestions of the heterogeneous constituent of complex heterogeneous–homogeneous reactions, the intermediates formed may be radicals, radical-like species, excited atoms or molecules, and charged species (ions or ion-radicals), mainly in the form of surface complexes. The surface reaction may propagate as a radical or nonradical, equally as a chain-radical or non-chain-radical reaction. The experimental investigation and revelation of surface reactions in a bimodal reaction sequence are related to the overall problems of heterogeneous catalysis, although it has its own peculiarities.

In revelation of the nature of heterogeneous reactions in experimental investigations, complications may arise if the heterogeneous constituent of the heterogeneous–homogeneous process partially occurs through a chain-radical mechanism.[305] Often, the heterogeneous pathway may be inconspicuous, even in the case of long heterogeneous chains. For example, the oxidation of slight hydrocarbons, induced by the heterogeneous-radical decomposition of hydrogen peroxide, may occur by either chain-radical or nonradical pathways on the surface of metal oxides.[305] According to Elchyan and Grigoryan,[305] the amount of formed product (N) in a chain-radical reaction will be equal:

$$N = \alpha \ v[H_2O_2]_0$$

where α is a fraction of radicals, forming in decomposition of H_2O_2 and participating in the surface chain reaction, and v is the length of the surface chain. If $\alpha v < 1$, the amount of the product may be $N < [H_2O_2]_0$. Even when $v \gg 1$, again, the condition $\alpha v < 1$ may be determining if $\alpha \ll 1$. Thus, observation of the surface chain reaction by kinetic analysis of the final product is not always possible because of the low concentration of radicals. Usually, the surface chains, as shown in the case of the oxidation of aldehydes, are quite short (about $v = 30$–50 links[124 in Ch. 1, 280, 306]).

In investigations of complex heterogeneous–homogeneous and homogeneous–heterogeneous processes, the following fundamental question often arises. What is the contribution of each constituent (homogeneous and heterogeneous) in the overall rate of the process? We have seen some examples where the main products of reaction, in certain conditions, may be formed both in the fluid phases and on the solid surfaces by parallel reaction pathways (formation of H_2O in $H_2 + O_2$; carboxylic acid in aldehyde $+ O_2$ reactions). In the case of "classical" heterogeneous–homogeneous processes, considering only heterogeneously initiated homogeneous chain-radical reactions in the fluid phases, the rate of the primary homogeneous reaction depends on the ability of the heterogeneous phase(s) to generate and transfer radicals to the fluid phases. However, the progressing homogeneous process itself produces radicals, which not only develop homogeneous chains, but also reach to the surface of the solid substances, where they may sometimes initiate new heterogeneous processes. Previously, it was considered that the solid surface plays the role of "sink" for radicals diffusing from the fluid phases, in progressing homogeneous reaction. In some other cases, the radical decomposition of the metastable intermediate products (as peroxides) on the solid surface may become an additional pathway for the initiation of new heterogeneous reactions. Hence, distribution of the leading active species between two phases determines the rate and selectivity of the final products in this kind of complex processes. Obviously, then, the rates and quantities of the final products will be different for the homogeneous and heterogeneous pathways. Apparently, in certain cases, the homogeneous chains transfer to the solid surface, an example of which was the low-temperature bimodal oxidation of aldehydes (Section 3.5.4.2).

The simplest criterion (indeed, not the only one) of the probability (α) of the transfer of the active centers of the homogeneous reaction to the surface, in homogeneous–heterogeneous reactions, may be expressed by the following ratio

$$\alpha = t_2/t_1 = 2D/k_{ef}a^2[C]$$

where α is the ratio of the half-lifetimes of the main active species in the gas-phase reaction (t_2) and diffusion to the surface (t_1); a is the average distance of the gas gap between the solid surfaces in the reactor; and D is the diffusion coefficient in the given conditions and medium.[306]

In oxidation of propionic aldehyde at low temperatures[280] (T < 423 K), $\alpha = 1$ condition takes place when $a = 0.053$ cm, in a reaction vessel packed with the glass pieces and precoated by KCl, having an S/V ratio of about 50 cm^{-1}. Examples of the calculation of the diffusion coefficients of OH, H, HO$_2$, RO$_2$, in different experimental conditions, were given by Porter et al.[307]

An interesting comparison was made by Krishtopa et al.,[227] who investigated the ability of some metal oxides to generate the peroxy radicals (RO$_2$) by further transfer to the gas phase in the heterogeneous–homogeneous oxidation reactions of organic compounds with molecular oxygen. They found that in the oxidation reaction sequences:

Hydrocarbon → Alcohol → Aldehyde → Carboxylic acid

the quantities of radicals generated on the surface of the same catalysts and transferred to the gas phase from the surface, in comparable conditions, had the following ratios

$$1 : 10 : 5 : 5$$

According to other authors, after the heterogeneous initiation of the homogeneous reaction, at certain moments, the solid surface slows down the supply of radicals to the gas phase and the given concentration of radicals is a result only of their homogeneous generation.[208, 209, 240] One example of these is partial oxidation of methane on Pt-catalyst (1000–1300°C).[209] The reaction starts at 600°C, producing mainly H$_2$O, CO$_2$, and with the increase in temperature, also CO, H$_2$. At about 1100°C (upon "ignition"), the reaction takes on a heterogeneous–homogeneous character with the formation of products, mainly those known as products of the oxidative coupling of methane (OCM), related to the methyl radicals. The formation of methyl radicals was shown to be a result of the homogeneous reaction rather than of their generation on the surface of Pt with the further transfer to the gas phase, at elevated temperatures. According to this work,[209] the role of the heterogeneous reaction is related to the release of heat accelerating the homogeneous reaction rather than to the desorption of radicals from the surface. This conclusion was in agreement also with the experimental data obtained by Lunsford et al. in their study of the reaction of the oxidation of methane on the Pt single crystal and polycrystalline Pt supported on SiO$_2$ at 900°C.[208] Investigating the reaction also by application of the matrix isolation EPR method, they found that the generation of radicals from the surface did not "facilitate" the homogeneous reaction in the heterogeneous–homogeneous oxidation of methane with dioxygen. In other words, in both examples, the heterogeneous reaction was not, in certain conditions, the main supplier of methyl radicals to the gas phase. Analogous conclusions were drawn in the investigation of methanol oxidation on the Pt-catalyst supported on SiO$_2$ and Al$_2$O$_3$[240] (see Section 3.5.3). In general, however, the main factor determining the bimodal nature of processes lies in the chemical reactions in both phases, but not only heat transfer processes, as the heat exchange between phases is a consequence of the chemical reactions. In these cases, the quantity of radicals in the gas phase may remain unchanged, while the rate of the overall reaction in all the above-mentioned examples and many similar cases usually increases with time, as well as with temperature.[308, 309] Thus, at relatively elevated temperatures, as mentioned above, the occurring homogeneous reaction itself may initiate heterogeneous reactions, supplying radicals to the surface. In all of these cases, the complex reaction progresses by parallel heterogeneous and homogeneous reactions rather than only by homogeneous reaction.

A probable explanation of a strong dependence of the rate of heterogeneous–homogeneous reaction on the ratio of initial reactants, dioxygen/organic compounds was given by Ismagilov et al.[123 in Ch. 1] At the high ratio of the dioxygen/organic compound, the surface oxygen species in enough concentrations may oxidize organic compounds by the heterogeneous mechanism on the metal oxides. At the same time, at low concentration of oxygen, the probability of the concurring reactions of dehydrogenation, isomerization or others, and, especially, the desorption of intermediate radicals to the gas-phase increases. Thus, "oxygen deficiency" may become an essential condition for the heterogeneous–homogeneous occurrence of the oxidation reaction.

As mentioned earlier, in oxidation reactions, the heterogeneous nature was relatively well pronounced at low temperatures. At low concentrations of reactant, as well as at low total pressure in a heterogeneous catalytic system, the reaction occurs rather by a "purely" heterogeneous than by a heterogeneous–homogeneous mechanism. In the latter case, the heterogeneous constituent in bimodal reaction sequences may be part of a process that occurs through the participation of radical-like species and, apparently, developing short chains, like chain reactions in the gas phase, but limited on the frameworks of the surface.

These mentioned peculiarities, however, are rather tendencies more, than general rules affecting oxidation by the heterogeneous–homogeneous mechanism. In fact, numerous other factors determine the pathway of the complex reaction.

In general, the examples provided in this chapter show that in complex reactions, having a bimodal sequence of occurrence, as van Santen describes it, "the heterogeneous catalytic and homogeneous catalytic process steps sometimes complement each other in the overall production process."[36] In this regard, they are single, unified processes that are essentially different from "purely" heterogeneous or "purely" homogeneous reactions.

REFERENCES

3.1

1. Breslow R. Oxidation by remote functionalization methods (1.3), p. 39–40. In: Trost B. M., Fleming I., Semmelhack M. F. (eds). Comprehensive Organic Syntheses, Strategy and Efficiency in Modern Organic Chemistry, v. 8, 1991, Elsevier, 1143 pages.

3.2

2. Hou A. Quantum Chemistry. N.Y. Research Press, USA, 2015, 214 pages.
3. Mulliken R. S. Interpretation of the atmospheric oxygen bands; Electronic levels of the oxygen molecule. Nature, 1928, v. 122, 3075, p. 505.
4. Mulliken R. S. Interpretation of the atmospheric bands of oxygen. Phys. Rev., 1928, v. 32, p. 880–887.
5. Krasnovsky A. A., Jr. Primary mechanisms of photoactivation of molecular oxygen. History of development and the modern status of research. Biochemistry (*Biokhimiya*, in Russian), 2007, v. 72, 10, p. 1065–1080, p. 1311–1329.
6. House E. J. Fundamentals of Quantum Chemistry, Academic Press, USA, 2004, 290 pages, p. 204.
7. Fraunhofer J. *Denkshriften der Munch. Akad. Wiss.* 1814–1815, v. 5, p. 193–226.

8. Ellis J. W., Kneser H. O. *Kombinationsbeziehungen im absorptionsspektrum des flussigen sauerstoffs, Z. Phys.,* 1933, v. 86, p. 583–591.
9. Herzberg G. Photography of infrared solar spectrum to wavelength 12 900Å. Nature (London), 1934, v. 133, p. 759.
10. Turro N. J., Ramamurthy V., Scaiano J. C. Modern Molecular Photochemistry of Organic Molecules. University Science Books, 2010, U.S. 1084 pages, p. 1006.
11. Tachikawa T., Majima T. Single-molecule detection of reactive oxygen species: Application to photocatalytic reactions. J. Fluoresc., 2007, v. 17, p. 727–738. DOI 10.1007/s10895-007-0181-5
12. Wiberg E., Wiberg N. Inorganic Chemistry. Academic Press, San Diego, USA, 2001, 1884 pages, p. 476.
13. Herzberg G. Electronic Spectra and Electronic Structure of Polyatomic Molecules. Van Nostrand, Princeton, 1966.
14. Herzberg G. Spectra of Diatomic Molecules, 2nd ed. Van Nostrand, Princeton, N.Y., 1950.
15. Hoffmann R. Geometry changes in excited states. Pure and Appl. Chem., 1970, v. 24, 3, p. 567–584.

3.3

16. Mainzer K. Symmetry and Complexity: The Spirit and Beauty of Nonlinear Science. 2005, World Scientific Publishing, USA, Series A, v. 51, 437 pages.
17. Wilczek F. A Beautiful Question: Finding Nature's Deep Design, 2015, Penguin Press, USA, 448 pages.
18. Wigner E. P., Witmer E. E. *Uber die Struktue der zweiatomogen Molekelspektren nach der Quantenmechanik, Zeitschrift fur Physik,* 1928, v. 51, p. 859–886. In: The Collected Works of Eugene Paul Wigner, 1993, Springer, Verlag Berlin Heidelberg, 727 pages. p. 168.
19. Hargittai M., Hargittai I. Symmetry through the Eyes of a Chemist. Chapter 7. Chemical Reactions, p. 1–58. 2009. Kluwer Academic Pub., 520 pages.
20. Hoffmann R., Woodward R. B. Orbital Symmetry control of chemical reactions. Science, 1970, v. 167, 3919, p. 825–831.
21. Blinder S. M. Introduction to Quantum Mechanics: In: Chemistry, Materials Science, and Biology. 2012, Academic Press; 1st edition, 319 pages.
22. Pearson R. G. Orbital symmetry rule and the mechanism of inorganic reactions. Pure and Applied Chemistry. 1971, v. 27, 1–2, p. 145–160. DOI: 10.1351/pac197127010145
23. Moore J. W., Pearson R. G. Kinetics and Mechanism. 3rd ed., 1981, Wiley, NY, 473 pages p. 186.
24. Tsukahara H., Ishida T., Mayumi M. Gas-phase oxidation of nitric oxide: chemical kinetics and rate constant. Nitric Oxide. 1999, v. 3, 3, p. 191–198.

3.4

3.4.1

3.4.1.1

25. Chiesa M., Giamello E., Che M. EPR Characterization and Reactivity of Surface-Localized Inorganic Radicals and Radical Ions. Chem. Rev., 2010, v. 110, 3, p. 1320–1347.
26. Taube H. Mechanism of oxidation with oxygen. J. Gen. Physiol., 1965, v. 49, 1, p. 29–50.
27. Lu W., Zhou L. Oxidation of C-H bonds. 2017. Wiley and Sons. NJ, USA, 508 pages.

3.4.1.2

28. Steininger H., Lehwald S., Ibach H. Adsorption of oxygen on Pt(111). Surface Science, 1982, v. 123, 1, p. 1–17.
29. Wang C., Bai C. Single Molecule Chemistry and Physics: An Introduction. Series: Nanoscience and Technology. Springer Berlin. 2006, 304 pages, p. 49.
30. Spasic A. M., Hsu J-P. Finely Dispersed Particles: Micro-, Nano-, and Atto-Engineering. Micro-, Nano- and Atto- Engineering. 130. CRC, Taylor and Francis, Broken, 2006, 913 pages. p. 164.
31. Bashlakov D. Interaction of oxygen and carbon monoxide with Pt(111) at intermediate pressure and temperature: revisiting the fruit fly of surface science. Leiden University dissertation, 2004, p. 38. http://hdl.handle.net/1887/29023.
32. Chen M. S., Cai Y., Yan Z., Gath K. K., Axnanda S., Goodman D. W. Surface Science, 2007, v. 601, p. 5326.
33. Li W-X., Stampfl C., Scheffler M. Oxygen adsorption on Ag(111): A density-functional theory investigation Physical Review B, 2002, v 65, 075407- p. 1–19.
34. Bielanski A., Haber J. Oxygen in Catalysis. Marcel Dekker Inc, New York, 1991, 473 pages, p. 67.
35. Ertl G. Reactions at Surfaces: from Atoms to Complexity. Nobel Lecture, December 8, 2007. In: https://www.nobelprize.org/nobel_prizes/chemistry/laureates/2007/ertl_lecture.pdf.

3.4.1.3

36. Van Santen R. A. Modern Heterogeneous Catalysis: An Introduction. 2017, Wiley-VCH, Weinheim, 564 pages, p. 478.
37. Netzer F. P., Fortunelli A. Oxide Materials at the Two-Dimensional Limit, 2016, Springer International Publishing Switzerland, 388 pages
38. Fierro J. L. G. Metal Oxides: Chemistry and Applications (Series: Chemical Industries), 2005, CRC Press, Taylor and Francis, Boca Raton, 808 pages.
39. Centi G., Cavani F., Trifiro F. Selective Oxidation by Heterogeneous Catalysis (Fundamental and Applied Catalysis), p. 388, 8.3; 1, Nature of the interaction between molecular oxygen and oxide surfaces and types of oxygen adspecies. 2001, Springer Science, N.Y., 505 pages.
40. Olmsted J., Williams G. M. Chemistry: The Molecular Science, Second edition, 1997, p. 918. Wm. C. Brown Publishers, Dubuque, 1048 pages.

3.4.2

41. Che M., Tench A. J. Characterisation and reactivity of molecular oxygen species on oxide surfaces, p. 2–206. In: Eley D. D., Pines H., Weisz P. B. (eds), Advances in Catalysis, v. 32, 1983, Academic Press, N.Y., 515 pages.
42. Misono M. Heterogeneous Catalysis of Mixed Oxides: Perovskite and Heteropoly Catalysts, 76, series: Surface Science and Catalysis, 2013, Elsevier, Amsterdam, 182 pages.
43. Pike O. J., Mackenroth F., Hill E. G., Rose S. J. A photon–photon collider in a vacuum hohlraum. Nature Photonics, 2014, v. 8, p. 434–436. DOI: 10.1038/nphoton.2014.95

3.4.2.1

44. Singh V. P. Gymnosperm (naked seeds plant): structure and development, Sarup and Sons, Delhi, First edition 2006. 683 pages, p. 21.

45. Ho R. Y. N., Liebman J. F., Valentine J. S. Overview of the Energetics and Reactivity of Oxygen. In: Active Oxygen in Chemistry (Editor: C. S. Foote), Kluwer, Blackie Academic and Professional, Chapman and Hall, 1995, USA, Chapter 1, 342 pages.

46. Zhang J-Y., Boyd I. W. Rapid photooxidation at room temperature using 126 nm ultraviolet radiation, Applied Surface Science, 2002, v. 186, p. 64–68.

47. McNesby J. R., Okabe H. In: Noyes W. A., Hammond G., Pitts J. N. (eds). Advances in Photochemistry, v. 3, Wiley, New York, 1966, p. 157.

48. Lugez C., Schriver A., Levant R., Schriver-Mazzuoli L. A matrix isolation infrared spectroscopic study of the reaction of methane and methanol with ozone. Chem. Phys. 1994, v. 181, p. 129–146.

49. Finlayson-Pitts B. J., Pitts J. N. Jr. Chemistry of the Upper and Lower Atmosphere: Theory, Experiments and Application. 2000. Academic Press. San Diego, 970 pages, p. 660.

50. Newman P. A., Stratospheric Photochemistry. In: Stratospheric Ozone (Chapter 5, Section 3), Electronic textbook written and edited by members of NASA's Goddard Space Flight Center Atmospheric Chemistry and Dynamics Branch (Code 916), May, 2003. Website: http://www.ccpo.odu.edu

51. Sethi M. S., Satake M. Chemical Bonding, Discovery Publishing House, Delhi, 2003, 149 pages, p. 22.

52. Volodin A. M., Malykhin S. E., Zhidomirov G. M. O⁻ radical anions on oxide catalysts: Formation, properties, and reactions, *Kinetika i Kataliz* (in Russian), 2011, v. 52, 4, p. 615–629; Kinetics and Catalysis, 2011, v. 52, 4, p. 605–619 (in English).

53. Kazanski V. B. About possibility of the initiation of the chains in the reactions of the catalytic oxidation on the oxides. In: Problems of the Chemical Kinetics, edited by Kondratev, 1979, Moscow, Publishing House "*Nauka*", p. 232–244.

54. Lee J., Grabowsky J. J. Reactions of atomic oxygen radical anion and the syntheses of organic reactive intermediates. Chem. Rev. 1992, v. 92, p. 1611.

3.4.2.2

55. Chiesa M. G., Giamello E., Paganini C., Sojka Z., Murphy D. M. Continuous wave electron paramagnetic resonance investigation of the hyperfine structure of $^{17}O_2^-$ adsorbed on the MgO surface. Journal of Chemical Physics. 2002, v. 116, 10, p. 4266–4274.

56. Il'ichev A. N., Konin G. A., Matyshak V. A., Kulizade A. M., Korchak V. N., Yan Yu. B. Formation mechanism of O_2^- radical anions in the adsorption of $NO + O_2$ and $NO_2 + O_2$ mixtures on ZrO_2 according to EPR and TPD data. *Kinetika i Kataiz* (Kinetics and Catalysis, in Russian), 2002, v. 43, 2, p. 214–222.

57. Anpo M., Che M., Fubinic B., Garronec E., Giamello E., Paganinic M. C. Generation of superoxide ions at oxide surfaces. Topics in Catalysis, 1999, v. 8, p. 189–198.

58. Li Y-F., Aschauer U., Chen J., Selloni A. Adsorption and reactions of O_2 on anatase TiO_2. Acc. Chem. Res., 2014, v. 47, 11, p. 3361–3368.

59. Panov G. I., Dubkov K. A., Starokon E. V. Active oxygen in selective oxidation catalysis. Catal. Today, 2006, v. 117, 1–3, p. 148–155.

60. Che M., Giamello E. Electron paramagnetic resonance (Chapter 5), Mobility of adsorbed species (5.3.6), p. 314. In: Fierro J. L. G. Spectroscopic Characterization of Heterogeneous Catalysts, Part B: Chemisorption of Probe Molecules (Studies in Surface Science and Catalysis) 57, 1990, Elsevier Sci., 408 pages.

61. Riley D., Ebner J. Catalytic oxidation with dioxygen. An industrial perspective. 6. p. 204–248. In: Foote C. S., Greenberg A., Valentine J. S., Liebman J. F. Active Oxygen in Chemistry. v. 2, 1995, Blackie Academic and Professionals, London, 341 pages, p. 321.

62. Puigdollers A. R., Schlexer P., Tosoni S., Pacchioni G. Increasing oxide reducibility: The role of metal/oxide interfaces in the formation of oxygen vacancies. ACS Catal., 2017, v. 7, p. 6493–6513. DOI: 10.1021/acscatal.7b01913

63. Tsuji H., Hattori H. Oxide surfaces that catalyse an acid-base reaction with surface lattice oxygen exchange: evidence of nucleophilicity of oxide surfaces. Chemphyschem. 2004, v. 17, 5, 5, p. 733–736.

3.4.2.3

64. Minero C. Surface modified photocatalyst. In: Bahnemann D. W., Robertson P. K. J. (eds). Environmental Photochemistry, Part 3, 2015, Springer, Berlin, 346 pages, p. 29.

65. Hernández-Ramírez A., Medina-Ramírez I. Semiconducting materials. In: Hernández-Ramírez A., Medina-Ramírez I. (eds). Photocatalytic Semiconductors: Synthesis, Characterization, and Environmental Application. Springer, 2015, 289 pages, p. 7.

66. Apak R. Adsorption of heavy metal ions on soil surfaces and similar substances. Surface functional groups, p. 387. In: A. T. Hubbard (ed.), Encyclopedia of Surface and Colloid Science, volume 1. Marcel Dekker, N.Y., 2002, 1472 pages.

67. Nosaka Y., Nosaka A. Y. Identification and roles of the active species generated on various photocatalysts, p. 3–24. In: Lu M., Pichat P. (eds). Photocatalysis and Water Purification: From Fundamentals to Recent Applications, Part I. First Edition. 2013, Wiley-VCH Verlag GmbH and Co. KGaA, 438 pages.

68. Fisher G. B., Sexton B. A. Identification of an Adsorbed Hydroxyl Species on the Pt(111) Surface. Phys. Rev. Lett., 1980, v. 44, p. 683.

69. Talley L. D., Lin M.C. Energetics and mechanism for hydroxyl radical production from the Pt-catalyzed decomposition of water. Chemical Physics. Volume 61, 3, 15, 1981, p. 249–255.

70. Hu Z-M., Nakatsuji H. Adsorption and disproportionation reaction of OH on Ag surfaces: dipped adcluster model study. Surface Science, 1999, v. 425, p. 296–312.

71. Talley L. D., Tevault D. E., Lin, M. C. Laser diagnostic of matrix-isolated OH radicals from oxidation of H_2 on platinum. Chemical Physics Letters, 1979, v. 66, p. 584–586.

72. Talley L. D., Lin M. C. Energetics and mechanism for hydroxyl radical production from the Pt-catalyzed decomposition of water. Internal energy of hydroxyl radicals desorbing from polycrystalline Pt surfaces Chemical Physics. 1981, v. 61, p. 249–255.

73. Talley L. D., Sanders W. A., Bogan D. J. Internal energy of hydroxyl radicals desorbing from polycrystalline Pt surfaces. Chemical Physics Letters. 1981, v. 78, p. 500–503.

3.4.2.4

74. Marshall A. L. Photosensitization by optically excited mercury atoms. The hydrogen oxygen reaction. J. Phys. Chem., 1926, v. 30, 34, p. 34–46.

75. Foner S. N., Hudson R. L. Detection of the HO_2 radical by mass spectrometry. J. Chem. Phys., 1953, v. 21, p. 1608–1609.

76. Hucknall D. Chemistry of Hydrocarbon Combustion (Hydroperoxy radical HO_2), Chapter 4, p. 129, 1985, Chapman and Hall, London, 414 pages.

77. Cox R. A., Burrows J. P. Kinetics and mechanism of the disproportionation of hydroperoxyl radical in the gas phase. J. Phys. Chem., 1979, v. 83, 20, p. 2560–2568.

78. Giguère P. A. Thermochemistry of the hydrogen polyoxides H_2O_3 and H_2O_4. Transactions of the New York Academy of Sciences, 1972, v. 34, 4, p. 334–343.

79. Arutyunyan A. Z., Grigoryan G. L., Nalbandyan A. B., Decomposition of hydrogen peroxide vapor on oxide catalysts. *Kinetika i Kataliz* (Kinetics and Catalysis, in Russian), 1986, v. 27, p. 1173–1178.

80. Carlier M., Sahetchian K., Sochet R-L. ESR evidence for halogen oxy-radical forma-
    tion in the study of the heterogeneous decomposition of gaseous hydrogen peroxide in
    halides. Chemical Physics Letters, 1979, v. 66, 3, 15, p. 557–560.
81. Grigoryan G. L., Beglaryan H. A. Low temperature chemical transportation of zinc
    compounds by means of water vapors. Chem. J. Arm. 2009, v. 62, 1–2, p. 57–64.
82. Baldwin R. R. Walker R. W. Kinetics of hydrogen-oxygen and hydrocarbon-oxygen. Final
    Scientific Report 1977–81, Air Force Office of Scientific Research USA, 1981, 64 pages.
83. Taketani F., Kanaya Y., Akimoto H. Kinetics of heterogeneous reactions of HO_2 radi-
    cal at ambient concentration levels with $(NH_4)_2SO_4$ and NaCl aerosol particles. J. Phys.
    Chem. A, 2008, v. 112, 11, p. 2370–2377. DOI: 10.1021/jp0769936
84. Dissanayake D. P., Lunsford J. H. Evidence for the role of colloidal palladium in the
    catalytic formation of H_2O_2 from H_2 and O_2. J. Catal., 2002, v. 206, p. 173–176.
85. Sivadinarayana C., Choudhary T. V., Daemen L. L., Eckert J., Goodman D.W. The
    nature of the surface species formed on Au/TiO_2 during the reaction of H_2 and O_2: An
    inelastic neutron scattering study. J. Am. Chem. Soc. 2004, v. 126, p. 38–39.

3.4.2.5

86. Orlando J. J., Tyndall G. S. Laboratory studies of organic peroxy radical chemistry: an
    overview with emphasis on recent issues of atmospheric significance. Chem. Soc. Rev.,
    2012, 41, p. 6294–6317. DOI: 10.1039/C2CS35166H
87. Kuwata K. T. Computational treatment of peroxy radicals in combustion and atmo-
    spheric reactions. Ch. 15. In: J. F. Liebman, A. Greer, Z. Rappoport (eds). The Chemistry
    of Peroxides, Volume 3, Wiley, 2014, 1120 pages, p. 634.
88. Ingold K. U. Peroxy radicals. Acc. Chem. Res., 1969, v. 2, 1, p. 1–9.
89. Hucknall D. Chemistry of Hydrocarbon Combustion, 1985, Chapman et Hall, London,
    414 pages.
90. Attwood A. L., Edwards J. L., Rowlands C. C., Murphy D. M. Identification of a surface
    alkylperoxy radical in the photocatalytic oxidation of acetone/O_2 over TiO_2. J. Phys.
    Chem. A, 2003, v. 107, 11, p. 1779–1782. DOI: 10.1021/jp022448n
91. Bakhchadjyan R. H., Vardanyan I. A. Study of the radical decomposition of the vapors
    of peroxide compounds by the methods of EPR and electro conductivity. Khimcheskaya
    Phizika (Chemical Physics, in Russian), 1983, n. 11, p. 1536–1540.
92. Bakhchadjyan R. H., Gazaryan K. G., Vardanyan I. A. Formation of the acetylperoxy
    radicals on the solid surfaces. Khim. Phiz. (Chemical Physics, in Russian), 1991, v. 10,
    4, p. 659–664.
93. Bagdasaryan G. O., Vardanyan I. A., Nalbandyan A. B. OH radicals formation by
    methyl hydroperoxide on Pt surface. Arm. Chim. J. (Arm Chem. J. in Russian), 1979,
    v. 32, 2, p. 157–158.
94. Vardanyan I. A., Manucharova L. A., Jalali H. A., Tsarukyan S. V. Interaction of CH_3O_2
    radicals with CH_3CHO and CH_4 on TiO_2 surface. Chem. J. Armenia. 2012, v. 65, 1,
    p. 132–135.
95. Vardanyan I. A., Arustamyan A. M., Martirosyan A. S., Tsarukyan S. V. Interaction
    between peroxy radicals and acetaldehyde on solid surfaces and its role in the oxidation
    of aldehydes on NH_4NO_3 and NaCl. Russian Journal of Physical Chemistry A, 2016,
    v. 90, 4, p. 744–747.
96. Jalali H. A., Vardanyan I. A Modeling process of heterogeneous interaction of peroxy
    radicals with organic compound. Archivum Combustionis, 2010, v. 30, 4, p. 297–302.
97. Carter E., Carley A. F., Murphy D. M. Free-Radical pathways in the decomposition
    of ketones over polycrystalline TiO_2: The role of organoperoxy radicals. Chem. Phys.
    Chem., 2007, v. 8, p. 113.

3.4.2.6

98. Murphy D. EPR of polycrystalline oxide systems. In: Jackson S. D., Hargreaves J. S. J. (eds). Metal Oxide Catalysis. J. Metal Oxide Catalysis, Volume 1, Characterization of metal oxides. 2008, Wiley-VCH, Weinheim, 887 pages, p. 38.
99. Chiesa M., Giamello E. Carbon dioxide activation by surface excess electrons: An EPR study of the CO_2^- radical ion adsorbed on the surface of MgO. Chem. Eur. J., 2007, v. 13, p. 1261–1267.
100. Preda G., Pacchioni G., Chiesa M., Giamello E. Formation of CO_2^- Radical anions from CO_2 adsorption on an electron-rich MgO surface: A combined *ab initio* and pulse EPR study. J. Phys. Chem. C, 2008, v. 112, 49, p. 19568–19576. DOI: 10.1021/jp806049x
101. Todres Z. V. Ion-Radical Organic Chemistry: Principles and Applications, Second Edition, 2008, CRC Press, 496 pages, p. 59.
102. Mikheikin I. D., Zhidomirov G. M., Chuvylkin N. D., Kazanskii V. B. Parameters of the ESR spectra and structure of O_3^- radicals. Journal of Structural Chemistry, 1975, v. 15, 5, p. 678–681.

3.4.2.7

103. Kautsky H., de Bruijn H. *Die Aufklärung der Photoluminescenztilgung fluorescierender Systeme durch Sauerstoff: Die Bildung aktiver, diffusionsfähiger Sauerstoffmoleküle durch Sensibilisierung. Naturwissenschaften*, 1931, v. 19, 52, p. 1043–1043.
104. Kautsky H. Quenching of luminescence by oxygen. Trans. Farad. Soc., 1939, v. 35, p. 216–219.
105. Ho R. Y. N., Liebman J. F., Valentine J. S. Overview of the energetics and reactivity of oxygen. Chapter 1. In: Active Oxygen in Chemistry (Editor: C. S. Foote), Kluwer, Blackie Academic and Professional, Chapman and Hall, 1995, USA, 342 pages.
106. Jockusch S., Turro N. J., Thompson E. K., Gouterman M., Callis J. B. Khalil G. E. Singlet molecular oxygen by direct excitation. Photochem. Photobiol. Sci., 2008, v. 7, p. 235–239.
107. Wang W., Maa C., Chena H. Ji., Zhaoa J. Probing paramagnetic species in titania-based heterogeneous photocatalysis. Chemical Engineering Journal, 2011, v. 170, p. 353–362.
108. Timoshenko V. Y. Photoluminescence study of porous silicon as photosensitizer of singlet oxygen generation. Ukr. J. Phys., 2011. v. 56, 10, p. 1097–1109.
109. Portoles M. J. L., Gara P. M. D., Kotler M. L., Bertolotti S., Roman E. S., Rodriguez H. B., Gonzalez M. C. Silicon nanoparticle photophysics and singlet oxygen generation. Langmuir, 2010, v. 26, 13, p. 10953–10960 DOI: 10.1021/la100980x
110. Daimon, T., Nosaka Y. Formation and behavior of singlet molecular oxygen in TiO_2 photocatalysis studied by detection of near-infrared phosphorescence. J. Phys. Chem. C, 2007, v. 111, p. 4420–4424.
111. Centi G., Cavani F., Trifiro F. Selective Oxidation by Heterogeneous Catalysis (Fundamental and Applied Catalysis) 2001st edition Springer 505 pages, p. 389 (ref. within: Shvarts A. M., Kazanski V. B. Study of the state double bonded oxygen on the oxide catalysts by the luminescence method. Reports of the Academy of Sciences of the USSR (*Dokladi AN SSSR*, in Russian), 1975, v. 221, 4, p. 868–871.
112. Kiselev V. F., Krilov O. V. Adsorption and Catalysis on Transition Metals and Their Oxides. Technology and Engineering, 1989, Springer-Verlag, Berlin, 445 pages, p. 192.
113. Boikov E. V., Vishnetskaya M. V., Emel'yanov A. N., Rufov Yu. N., Shcherbakov N. V., Tomskii I. S. Singlet oxygen in the heterogeneous catalytic oxidation of benzene. Russian Journal of Physical Chemistry A, 2007, v. 81, 6, p. 861–865.

114. Carley A. F., Davies P., Hutchings G. J., Spencer M. S. (eds). Surface Chemistry and Catalysis. Springer 2002, v. 8, 381 pages, p. 234.

115. Vishnetskaya M. V., Tomskiy I. S. Role of singlet oxygen in the oxidation of toluene on vanadium. Molybdenum catalytic systems. Chemistry for Sustainable Development. 2011, v. 19, p. 321–325.

116. Udalova O. V., Khaula E. V., Rufov Yu. N. The heterogeneous catalytic oxidation of propylene in the presence of singlet oxygen in the reaction mixture. Russian Journal of Physical Chemistry A. 2007, v. 81, 9, p. 1511–1514.

117. Gohre K., Miller G. C. Photochemical generation of singlet oxygen on non-transition-metal oxide surfaces. J. Chem. Soc., Faraday Trans., 1985, v. 1, 81, p. 793–800.

118. Linsebigler A. L., Lu G., Yates J. T. J. Photocatalysis on TiO_2 surfaces: principles, mechanisms, and selected results. Chem. Rev., 1995, v. 95, p. 735–758.

119. Tachikawa T., Majima T. Single-molecule fluorescence imaging techniques for the detection of reactive oxygen species, p. 656. In: Méndez-Vilas A., Díaz J. (eds). Modern Research and Educational Topics in Microscopy. Formatex, 2007.

120. Tachikawa T., Majima T. Single-molecule, single-particle fluorescence imaging of TiO_2-based photocatalytic reactions (Critical Review). Chem. Soc. Rev., 2010, v. 39, p. 4802–4819. DOI: 10.1039/B919698F

3.5

3.5.1

3.5.1.1

121. Masel I. R. Principles of Adsorption and Reactions on the Solid Surfaces. 1996. John Wiley, N.Y.

122. Verheij L. K., Hugenschmidt M. B., Poelsema B., Comsa G. Influence of a very low density of defects on the deuterium—oxygen reaction on Pt(111). Chem. Phys. Lett., 1990, v. 174, p. 449–454.

123. Hollman A. F., Wiberg E. Inorganic Chemistry, 2001 Academic Press San Diego, 1780 pages, p. 246.

124. Virmani Y. P., Zimbrick J. D., Zeller E. J. ESR studies of hydrogen trapped in alkali halides by proton irradiation. Chemical Physics Letters, 1970, v. 6, 5, 1, p. 508–512.

125. Hasegawa A., Nishikida K., Williams F. ESR spectra of interstitial hydrogen atoms in dipotassium hexafluorosilicate. J. Phys. Chem., 1980, v. 84, 26, p. 3630–3633. DOI: 10.1021/j100463a027

126. Plonka A. Dispersive Kinetics. Chapter 4. Kinetics on Condensed Media, 2001, Springer Science, Berlin, 234 pages, p. 132.

127. Ammerlaan C. A. J. Hydrogen, hydrogen atoms and molecules, p. 263–269. In. Siffert P., Krimmel E. (eds). Silicon: Evolution and Future of a Technology. Springer, Berlin, 2004, 549 pages, p. 266.

128. Jackson S. D., Hargreaves J. S. J. Metal Oxide Catalysis. 2009, v. 1–2, Wiley-VCH; 847 pages, p. 35.

129. Chen W. P., He K. F., Wang Y., Chan H. L. W., Yan Z. Highly mobile and reactive state of hydrogen in metal oxide semiconductors at room temperature. Sci. Rep., 2013, v. 3, p. 3149.

130. Song Z., Xu H. Unusual dissociative adsorption of H_2 over stoichiometric MgO thin film supported on molybdenum. Applied Surface Science, 2016, v. 366, p. 166–172.

131. Bowker M. The surface structure of titania and the effect of reduction. Curr. Opin. Solid State Matter, 2006, v. 10, p. 153–162.

132. Züttel A. Interaction of hydrogen with solid surfaces, Properties of hydrogen, p. 94. In: Züttel A., Borgschulte A., Schlapbach L. (eds). Hydrogen as a Future Energy Carrier. 2008, Wiley VCH, Weinheim, 427 pages, p. 104.
133. Horch S., Lorensen H. T., Helveg S., Lægsgaard E., Stensgaard I., Jacobsen K. W., Norskov K. J. Enhancement of surface self-diffusion of platinum atoms by adsorbed hydrogen. Letters to Nature, Nature 398, 1999, p. 134–136.
134. Henrici-Olive G., Olive S. The Chemistry of the Catalyzed Hydrogenation of Carbon Monoxide. Springer-Verlag, Berlin, 1984, p. 19.
135. Pariiskii G. B., Kazanskii V. B. *Kinetika i Kataliz* (Kinetics and Catalysis, in Russian), 1964, v. 5, 96, p. 327.
136. Nalbandyan A. B., Shubina M. S. *Zhurnal Phiz. Khim.* (Journal of Physical Chemistry, in Russian), 1946, v. 20, p. 1249.

3.5.1.2

137. Roginsky S. Z. Introduction to volume 1. Active hydrogen, p. 1-XIV. In: Suits C. G. (Ed.). Low-Pressure Phenomena: With Contributions in Memoriam Including a Complete Bibliography of His Works (The collected works of Irving Langmuir), 2015, Pergamon Press, N.Y., 428 pages.
138. Azatyan V. V., Piloyan A. A., Baimuratova G. R., Masalova V. V. Accelerated diffusion of chain carriers and kinetic features of heterogeneous processes in gas-phase chain reactions. *Kinetika i Kataliz* (Kinetics and Catalysis, in Russian), 2008, v. 49, 2, p. 190–197.
139. Azatyan V. V. Problems of combustion, explosion and detonation in the gases in theory of the Low temperature reactions of chain processes. *Zhurnal Fiz. Khimii* (Journal of Physical Chemistry, in Russian), 2014, v. 88, 5, p. 759.
140. Azatyan V. V., Rubtsov N. M., Tsvetkov G. I., Chernysh V. I. The participation of preliminarily adsorbed hydrogen atoms in reaction chain propagation in the combustion of deuterium. Russian Journal of Physical Chemistry. 2005, v. 79, 3, p. 320–324.
141. Zwinkels F. M., Jaras S. G., Menon P. G. Catalytic Fuel Combustion on Honeycomb Monolithic Reactors p. 157. In: Cybulski A., Moulijn J. A. Structured Catalysts and Reactors. First edition 1998 Marcel Dakker N.Y., 671 pages, Second Edition 2005, CRC Press, 856 pages.
142. Hellsing B., Kasemo B., Zhdanov P. V. Kinetics of the hydrogen-oxygen reaction on platinum. Journal of Catalysis. 1991, v. 132, 1, p. 210–228.
143. Ellinger C., Stierle A., Robinson I. K., Nefedov A., Dosch H. Atmospheric pressure oxidation of Pt(111) Journal of Physics: Condensed Matter, 2008, v. 20, 18, 184013–18.
144. Fridell E., Elg A-P., Rosén A., Kasemo B. A laser-induced fluorescence study of OH desorption from Pt(111) during oxidation of hydrogen in O_2 and decomposition of water. Journal of Chemical Physics, 1995, v. 102, 14, p. 5827.
145. Hsu D. S. Y. Hoffbauer M. A., Lin M. C. Dynamics of OH desorption from single crystal Pt(111) and polycrystalline Pt foil surfaces. Surface Science, 1987, v. 184, 1–2, p. 25–56.
146. Germer T. A., Ho W. Direct characterization of the hydroxyl intermediate during reduction of oxygen on Pt(111) by time-resolved electron energy loss spectroscopy. Chemical Physics Letters. 1989, v. 163, 4–5, p. 449–454.
147. Qi L., Yu J., Lia J. Coverage dependence and hydroperoxyl-mediated pathway of catalytic water formation on Pt (111) surface. J. Chemical Physics 2006, v. 125, 054701 (1–8).
148. Fisher G. B., Gland J. L. The Interaction of water with the Pt(111) surface. Surface Science. 1980, v. 94, p. 446–455.

149. Bowker M. Davies P. (eds) Scanning Tunneling Microscopy in Surface Science, Wiley-VCH Verlage, Weinheim, 2010, 258 pages, p. 55–95.

150. Sachs C., Hildebran M., Völkening S., Wintterlin J., Ertl G. Reaction fronts in the oxidation of hydrogen on Pt(111): Scanning tunneling microscopy experiments and reaction–diffusion modeling. The Journal of Chemical Physics, 2002, v. 116, p. 5759.

151. Völkening S., Bedürftig K., Jacobi K., Wintterlin J., Ertl G. Dual-path mechanism for catalytic oxidation of hydrogen on platinum surfaces. Phys. Rev. Lett. 1999, v. 83, p. 2672.

152. Verheij L. K. Kinetic modeling of the hydrogen-oxygen reaction on Pt(111) at low-temperature (less-than-170 K), Surface science. 1997, v. 371, 1, p. 100–110.

153. Verheij L. K., Hugenschmidt M. B. On the mechanism of the hydrogen-oxygen reaction on Pt(111), Surface science, 1998, v 416, 1–2, p. 37–58.

154. Zhu Z., Melaet G., Axnanda S., Alayoglu S., Liu Z., Salmeron M., Somorjai G. A. Structure and chemical state of the Pt(557) surface during hydrogen oxidation reaction studied by in situ scanning tunneling microscopy and X-ray photoelectron spectroscopy. Journal of the American Chemical Society. 2013, v. 135, 34, p. 12560–12563 DOI: 10.1021/ja406497s

155. L'vov B. V., Galwey A. K. Catalytic oxidation of hydrogen on platinum. Thermochemical approach. Journal of Thermal Analysis and Calorimetry. 2013, v. 12, 2, p. 815–822.

156. Acres G. J. K. The reaction between hydrogen and oxygen on platinum. Progress in establishing kinetics and mechanisms. Platinum Metals Rev., 1966, v. 10, 2, p. 60–64.

157. Matthiesen J., Wendt S., Hansen J. O., Madsen G. K. H., Lira E., Galliker P., Vestergaard E. K., Schaub R., Lægsgaard E., Hammer B., Besenbacher F. Observation of all the intermediate steps of a chemical reaction on an oxide surface by scanning tunneling microscopy. ACS Nano., 2009, v. 3, 3, p. 517–526.

3.5.2

158. Arutyunov V. Direct Methane to Methanol: Foundations and Prospects of the Process. Elsevier, Amsterdam, 2014, 310 pages, p. 90.

159. Takanabe K., Shahid S. Dehydrogenation of ethane to ethylene via radical pathways enhanced by alkali metal based catalyst in oxysteam condition. AIChE Journal, 2016. DOI: 10.1002/aic.15447

160. Donazzi A., Livio D., Maestri M., Beretta A., Groppi G., Tronconi E., Forzatti P., Synergy of homogeneous and heterogeneous chemistry probed by in situ spatially resolved measurements of temperature and composition. Angewandte Chemie, International Edition, 2011, v. 50, 17, p. 3943–3946.

161. Shilov A. E., Shul'pin G. B. Activation and Catalytic Reactions of Saturated Hydrocarbons in the Presence of Metal Complexes. Chapter III. 6. A. Oxidation with molecular oxygen, p. 90, Kluwer Academic Publishers, N.Y. 2002, 536 pages.

162. Margulis L. Ya. Catalytic oxidation of hydrocarbons, p. 429–501. In: Advances in Catalysis and Related Subjects (ed. Eley D. D.), v. 14. Academic Press, N. Y., London, 1963. 522 pages.

163. Emanuel M. N. Chemical and biological kinetics, p. 447. In: Science, Technology and the Future (eds E. P. Velikhov, J. M. Gvishiani, S. R. Mikulinsky), Pergamon Press, Oxford, 1980, 480 pages.

3.5.2.1

164. Stern V. Ya. The Gas Phase Oxidation of Hydrocarbons, 1964, Pergamon Press, Oxford, 710 pages.

165. Olah G. A., Molnar A. Hydrocarbon Chemistry, 2nd ed., 2003, John Wiley and Sons, Hoboken, 870 pages.

166. Satterfield C. N., Reid R. C. Effects of surfaces on the products formed in the oxidation of propane. J. Chem. Eng. Data, 1961, v. 6, 2, p. 302–304. DOI: 10.1021/je60010a035

167. Driscoll D. J., Lunsford J. H. Gas phase radical formation during the reaction of methane, ethane, ethylene and propylene over selected oxide catalysts. J. Phys. Chem., 1985, v. 89, p. 4415–4418.

168. Driscoll D. J., Cambell K. D., Lunsford J. H. Surface generated gas phase radicals: formation detection, and role in catalysis, p. 139–186. In: Eley D. D., Pines H., Weiz P. B. (ed). Advances in Catalysis. 1985, Academic Press, N.Y.

169. Ismagilov R. Z., Pak S. N., Yermolaev V. K., Zamaraev K. I. Heterogeneous-homogeneous reactions involving free radicals in processes of total catalytic oxidation. Journal of Catalysis, 1992, v. 136, p. 197–201 and Doklady Akadrmii Nauk SSSR (Reports of AN USSR), v. 298, N 3, p. 637.

170. Magomedov R., Proshina A. Yu., Peshnev B. V., Arutyunov V. S. Effects of the gas medium and heterogeneous factors on the gas-phase oxidative cracking of ethane. Kinetika I Kataliz (Kinetics and Catalysis, in Russian), 2013, v. 54, 4, p. 394–399.

171. Labinger J. A., Bercaw J. E. Understanding and exploiting C–H bond activation (Review). Nature, 30 May 2002, v. 417, p. 507–514. DOI: 10.1038/417507a

172. Haber J. Molecular mechanism of the heterogeneous oxidation organic and solid-state chemists views, p. 5. In: Oyama S. T., Gaffney A. M., Lyons J. E., Grasselli R. K. (eds). Third World Congress on Oxidation Catalysis. Studies in Surface Science and Catalysis, v. 110, 1997, Elsevier Science, Amsterdam, 1247 pages.

173. Witko M., Hermann K. Selective hydrocarbon oxidation on vanadium pentoxide surfaces: Ab initio cluster model studies, p. 75. In: New Developments in Selective Oxidation II, volume 82, first edition (eds: Corberán V. C., Bellon S. V.), 1997, Elsevier Science, 883 pages.

174. Hagen J. Industrial Catalysis: A Practical Approach. Wiley-VCH, 2015, Weinheim (Germany), 522 pages, p. 128.

175. Boudart M. Turnover rates in heterogeneous catalysis. Chem. Rev., 1995, v. 95, 3, p. 661–666.

176. Balandin A. A. The multiplet theory of catalysis, structural factors in catalysis. Russian Chemical Reviews, 1962, v. 31, 11, p. 589–614.

177. Tomas J. M., Tomas W. J. Principles and Practice of Heterogeneous Catalysis, second edition, Wiley-VCH 2015, Chapter 1, p. 38, 688 pages.

178. Védrine J. C., Fechete I. Heterogeneous partial oxidation catalysis on metal oxides. Comptes Rendus Chimie, 2016, v. 19, 10, p. 1203–1225.

179. Sinev M. Yu. Free radicals in catalytic oxidation of light alkanes: kinetic and thermochemical aspects. Journal of Catalysis, 2003, v. 216, p. 468–476. www.elsevier.com/locate/jcat

180. Kondratiev V. N. Chemical Bonding Energies, Enthalpy of Chemical Processes, Ionization Potentials and Electron Affinities, 1974, Nauka, Moscow (Handbook, in Russian).

181. Conley M. P., Delley M. F., Nunez-Zarur F., Comas-Vives A., Copéret C. Heterolytic activation of C-H bonds on Cr(III)-O surface sites is a key step in catalytic polymerization of ethylene and dehydrogenation of propane. Inorg. Chem., 2015, v. 1, 54 (11), p. 5065–78. DOI: 10.1021/ic502696n

182. Kung H. H. Desirable catalyst properties in selective oxidation reactions. Eng. Chem. Prog. Res. Dev., 1986, v. 25, p. 171–178. DOI: 10.1021/i300022a009

183. Trimm D. L., Brown D. M. Homogeneous–heterogeneous interaction during catalytic oxidation of benzene to maleic anhydride. Nature, Physical Science, 1971 (14 June), v. 231, p. 156–157. DOI: 10.1038/physci231156a0

184. Elchyan A. M., Grigoryan G. L. Chain regime in the oxidation of hydrocarbons with hydrogen peroxide on a SiO$_2$. *Khim. Phys.* (Chemical Physics, in Russian), 1990, v. 9, 5, p. 639–643.
185. Deo G., Cherian M., Rao T. V. M. Oxidative dehydrogenation of alkanes on the metal oxides, p. 491. In: Fierro J. L. G. Metal Oxides: Chemistry and Applications. 2006, CRC, Taylor and Francis.

3.5.2.2

186. Khcheyan Kh. E., Shatalova A. V., Temkin O. N., Revenko O. M. *Neftekhimiya* (in Russian), 1981, v. 21, 2, p. 303; Khcheyan Kh. E., Shatalova A. V., Revenko O. M., Agrinskaya L. I. *Neftekhimiya* (in Russian), 1980, v. 20, 6, p. 876.
187. Shilov A. E., Shul'pin G. B. Activation and Catalytic Reactions of Saturated Hydrocarbons in the Presence of Metal Complexes. Kluwer Academic Publishers, 2001, Dordrecht, 539 pages, p. 109.
188. Nersissyan L. A. Study of the mechanism of the heterogeneous-homogeneous oxidation of methane. Candidate's Thesis in Chemical Sciences. 1979, Institute of Chemical Physics, Academy of Sciences of the Armenian SSR, 27 pages.
189. Sinev M. Yu., Fattakhova Z. T., Lomonosov V. I., Gordienko Yu. A. Kinetics of oxidative coupling of methane: bridging the gap between comprehension and description. Journal of Natural Gas Chemistry, 2009, v. 18, 3, p. 273–287.
190. Nersissyan L. A., Vardanyan I. A., Kegheyan E. M. Margolis L. Y., Nalbandyan A. B. Heterogeneous-homogeneous oxidation of methane. *Dokladi* of the Academy of Sciences of USSR (in Russian), 1975, v. 220, 1, p. 605–607.
191. Mitchell H. L., Waghorne R. H. Patent USA, 4 205 194, 1980.
192. Zavyalova U., Holena M., Schlögl R., Baerns M. Statistical analysis of past catalytic data on oxidative methane coupling for new insights into the composition of high-performance catalysts. ChemCatChem. 2011, v. 3, 12, p. 1935–1947, DOI: 10.1002/cctc.201100186.
193. Kondratenko E. V., Baerns M. Catalysis of oxidative methane conversions. Chapter 3.2. In: Hess C., Schlögl R. (eds). Nanostructured Catalysts. Selective Oxidations. RSC Publishing, Cambridge, 2011. 452 pages
194. Bhasin D. W., Campbell K. D. Oxidative coupling of methane – a progress report, p. 3–18. In: Bhasin D. W. Slocum M. M. (eds). Methane and Alkane Conversion Chemistry. Springer Science, 1995, N.Y. 349 pages.
195. Margolis L. Ya. Oxidation of Hydrocarbons on Heterogeneous Catalysts, 1977, Moscow, Khimiya, 328 pages, p. 109.
196. Bhasin M. M., Slocum D. W. Methane and Alkane Conversion Chemistry. 2012, Springer Science and Business Media, 349 pages.
197. Horn R., Schlögl R. Methane Activation by Heterogeneous Catalysis. Catal. Lett., 2015, v. 145, p. 23–39. DOI: 10.1007/s10562-014-1417-z
198. Deo G., Cherian M., Rao T. V. M. Oxidative dehydrogenation of alkanes on the metal oxides, p. 491; In: Fierro J. L. G. (ed). Metal Oxides: Chemistry and Applications. 2006, CRC, Taylor and Francis, Boca Raton, 808 pages.
199. Kurzina A., Kurina L. N. Tin-containing catalysts, modified with alkaline earth metals in the oxidative dimerization of methane. Theoretical and Experimental Chemistry, 2003, v. 39, 1, p. 55–59.
200. Arndt S., Laugel G., Levchenko S., Horn R., Baerns M., Scheffler M., Schlögl R., Schomäcker R. A critical assessment of Li/MgO-based catalysts for the oxidative coupling of methane catalysis reviews. Journal Catalysis Reviews Science and Engineering, 2011, v. 53, 4, p. 424–514.

201. Moro-oka Y. The role of acidic properties of metal oxide catalysts in the catalytic oxidation. Applied Surface Catal. A, 1999, v. 181, 2, p. 323–329.
202. Xu M. T., Ballinger T. H., Lunsford J. H. Quantitative studies of methyl radicals reacting with metal-oxides. Journal of Physical Chemistry, 1995, v. 99, 39, p. 14494–14499.
203. Lio L., Tang X., Wang W., Wang Y., Sun S., Qi F., Huang W. Methyl radicals in oxidative coupling of methane directly confirmed by synchrotron VUV photoionization mass-spectrometry. Scientific Reports, 2013, Article 1625. DOI: 10.1036/srep01625
204. Sinev M. Yu. Reactions of free radicals in the processes of the catalytic oxidation of light alkanes (in Russian). Dissertation Autoabstract on the degree of the Doctor of Chemical Sciences. Semenov Institute of chemical Physics, Moscow 2011, 61 pages. p. 19.
205. Horn R., Schlögl R. Methane activation by heterogeneous catalysis. Catal. Lett., 2015, v. 145, p. 23–39. DOI: 10.1007/s10562-014-1417-z
206. Cavani F, Trifirò F. Selected reactions in heterogeneous catalysis (II), Partial oxidation of C_2 to C_4 paraffins, p. 19–84; p. 25–26. In: Baerns M. (ed.). Basic Principles in Applied Catalysis 2004 Springer, Berlin, Heidelberg, 557 pages.
207. Bol C. W. J., Friend C. M. C-O bond formation by direct addition of methyl radicals to surface oxygen on Rh(111). J. Am. Chem. Soc., 1995, v. 117, 30, p. 8053–8054.
208. Walter E. J., Rappe A. M. Co-adsorption of methyl radicals and oxygen on Rh(111). Surface Science, 2004, v. 549, p. 265–272.
209. Labinger J. A., Bercaw J. E. Understanding and exploiting C–H bond activation. Nature. 2002, 417 (6888), p. 507–514.
210. Liu Yu., Lin J., Tan K. L. Conversion of methane to styrene over the mixed catalyst La_2O_3 + MoO_3/HZSM-5. Catalysis Letters, 1998, v. 50, 3, p. 165–168.
211. Gao J., Zheng Y., Fitzgerald G. B., de Joannis J., Tang Y., Wachs I. E., Podkolzin S. G. Structure of Mo_2C_x and Mo_4C_x molybdenum carbide nanoparticles and their anchoring sites on ZSM-5 zeolites. The Journal of Physical Chemistry C, 2014, v. 118, 9, p. 4670–4679.
212. Gao J., Zheng Y., Jehng J. M., Tang Y., Wachs I. E., Podkolzin S. G. Catalysis. Identification of molybdenum oxide nanostructures on zeolites for natural gas conversion. Science, 2015, 348(6235), p. 686–690.
213. Nguyen K. T., Kung H. H. Heterogeneous-homogeneous pathway in catalytic oxidative dehydrogenation of propane, p. 285–306. In: Albright L. F., Crynes B. L., Nowal S. (eds). Novel Production Methods for Ethylene, Light Hydrocarbons, and Aromatics, 1992, Marcel Dekker, New York, 549 pages.

3.5.2.3

214. Keulks G. W., Rosynek M. P., Daniel C. Bismuth molybdate catalysts. Kinetics and mechanism of propylene oxidation. Ind. Eng. Chem. Prod. Res. Dev., 1971, v. 10, 2, p. 138–142.
215. Sancier K. M., Wentrcek P. R., Wise H. Role of sorbed and lattice oxygen in propylene oxidation catalyzed by silica-supported bismuth molybdate. Journal Catalysis, 1975, v. 39, 1, p. 141–147.
216. Misono M. Heterogeneous Catalysis of Mixed Oxides: Perovskite and Heteropoly Catalysts, Perovskite and Heteropoly Catalysts, Series: Studies in Surface Science and Catalysis, first edition, v. 176, 2013, Elsevier, Amsterdam, 182 pages, p. 57.
217. Adams C. R., Jennings T. J. Investigation of the mechanism of catalytic oxidation of propylene to acrolein and acrylonitrile. Journal of Catalysis, 1963, v. 2, 1, p. 63–68.

218. Pudar S., Oxgaard J., Chenoweth K., van Duin A. C. T., Goddard W. A. Mechanism of selective oxidation of propene to acrolein on bismuth molybdates from quantum mechanical calculations. J. Phys. Chem. C, 2007, v. 111, p. 16405–16415.

219. Dozono T., Thomas D. W., Wise H. Mechanism of [3-^{13}C] propene ammoxidation on bismuth molybdate catalyst. J. Chem. Soc., Faraday Trans. 1, 1973, v. 69, p. 620–623. DOI: 10.1039/F19736900620

220. Sinfelt J. H., Cusumano J. A bimetalic catalysts, 1–29 pages, p. 26. In: Burton J. J., Garten R. L. (eds). Advanced Materials in Catalysis, 1977, Academic Press, N.Y., 329 pages.

221. Keulks G. W. Selective oxidation of propylene, p. 109–118. In: Smith G. V. (ed). Catalysis in Organic syntheses, 1977, Academic Press, N.Y., 295 pages.

222. Dolejsek Z., Nováková J. Mass spectrometric observation of allyl radicals during the interaction of propene with some oxides at low pressures. Journal of Catalysis. 1975, v. 37, 3, 6, p. 540–543.

223. Martir W., Lunsford J. H. The formation of gas-phase π-allyl radicals from propylene over bismuth oxide and γ-bismuth molybdate catalysts. Journal of the American Chemical Society, 1981, v. 103, 13, p. 3728–3732.

224. Getsoian A., Shapovalov V., Bell A. T. DFT+U investigation of propene oxidation over bismuth molybdate: active sites, Reaction intermediates, and the role of bismuth. J. Phys. Chem., C, 2013, v. 117, 14, p. 7123–7137. DOI: 10.1021/jp400440p

225. Bruckner A. Electron Paramagnetic Resonance. A powerful tool for monitoring working catalyst, p. 266–309, p. 294. In: Gates B. C., Knozinger H. (eds). Advances in Catalysis, Volume 51, 2007, Elsevier, Amsterdam, 389 pages.

3.5.2.4

226. Lin J. L., Bent B. E. Two mechanisms for formation of methyl radicals during the thermal decomposition of methyl iodide on a copper surface. J. Phys. Chem., 1993, v. 97, 38, p. 9713–9718. DOI: 10.1021/j100140a030 (see also Chapter II, Section: Methyl iodide decomposition)

227. Krishtopa L. G., Yermolayev V. K., Isrriagilov Z. S., Zamaraev K. I. Desorption rate of RO$_2$-radicals formed during reaction of n-propanol with over CuO and its dependence oil the surface reduction studied by "in situ" gravimetry, XBD and X-ray microbeam method. Proc. Vlll-th Soviet-French Seminar. Novosibirsk, 1990, p. 96–98.

3.5.3

228. Ismagilov R. Z., Pak S. N., Krishtopa L. G., Yermolaev K. E. Role of the free radicals in heterogeneous complete oxidation of organic compounds over IV period transition metal oxides, p. 231–243. In: Guczi L., Solymosi F., Tétényi P. (eds). New Frontiers in Catalysis, Parts A-C, Studies in Surface Science and Catalysis, 75, Proceedings of the 10th international Congress on Catalyst, 19–24 July, 1992, Budapest, Hungary, 1993, Elsevier, Amsterdam, 1480 pages.

229. Mellat T., Baiker A. Oxidation of alcohols with molecular oxygen on solid catalysts (a review). Chem. Rev., 2004, v. 104, p. 3037–3058.

230. Yu Y., Huang J., Ishida T., Haruta M. Unique catalytic performance of supported gold nanoparticles in oxidation, p. 98. In: Mizuno N. (ed). Modern Heterogeneous Oxidation Catalysis: Design, Reactions and Characterization, 2009, Wiley VCH, Weinheim, 340 pages.

231. Sapi A., Liu F., Cai X., Thompson C. M., Wang H., An K., Krier J. M., Somorjai G. A. Comparing the catalytic oxidation of ethanol at the solid-gas and solid-liquid interfaces over size-controlled Pt nanoparticles – striking differences in kinetics and mechanism, Nano Lett., 2014, v. 14, 11, p. 6727–6730.

232. Wang H., Sapi A., Thompson C. M., Liu F., Zherebetskyy D., Krier J. M., Carl L. M., Cai X., Wang L.W., Somorjai G. A. Dramatically different kinetics and mechanism at solid/liquid and solid/gas interfaces for catalytic isopropanol oxidation over size-controlled platinum nanoparticles. J. Am. Chem. Soc., 2014, v. 136, 29, p. 10515–10520. DOI: 10.1021/ja505641r. Epub 2014 Jul 15

3.5.3.1

233. Hofmann A. W. *Zur Kenntnis des Methylaldehyds. Monatsbericht der Königlich Preussischen Akademie der Wissenschaften zu Berlin*, 1867, vol. 8, p. 665–669.
234. Bazilio C. A., Thomas W. J., Ullah U., Hayes K. E. Catalytic Oxidation of Methanol, Proceedings of the Royal Society of London. Series A, Mathematical and Physical Sciences, 1985, p. 181–194.
235. Wachs I. E., Madix R. The oxidation of methanol on a silver (110) catalyst. Surface Science, 1978, v. 76, p. 531–558.
236. N'dollo M., Moussounda P. S., Dintzer T., Garin F. A Density Functional theory study of methoxy and atomic hydrogen chemisorption on Au(100) surface. Journal of Modern Physics, 2013, v. 4, p. 409–417.
237. Benson T. J., Daggolu P.R., Hernandez R.A., Liu S., White M. G. Catalytic deoxygenation chemistry: Upgrading of liquids derived from biomass processing, p. 189–344 (5. Alcohols, 5.1. Alcohols in Metals, p. 212–214). In: Gates B. C., Jentoft F. C. (eds). Advances in Catalysis, v. 56. First Edition, 2013, Elsevier, Amsterdam. 381 pages.
238. Baumer M., Libuda J., Neyman K. M., Rosch N., Rupprechter G., Freund H-J. Adsorption and reaction of methanol on supported palladium catalysts: microscopic-level studies from ultrahigh vacuum to ambient pressure conditions. Phys. Chem. Chem. Phys., 2007, v. 9, p. 3541–3558.
239. McCabe R. W. D., McCready F. J. Kinetics and reaction pathways of methanol oxidation on platinum. Phys. Chem., 1986, v. 90, 7, p. 1428–1435.

3.5.3.2

240. Malakhova I. V., Ermolaev V. K., Danilova I. G., Paukshtis E. A., Zolotarskii I. A. Kinetics of free radical generation in the catalytic oxidation of methanol. *Kinetika i Kataliz* (Kinetics and Catalysis, in Russian), 2003, v. 44, 4, p. 536–546.
241. Malakhova I. V., Ermolaev V. K. Identification of matrix-isolated methoxyl radicals by recording EPR spectra under photolysis. Mendeleev Communications, 1998, v. 8, 3, p. 83–84.
242. Matyshak V. A., Khomenko T. I., Lin G. I., Zavalishin I. N., Rozovskii A. Ya. Surface species in the methyl formate-methanol-dimethyl ether-γ-Al$_2$O$_3$ system studied by in situ IR spectroscopy. *Kinetika i Kataliz* (Kinetics and Catalysis, in Russian), 1999, v. 40, 2, p. 295; *Kinet. Catal.* (Engl. Transl.), 1999, v. 40, 2, p. 269.
243. Pak S. N. Cand. Sci. (in Chemistry) Dissertation, Novosibirsk Institute of Catalysis, 1990.
244. Lunina V. V. *Kataliz. Fundamental'nye i prikladnye issledovaniya* (In: Petrii O. A., Lunina V. V. (eds), Catalysis: Fundamental and Applied Studies, in Russian), Moscow: Moskov. Gos. Univ., 1987, p. 262.
245. Dickens K. A., Stair P. C. A Study of the adsorption of methyl radicals on clean and oxygen-modified Ni(100). Langmuir, 1998, v. 14, 6, p. 1444–1450. DOI: 10.1021/la970715o
246. Sexton B. A. Surface vibrations of adsorbed intermediates in the reaction of alcohols with Cu(100). Surface Sci., 1979, v. 88, p. 299–318.

3.5.4

247. Liebig J. Ueber die Producte der Oxydation des Alkohols. Annalen Der Pharmacie, 1835, v. 14, 2, p. 133–167.
248. Dixson D. J., Scirrow G. Gas phase combustion of aldehydes, Chapter 3, p. 369–435. In: Bamford C. H. Tipper C. F. H. (ed). Gas Phase Combustion, Elsevier, 1977, Amsterdam, 519 pages.
249. Sajus L., Seree fe Roche. The liquid phase oxidation of aldehydes. p. 89–122. In: Bamford C. H., Compton R. G., Tipper C. F. H. (eds). Comprehensive Chemical Kinetics, v. 16. Liquid Phase Oxidation. Elsevier, 1980, Amsterdam, 263 pages.

3.5.4.1

250. Maslov S. A., Blyumberg E. A. Liquid-phase Oxidation of Aldehydes. Russian Chemical Reviews, v 45, 2, 1976, p. 155–167.
251. Horner E. C. A., Style D. W. G. The oxidation of formaldehyde. Part 1. The photochemical reaction and the influence of surface. Transactions of the Faraday Society, v. 50, 1954, p.1197 DOI: 10.1039/tf9545001197
252. Horner E. C. A., Style D. W. G., Summers D. The oxidation of formaldehyde. Part 2.—General discussion and mechanism of the reaction. Trans. Faraday Soc, v. 50, 1954, p. 1201. DOI: 10.1039/TF9545001201
253. Lewis B., Von Elbe L. Combustion, Flames and Explosion of Gases Chapter 4. The reaction between hydrocarbon and oxygen, Methane and formaldehyde, p. 90–112, second edition (1961), 730 pages, third edition (1987) 739 pages, Academic Press, N.Y, p. 127–133.
254. Fort R., Hinshelwood C. N., Further investigations on the kinetics of gaseous oxidation reactions. Proc. Roy. Soc., A, 1930, v. 129, p. 284.
255. Askey P. J. The oxidation of benzaldehyde and formaldehyde in the gaseous phase. J. Amer. Chem. Soc., 1930, v. 52, p. 974.
256. Bone W. A., Gardner J. B. Comparative studies of the slow combustion of methane, methyl alcohol, formaldehyde and formic acid. Proc. Roy. Soc., A, 1936, v. 154, p. 297.
257. Axford D. W. E., Norrish R. G. W. The oxidation of gaseous formaldehyde. Proc. Roy. Soc. (London) A. 1948, v. 192, p. 518.
258. Nie L., Yu J., Jaroniec M., Tao F. F. Room-temperature catalytic oxidation of formaldehyde on catalysts Catal. Sci. Technol., 2016, v. 6, p. 3649–3669.
259. Bai B., Qiao Q., Li J., Hao J. Progress in research on catalysts for catalytic oxidation of formaldehyde. Chinese Journal of Catalysis, 2016, v. 37, p. 102–122.
260. Medlin J. W. Surface science studies relevant for metal-catalyzed biorefining reactions p. 33–63. In: Simmons B. A., Schüth F., Peter L. (eds). Chemical and Biochemical Catalysis for Next Generation Biofuels. 2011, RSC Publishers, 205 pages, p. 43.
261. Torres J. Q., Royer S., Bellat J-P., Giraudon J-M., Lamonier J-F. Formaldehyde: catalytic oxidation as a promising soft way of elimination. ChemSusChem, 2013, v. 6, 4, p. 578–92. DOI: 10.1002/cssc.201200809
262. Gomes J. R. B., Gomes J. A. N. F. Theoretical study of dioxymethylene, proposed as intermediate in the oxidation of formaldehyde to formate over copper. Surface Science, 2000, v. 446, 3, p. 283–293.
263. Schlörer N. E., Berger S. First spectroscopical evidence of a dioxomethylene intermediate in the reaction of CO_2 with $Cp_2Zr(H)Cl$: A ^{13}C NMR study. Organometallics, 2001, v. 20, 8, p. 1703–1704.
264. Zhang Z., Tang M., Wang Z-T., Ke Z., Xia Y., Park K. T., Lyubinetsky I., Dohnálek Z., Ge Q. Imaging of Formaldehyde Adsorption and Diffusion on $TiO_2(110)$. Topics in Catalysis, 2015, v. 58, 2, p. 103–113.

265. Gomes R. B., Gomes J. A. N. F. Adsorption of the formyl species on transition metal surfaces J. Electroanalytical Chem., 2000, v. 483, p. 180–187.
266. Gomes J. R. B., Gomes J. A. N. F., Illas F. First-principles study of the adsorption of formaldehyde on the clean and atomic oxygen covered Cu(111) surface. Journal of Molecular Catalysis A: Chemical, 2001, v. 170, 187–193.
267. Diebold U. The surface science of titanium dioxide. Surface Science Reports. 2003, v. 48, p. 53–229.

3.5.4.2

268. Cairns G. T., Waddington D. J. Gas-phase oxidation of aldehydes in the low-temperature region: I—Propanal and propanal-2,2-d$_2$. Combustion and Flame, v. 31, 1978, p. 25–35.
269. Pease R. N. The thermal reaction between acetaldehyde vapor and oxygen. J. Amer. Chem. Soc., 1933, v. 55, p. 2753–2761.
270. Steacie E. W., Hatcher W. A., Rosenberg S. The kinetics of oxidation of gaseous propionaldehyde. J. Phys. Chem., 1934, v. 38, 9, p. 1189–1200.
271. Emanuel N. M. Kinetics of Aldehyde Oxidation in the Gas Phase, Kinetics of Oxidation Chain Reactions. Collected Works, 1950, Moscow, Leningrad, p. 185–232 (in Russian).
272. Vardanyan I. A., Nalbandyan A. B. Advances in the study of the mechanism of the gas-phase oxidation of aldehydes, *Usp. Khim.*, 1985, v. 54, 6, p. 903–922.
273. Nalbandyan A. B., Mantashyan A. H. Elementary Reactions in Slow Gas Phase Reactions, 1975, *Izdatelstvo AH Arm. SSR*, Yerevan.
274. Bakhchadjyan R. H., Vardanian I. A., Nalbandyan A. B. Influence of the heterogeneous factors in gas phase oxidation of propionaldehyde. *Armianski Khimicheski Journal* (Armenian Chemical Journal, in Russian), 1982, v. 35, 4, p. 209–212.
275. Bakhchadjyan R. H. Kinetic Peculiarities of Low Temperature Oxidation of Propionaldehyde in Reaction Vessel with Developed Surface. Dissertation for Candidate of Chemical Sciences degree (in manuscript). Institute of Chemical Physics, National Academy of Sciences of Armenia, Yerevan, 1986, p. 1–129.
276. Bakhchadjyan R. H., Vardanian I. A., Nalbandyan A. B. Studies of the reaction between the adsorbed propionaldehyde and peroxy radicals by the method of IR-spectroscopy. *Khimicheskaya Fizika* (Chemical Physics, in Russian), 1986, v. 5, 3, p. 393–396.
277. Bakhchadjyan R. H., Oganessyan E. A., Babertsyan L. P., Vardanyan I. A. Heterogeneous initiation of the oxidation and ignition processes by the radicals. Archivum Combustionis. 1990, v. 10, n. 1–4, p. 201–209.
278. Bakhchadjyan R. H., Gazaryan K. G., Vardanyan I. A. Formation and reactions of acetylperoxy radicals on the solid surface. *Khimicheskaya Fizika* (Chemical Physics, in Russian), 1991, v. 8, 5, p. 659–663.
279. Vardanyan I. A., Manucharova L. A., Jalali H. A., Tsarukyan S. V. The reaction of CH$_3$O$_2$ radicals with CH$_4$ and CH$_3$CHO on TiO$_2$ (rutile). Chemical Journal of Armenia. 2012, v. 65, 1, p. 132–136.
280. Bakhtchadjyan R. H. Appraisal of length of heterogeneous chains in reaction of low temperature aldehydes oxidation. Applied Surface Science. 2001, v. 182, n. 1–2, p. 20–24.
281. Oganessyan E. A., Lusparyan A. P., Vardanyan I. A., Nalbandyan A. B. Role of heterogeneous factors in the region of negative temperature coefficient of maximal rate of reaction. *Arm. Khim Zh.* (Arm. Chem. Journal, in Russian), 1988, v. 41, 1–2, p. 90–94.
282. Lusparyan A. P., Oganesyan E. A., Vardanyan I. A., Nalbandyan A. B. Heterogeneous formation of peroxy compounds in reaction of gas phase oxidation of C$_2$H$_5$CHO. *Arm. Khim Zh.* (Arm. Chem. Journal, in Russian), 1985, v. 38, 5, p. 333–335.

283. Jenkins C. A., Murphy D. M. Thermal and photoreactivity of TiO$_2$ at the gas–solid interface with aliphatic and aromatic aldehydes. Phys. Chem. B., 1999, v. 103, 6, p. 1019–1026. DOI: 10.1021/jp982690n

3.5.4.3

284. Medlin J. W. Surface Science Studies Relevant for Metal-catalyzed Biorefining Reactions, p. 33–63. In: Simmons B. A., Schüth F., Peter L. (eds). Chemical and Biochemical Catalysis for Next Generation Biofuels. 2011, RSC Publishers, 205 pages, p. 43.
285. Wu G., Stacchiola D., Collins M., Tysoe W. T. The adsorption and reaction of acetaldehyde on clean Ag(111). Surf. Rev. Lett., 2000, v. 7, 3, p. 271–275.
286. Davis J. L., Barteau M. A. The interactions of oxygen with aldehydes on the Pd(111) surface. Surface Science. 1992, v. 268, 1–3, p. 11–24. DOI: 10.1016/0039-6028(92)90946-4
287. Bowker M. Holroyd R., Perkins N., Bhantoo J., Counsell J., Carley A., Morgan C. Acetaldehyde adsorption and catalytic decomposition on Pd(110) and the dissolution of carbon. Surface Science, 2007, v. 601, p. 3651–3660.
288. Plata J. J., Collico V., Márquez A. M. Understanding Acetaldehyde Thermal Chemistry on the TiO$_2$ (110) Rutile Surface: From Adsorption to Reactivity. The Journal of Physical Chemistry C. 2011, v. 115, 6, p. 2819–2825. DOI: 10.1021/jp110696f
289. János R., Kiss J. Adsorption and surface reactions of acetaldehyde on TiO$_2$, CeO$_2$ and Al$_2$O$_3$. Applied Catalysis A General, 2005, v. 287, 2, p. 252–260. DOI: 10.1016/j.apcata.2005.04.003
290. Natal-Santiago M. A., Hill J. M. Studies of the adsorption of acetaldehyde, methyl acetate, ethyl acetate, and methyl trifluoroacetate on silica. Journal of Molecular Cat. A. Chemical. 1999, v. 140, 2, p. 199–214.
291. Role N. K., Glague A. D. H., Wilson A. E. The study of weakly adsorbed Molecules by ^{13}C NMR. Reaction propanol on silica. Appl. Surface Science, 1982, v. 10, p. 331–341.
292. Idriss H. Surface reactions of oxygen containing compounds on metal (TiO$_2$ and UO$_2$) oxide single crystal, p. 133–154. p. 146. In: Rioux R. (ed.). Model Systems in Catalysis: Single Crystals to Supported Enzyme Mimics. 2010, Springer Sci., N.Y., 525 pages.
293. Bakhchadjyan R. H., Vardanyan I. A. Catalyst for synthesis of 2-methyl-2-penten-1-al by crotonic condensation of propionaldehyde. Authorship Certificate USSR, n° 1532070; 1989, *Bull. Otkritiya i Izobreteniya v SSSR* (Bulletin of Discoveries and Inventions in USSR), 1989, n. 48, p. 43.
294. Suprun W. Y., Machold T., Schädlich H-K., Papp H. Oxidation of acetaldehyde and propionaldehyde on a VO$_x$/TiO$_2$ catalyst in the presence of water vapor. Chemical Engineering and Technology, 2006, v. 29, 11, p. 1376–1380.
295. Demydov D. V. Nanosized alkaline earth metal titanates: Effects of size on photocatalytic and dielectric V$_2$O$_5$-TiO$_2$ catalyst. *Zeitschrift fur physikalische Chemie*, 2006, v. 220, 12, p. 1575–1588.
296. Plata J. J., Collico V., Marquez A. M., Sanz J. F. Understanding acetaldehyde thermal chemistry on the TiO$_2$ (110) rutile surface: From adsorption to reactivity. J. Phys. Chem. C, 2011, v. 115, 6, p. 2819–2825. DOI: 10.1021/jp110696f
297. Iuga C., Ignacio Sainz-Díaz C., Vivier-Bunge A. On the OH initiated oxidation of C$_2$-C$_5$ aliphatic aldehydes in the presence of mineral aerosols. *Geochimica et Cosmochimica Acta*, 2010, v. 74, 12, p. 3587–3597.
298. Jenkins C. A., Murphy D. M. Thermal and photoreactivity of TiO$_2$ at the gas-solid interface with aliphatic and aromatic aldehydes. J. Phys. Chem. B, 1999, v. 103, 6, p. 1019–1026. DOI: 10.1021/jp982690n

299. Murphy D. M., Chiesa M. EPR of paramagnetic centers of solid surfaces, p. 105–130, p. 111. In: Gilbert B. C., Davies M. J., Murphy D. M. (eds). Electron Paramagnetic Resonance (EPR). Volume 21, 2008, Cambridge, RCS Publishers, 233 pages.

300. Carter E., Carley A. F., Murphy D. M. Evidence for O_2^- radical stabilization at surface oxygen vacancies on polycrystalline TiO_2. Journal of Physical Chemistry C, 2007, v. 111, 28, p. 10630–10638. (10.1021/jp0729516)

301. Carter E. EPR investigation of stable and transient oxygen centered radicals over polycrystalline titanium dioxide. PhD Thesis, 2007, Cardiff University.

302. Rajakumar B., Flad J. E., Gierczak T., Ravishankara A. R., Burkholder J. B. Visible absorption spectrum of the CH_3CO radical. J. Phys. Chem. A, 2007, v. 111, 37, p. 8950–8958.

303. Zhao H., Kim J., Koel B. E. Adsorption and reaction of acetaldehyde on Pt(111) and Sn/Pt(111) surface alloys. Surface Science, 2003, v. 538, 3, p. 147–159.

304. Xu C., Koel B. E. Adsorption and reaction of CH_3COOH and CD_3COOD on the MgO(100) surface: A Fourier transform infrared and temperature programmed desorption study. J. Chem. Phys., 1995, v. 102, p. 8158.

3.7

305. Elchyan A. M., Grigoryan G. L. Chain regime in oxidation of hydrocarbons by hydrogen peroxide on the surface of a SiO_2, *Khim. Phizika* (Chemical Physics, in Russian), 1990, v. 9, 5, p. 639–643.

306. Bakhchadjian R. H. Transfer of the chains to the solid surface in heterogeneous-homogeneous reaction. 16th International Symposium on Gas Kinetics. Cambridge, July 23rd-27th, 2000. Book of Abstracts, PE I, 2 pages.

307. Porter R., Glaude P-A., Buda F., Battin-Leclerc F. A tentative modeling study of the effect of wall reactions on oxidation phenomena. Energy and Fuels, 2008, v. 22, p. 3736–3743.

308. Berlowitz P., Driscoll D. J., Lunsford J. H., Butte J. B., Kung H. H. Does platinum generate gas phase methyl radicals in the catalytic combustion of methane? Combustion Science and Technology, 1984, v. 40, 5–6, p. 317–321 DOI: 10.1080/00102208408923815

309. Geske M., Pelzer K., Horn R., Jentoft F. C., Schlögl R. In-situ investigation of gas phase radical chemistry in the Catalytic partial oxidation of methane on Pt. Catalysis Today. 2008, v. 142, p. 61–69.

4 Bimodal Reaction Sequences in Oxidation with Dioxygen in Photocatalysis

Interactions of photons with components of the chemical system, containing also photosensitive solid substances, can lead to the generation of active intermediate species both in the fluid phase and on the surface of solid substance. If the active intermediates are able to be transferred from one phase to another by continuing the chemical reaction, the overall photochemical process obtains the character of the bimodal reaction sequences. As it appeared by the investigations in recent decades, this reaction mechanism may be suggested for a great number of photochemical oxidation processes of organic compounds on semiconductor catalysts or other photosensitive materials.

In this chapter some differences, as well as similarities of bimodal reaction sequences between the photochemical and thermal oxidation reactions of organic compounds have been revealed.

4.1 FROM OLDEN TO MODERN PHOTOCHEMISTRY

Application of light in the chemical transformation of organic compounds has a long history.[1, 2] Some of the reactions occurring under light, such as those involved in the decomposition of silver salts, were utilized in nineteenth-century photography.[1] However, the history of modern photochemistry and green chemistry, including the photochemical oxidation of organic substances with molecular oxygen, begins with the works of the Italian-Armenian chemist G. L. Ciamician (1857–1922),[2, 3] known as a pioneer in this field. His discovery of new photochemical reactions opened a wide area for the use of light, including sunlight, not only for chemical synthesis, but also for solar energy transformation into chemical energy. Ciamician predicted the future development of photochemistry, presently related to sustainable and green chemistry.[4] According to Schroll,[5] Ciamician, together with Silber, also discovered a number of photochemical oxidation reactions involving molecular oxygen. For instance, stilbene was transferred to benzaldehyde by the addition of molecular oxygen that reacted with the double bond of stilbene under light irradiation. The formed dioxetane-like intermediate, a four-member ring, underwent further decomposition giving benzaldehyde:

$$Ph\text{-}CH=CH\text{-}Ph \xrightarrow{\ O_2\,,\ h\nu\ } Ph\text{-}\underset{O-O}{\overset{H\quad H}{C-C}}\text{-}Ph \rightarrow 2\,Ph\text{-}CHO$$

The first notable work of Ciamcian was Spectra of Chemical Elements and Their Compounds, *in which he showed the similarities between various emission spectra of elements in the same groups of the periodic table of chemical elements. Dmitri Mendeleev (1834–1907), discoverer of the periodic law and creator of the periodic table, in his two famous works,* The Principles of Chemistry *(1891, German version) and* The Periodic Law of the Chemical Elements *(1889, J. Chem. Soc.), acknowledged the contribution of Ciamician in best confirmation of the periodic law.[6] However, the main contribution of Ciamician was in the photochemistry. He discovered a large number of new reaction types, including geometrical isomerization of C=C, C=N, and N=N bonds; [2+2] cycloaddition involving olefins, α,β-unsaturated ketones, and esters; reactions of ketones, such as α-cleavage and intermolecular hydrogen abstraction from alcohols, leading to reduction or coupling; inter- and intramolecular reactions of nitro compounds, etc. He was the first scientist who proposed "to fix the solar energy through suitable photochemical reactions with new compounds that master the photochemical process that hitherto have been the guarded secret of the plant."[3] For this reason, he is considered the father of solar panels. He is known to have used a solar-powered light bulb to illuminate his laboratory.[2] For his work in photochemistry, he was a nine-time nominee for the Nobel Prize but never received it, while the five scientists whom he nominated for Nobel Prizes in Chemistry and Physics did receive these awards.[5, 7]*

Ciamcian's ultramodern idea that "mankind should have learnt to use directly solar light, thus having at its disposal not only a boundless resource, but also a clean environment, abandoning the dirty world based on burning fuels"[2] is the basis of modern photocatalysis.

Unfortunately, however, only in the second half of the twentieth century, investigations in photocatalysis were directed toward knowledge of the chemical transformation of photon energy into other forms of energy. Presently, photocatalysis is widely used for the oxidative degradation of pollutants in both liquid and gas phases; one of the commercialized versions of photocatalysis is known as the advanced oxidation process (AOP). It is based on the semiconductor photocatalyst (mainly of TiO_2). On the other hand, photocatalysis is an excellent means of chemical synthesis in very "mild" conditions (room temperature, atmospheric pressure, inexpensive catalysts, and visible light).

Photocatalytic reactions, particularly heterogeneous photocatalytic reactions, are environmentally benign and economically the most suitable processes for modern and future developments in chemical industry. In this context, understanding of the mechanism of the heterogeneous photocatalytic processes attained a novel importance in recent decade.

As this chapter will show, a great number of photocatalytic processes of oxidation, known as heterogeneous reactions, are heterogeneous–homogeneous processes. They mostly through the participation of electronically excited species, subsequently forming radicals, which are the leading active intermediates of further chemical processes. In rare cases, the mechanism suggestions in these reactions are based only on electronically excited species, as the main intermediates leading directly to the formation of products. Although, in our knowledge, nearly nowhere in the chemical literature can the reader find the combination of the terms *heterogeneous–homogeneous* and *photocatalysis*, however, the reader can find reaction schemes in the heterogeneous photocatalysis, where part of radicals, generated on the surface, "escape" to the fluid phases and participate in homogeneous oxidation processes. In turn, the homogeneous constituent of the complex reaction interacts with the heterogeneous one, in photocatalytic oxidation.

4.2 PRELIMINARY NOTES ABOUT PHOTOCATALYTIC REACTIONS

Photochemical reaction takes place when the reactants or other components of the reaction media (atoms, molecules, ions, radicals, excited species, solid-phase substances, etc.) absorb the energy of light (photons) and use it for the chemical transformations. Light, being one of the forms of electromagnetic irradiation, is characterized by the wavelength (λ) and frequency (v). On the scale of electromagnetic waves, it occupies the diapason of the wavelength about 100–2500 nm. In turn, it is divided into diapasons of UV irradiation (100–400 nm), visible light (400–750 nm), and infrared irradiation (750–2500 nm). It is obvious that the energy of light decreases with increases in wavelength, as $E = hv$ and $v = 1/\lambda$. Correspondingly, in the 150–800 nm range, light possesses 800–150 kJ/mol energy that is comparable to energies sufficient to break the chemical bonds.[8] For example, molecular oxygen absorbing the energy of photons below $\lambda = 200$ nm that corresponds to the energy 143 kcal/mol (598 kJ/mole), can be dissociated into oxygen atoms, as the energy of O=O bond is only 119 kcal/mol (498 kJ/mol).

Photocatalysis is divided into two branches: homogeneous and heterogeneous photochemical catalytic processes. In homogeneous photocatalytic reactions, all "actors" of the reaction are in the same phase state. In heterogeneous photocatalysis at least, the catalyst or sensitizer is in the solid state, and other initial reaction components may be in the gas or liquid phase. In both cases, the photocatalytic process begins with absorption of the energy of photons (hv) by one or more reaction components named photoexcitation. Photoexcited species can interact with other species, transferring either energy or electron, proton, atom of hydrogen, oxygen, etc. On the other hand, the energy of light, absorbed by the components of the chemical system, may be transformed into heat and dissipated in the environment, without undergoing any chemical transformation.

The glossary of the photocatalysis is rich. In the scientific literature of physics, chemistry, and biology one encounters different terms with similar meanings, such as photogenerated, photocatalytic, photoinduced, photoassisted, phosensitized, and photoinitiated reaction.[9] According to the classification of photochemical reactions, given by Hennig et al.,[10] the main types of heterogeneous photocatalysis and homogeneous photocatalysis may be differentiated on the basis of the quantum yields of the processes (Φ). The quantum yield of the consumption of substrate Φ_S, in the presence of the catalyst, is a ratio of the number of substrate molecules, transformed to product(s), to the number of photons absorbed.

$$\Phi_S = \frac{\Delta[substrate]}{number\ of\ photons\ absorbed\ by\ system}$$

Analogously, if the quantum yield, in the same conditions but in the absence of the photocatalyst, is Φ_U, the photocatalytic reaction is defined as a process that corresponds to the condition

$$\Phi_S > \Phi_U \tag{1}$$

If the photocatalytic reaction occurs in the presence of the initiator or nominal catalyst, they become chemically active only after the absorption of light energy. The following condition may take place

$$\Phi_S > \Phi_C \tag{2}$$

where Φ_C is the quantum yield of the photochemical reaction in the presence of an initiator or nominal catalyst (precursor of catalyst). If $\Phi_S > \Phi_C$, the photochemical reaction is referred to as a photoinduced reaction. If

$$\Phi_S < \Phi_C \tag{3}$$

the process is named photoassisted. In this regard, the photosensitized reaction also is a photoassisted process.[9]

Thus, according to conditions (1)–(3), the photocatalytic processes may be divided into two main types: photoinduced and photoassisted reactions. In the photoinduced oxidation reaction, absorbing the energy of light, the precursor of the catalyst becomes a real catalyst. Creating a real catalyst, the photoinduced reaction may perform an unlimited number of catalytic cycles. Theoretically, in photoinduced reactions, one or two photons are sufficient for the complete transformation of the initial reactants into final products. In this categorization, a chain-radical photochemical reaction is also a photoinduced catalytic reaction.[10] In a photoassisted reaction, theoretically, every catalytic cycle takes place through the participation of photons. Photons, in this case, become an initial reactant.

Photosensitizers involve many classes of organic compounds, including the metal organic complexes, dyes, and substances of biological origin, such as porphyrins, chlorophylls, and organic dyes. Absorbing the light energy and being transferred to the excited state, the sensitizers activate other molecules, transforming them into excited states.

In 1900, in Germany, Hermann von Tappeiner[11] (1847–1927) and his student Raab[12] discovered an interesting phenomenon, which was later named photodynamic action. They observed that sunlight killed cells stained with a light-sensitive dye. This observation may be explained as follows. Absorbing the sunlight in a system composed from dye, oxygen, and substrate, a dye, or the oxygen and dye together, transfer into an excited state. Then, the photoexcited species initiates the oxygenation reactions of the biological components in the cell. The process is usually accompanied by the accumulation of peroxides in the system. Oxygenation is a combination of organic compounds with oxygen or its addition to the composition of the biological substrate. The photodynamic action is schematically presented in Figure 4.1, which is based on Krasnovsky's work.[5 in Ch. 3] Historically, investigations of photodynamic action played an important role in the discovery of singlet oxygen and in the knowledge of the photocatalytic oxidation reactions using photosensitive materials.[5 in Ch. 3] Subsequently, this phenomenon was widely applied in investigations of photobiology and laser medicine.

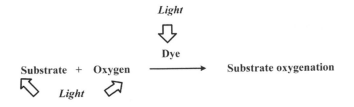

FIGURE 4.1 Scheme of the photodynamic action according to Krasnovsky's presentation. (Ref.[5 in Ch. 3])

Unfortunately, wide application of them in practice is often limited because of the low structural stability of dyes under light irradiation or when used in different solvents. One of the ways to perfect the efficacy of the sensitizer action in the photochemical process is through their heterogenization by anchoring the dyes on different supports (polymeric or composite materials, metal oxides, zeolites, etc.).[13] In 1973, Blossey et al.[14] were the first to report the synthesis of the heterogeneous photosensitizer, composed from Rose Bengal (a dye), anchored on a polymer material (chloromethylated polystyrene) as carboxylate ester. It was shown that new material was a more efficient photosensitizer of oxygen than suspension of Bengal Rose in CH_2Cl_2 in oxidation of the organic compounds under irradiation $\lambda = 563$ nm.[14]

The overall complex photochemical process may be divided into three main stages.[15] The first stage begins with adsorption of light and electronic excitation of one or more reaction components. The chemical transformation, named primary reaction, takes place in the second stage, when electronically excited species transform into product(s) or interact with other components of the reaction medium, forming other intermediate(s). In the third stage, the intermediates continue the reactions, named secondary or "dark" reaction. In this context, the above-mentioned first stage, photoexcitation, is only a physical process. In the timescale, the fastest chemical process may occur in a part of seconds having an order 10^{-16}–10^{-15} sec (an approximate time interval corresponding to the movement of the electron about 10 angstrom distances in a chemical reaction). The absorption of the photon and electronic excitation usually occur in attoseconds (10^{-18} sec). Therefore, the chemical reaction is the slowest stage of the photochemical process.

Albini and Fagnoni[16] state that the main difference in the formation of the reactive intermediates between thermal and photochemical reactions, making possible the occurrence of later under more "mild" conditions, lies in the different pathways of their formation. Although the same intermediate may be formed by both thermal and photochemical pathways, the energetic barrier of the intermediate formation in photochemical reactions, in general, is lower than that in thermal reactions. Intermediates in photochemical reaction are formed through energetically higher states (excited state), and they "descend" to an energetically low level (corresponding to an intermediate formation), while thermal oxidation happens through energy consumption, from the ground state to the same level. Moreover, the photochemical pathway may often be favorable to formation of intermediates corresponding to high energetic levels, which are not available for ground state molecules by the thermal reaction way. Schematically, this is obvious from Figure 4.2.

That is why the photochemical reaction may sometimes open a pathway that is not favorable in the same conditions by the thermal reaction pathway. This also opens up new perspectives for chemical synthesis under light, via intermediates, which are practically nonexistent in thermal reactions.

4.2.1 Homogeneous Photocatalysis

In homogeneous photocatalysis,[17–20] usually, catalysts are transition metal ions or their various complexes, including metal-organic coordination complexes of different

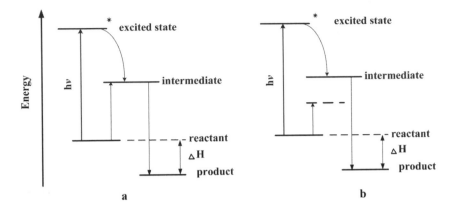

FIGURE 4.2   Scheme of the formation of intermediates: (a) when the same interme-diate forms in photochemical and thermal reactions; (b) when the heat in thermal reaction is insufficient for formation of the same intermediate than that in photochemical reaction. (Adapted from Ref.[16])

structure. As an example of the homogeneous photocatalysis may serve the photo-chemical analog of the well-known reaction with Fenton reagent $(M^{n+} + H_2O_2)$[18]

$$Fe^{3+} + H_2O + h\nu \rightarrow Fe^{2+} + HO^{\bullet} + H^{+}$$

$$H_2O_2 + h\nu \rightarrow HO^{\bullet} + HO^{\bullet}$$

The solution containing metal ions of the variable valence (M^{n+}) and hydrogen peroxide H_2O_2), named the Fenton reagent, is a strong oxidant agent. The reaction has an important practical application in wastewater treatment technologies.[21]

If the catalyst is a transition-metal organic compound[18] with aromatic rings, the exci-tation occurs through metal-to-ligand charge transfer (usually from the d orbital of metal to the delocalized π-orbital of the aromatic ring). Relaxation of the excited state may occur via the so-called internal conversion or radiationless de-excitation mechanism, when the energy of the excited molecule dissipates in the form of heat or transforms to vibrational energy. There are two ways for relaxation of the species in an excited state: (i) radiation of the photon and return to the ground state (named fluorescence) and (ii) movement to the lower excited triplet state, called intersystem crossing. Radiation return from the triplet state to the singlet state is called phos-phorescence, and it occurs by means of spin inversion. All these events, represented schematically in Figure 4.3, were first proposed by Polish physicist A. Jablonski (1898–1980) in 1933.[18]

   Electron transfers between the catalyst and reactants occur by an outer-sphere electron transfer mechanism (for coordination-saturated complexes, the inner-sphere electron transfer cannot take place, because the formation of a new chemical bond through a ligand, bridging other (one more) metal center, is impossible). Note that the majority of the metalorganic complex photocatalysts in triplet state have a lifetime that is long enough for electron transfer rather than relaxing and returning to their ground state. In the case of homogeneous photocatalysis by complexes of different transition metals, the mechanism suggestions are reminiscent of the perceptions of

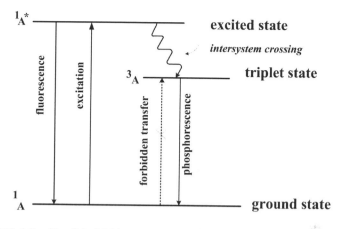

FIGURE 4.3   Simplified Jablonski diagram. ^1A: ground state (singlet); ^3A: triplet state; and $^1A^*$: excited state (singlet). (Adapted from Ref.[18])

redox catalysis. Thus, in homogeneous photocatalysis, the transition metal ion mediates the electron transfer from the oxidant to reductant compound.

Presently, the homogeneous photoredox catalysis is a rapidly developing area permitting to carry out many environmentally benign syntheses of organic compounds under "mild" conditions.[20] As a rule, they are distinguished by their high selectivity and rapidity. However, some limitations, mentioned in the above section for thermal oxidation reactions in homogeneous conditions (Section 3.1), also exist in photochemical conditions for their wide application in industrial scales. As already mentioned, one way to overcome these limitations is through heterogenization of the homogeneous photocatalysts in selective oxidation processes, some outlines of which were presented in Section 4.2.

The important achievements in homogeneous photocatalysis[17, 18, 20 and ref. within] are well documented in modern monographs, articles, and reviews, which also shed light on knowledge of heterogeneous photocatalysis, including photochemical reactions exhibiting a heterogeneous–homogeneous character.

4.2.2   HETEROGENEOUS PHOTOCATALYSIS

Although some observations of the occurrence of photoinitiated heterogeneous reactions may be found in the works of investigators at the end of the XIX and beginning of the XX centuries,[22, 23] the heterogeneous chemical processes under light did not become a subject of wide and systematic investigations until the 1960–1970s.[24, 25] The majority of authors relate the primary development of heterogeneous photocatalysis to works on the photoinduced electrochemical splitting of water into hydrogen and oxygen on TiO_2 under UV irradiation, discovered by Fujishima and Honda[26] in 1972. However, the historical roots of the modern theory of heterogeneous photocatalysis go also back to the 1960s and the concept of the so-called electronic theory of catalysis on semiconductors, created by Volkenstain.[27] At first glance, the adjective "electronic" does not add anything new in characterizing a theory of catalysis, as, *a priori,* every chemical reaction indispensably includes electronic processes. According to Volkenstain,[27] the electronic theory of catalysis is connected with two different, but mutually interdependent, properties of materials: electrophysical properties and catalytic activity. That is why, at that time, theoretical perceptions were named the "electronic theory" of catalysis. Indeed, the physical properties of materials in bulk differ from those on surfaces. Note that this theory is one of the bases for different modern interpretations in heterogeneous photocatalysis.

A period of the more intensive scientific investigations in this branch of chemistry may be considered the recent decades. This development is apparently related to some findings from heterogeneous photocatalysis, which presently are experiencing wide practical applications (advanced oxidation technologies (AOPs), treatment of wastewater, based primarily on the photocatalytic reactions of semiconductor materials, particularly materials based on TiO_2). Another important domain in heterogeneous photocatalysis is solar energy storage through synthesis of solar fuels.[28]

The main actors in heterogeneous photocatalysis are solid-phase catalysts, visible light or UV irradiation, substrate, oxidizing agents, and, often, solvents. The heterogeneous photocatalysts include semiconductor materials, metal oxides, polyoxometalates, and photosensitive composite materials.[29] Among semiconductors, TiO_2 is known for its catalytic activity in thermal oxidation reactions, in both gas and liquid phases, with the components of organic nature.[29] In photochemical reaction conditions, TiO_2 often exhibits exceptional catalytic activity. TiO_2 mainly is known as n-type photoactive material with bandgap 3,2eV, photocatalytically active under UV irradiation ($\lambda < 390$ nm).[29]

Recently, an international group of authors reported the synthesis of so-called titan terminated anatase with reduced bandgap 2eV, operating in the visible region.[30]

That is why the majority of investigations in heterogeneous photocatalysis focus on TiO_2. Commercially available and mainly used TiO_2 is TiO_2-Degussa P25, containing two polymorphs, crystalline structures as a mixture of anatase and rutile phases. Degussa P25 is known as a more active photocatalyst material than anatase or rutile, when taken separately.

Although not all details of the reaction mechanism under light are completely clear in oxidation photocatalysis, in the majority of cases, the main intermediates of these reactions have been identified using different experimental methods, including the "arsenal" of methods employed by surface science. In general, they are radicals or radical-like species, formed via the excited species on the surface of the solid photocatalyst, the reactions of which have many analogies with thermal oxidation reactions. Only in rare cases the distinctive role of singlet oxygen or other electronically excited species in photocatalysis was known. (Some examples were given in Section 3.4.2.7.) Here, note again that the physical process temporary is well separated from the genuine chemical reaction. Like thermal oxidation reactions with dioxygen, in certain conditions, the photoinduced oxidation on the catalyst may occur as a complex heterogeneous or heterogeneous–homogeneous reaction.

As different aspects of the problems of photocatalysis are well documented, the following discussion will be devoted to the heterogeneous-radical reactions leading to exhibition of heterogeneous–homogeneous mechanisms.

4.3 HETEROGENEOUS PHOTOGENERATION OF RADICALS

Gas- or liquid-phase thermal oxidation with molecular oxygen occurs mainly by the radical mechanism. Radicals also form in photochemical oxidation reactions, although the primary action of light of a certain frequency does not directly result in the formation of radicals, but it does generate excited state atoms and molecules. The formation of radicals occurs in the second stage of the photochemical process that was named the primary chemical reaction (see Section 4.2). Here, we return to the Kautsky experiment of the discovery of singlet oxygen (Section 3.4.2.7). The following question arises related to the mechanism of the substrate oxidation, in conditions generating singlet oxygen. Does singlet oxygen directly react with substrate passing through the gas phase, or does it transform to another intermediate? Weis[31] was the first (followed by Franck and Livingston[32]), to attempt to explain Kautsky's experiment by the formation of radicals and their further reactions, suggesting that

$$\left(\text{Photosensitive Dye}\right)^* + O_2 \rightarrow \left(\text{Photosensitive Dye}\right)^+ + O_2^-$$

$$H^+ + O_2^- \rightarrow HO_2$$

(Photosensitive Dye)* is a molecule in the excited state. As O_2^- and HO_2 are oxidizing radicals, further oxidation of substrate may occur by radical reactions. Note that the oxidation of substrate may take place if the intermediate, generated on the surface of a photosensitizer, passes through the gas phase, reaching to the surface of other silica gel grain, supporting substrate. It is evident that the overall process consists of two consecutive stages: (i) generation of the active species on the surface and their transfer to the gas phase (the heterogeneous–homogeneous process) and (ii) diffusion from the gas phase to the solid surface and reaction on the surface (homogeneous–heterogeneous processes). Unfortunately, at that time, this suggestion was only a working hypothesis, and the absence of experimental methods for direct detection of the radicals did not permit to verify it. Only some decades ago, after the creation and development of new methods such as EPR spectroscopy, did it become possible to prove the relation between the formation of singlet oxygen and free radicals in organic photochemistry.[5 in Ch.3]

Description of the formation of radicals on semiconductor materials, given mainly for the surface of TiO_2, may now be found in nearly every textbook, handbook, and monograph related to the problems of modern heterogeneous photocatalysis. The schematic presentation of the generation of holes and electron pairs, as well as free radicals, on the surface of the TiO_2 semiconductor under UV irradiation ($\lambda < 390$ nm) is illustrated in Figure 4.4.

The formation of the charge carriers (hole-electron pair) under UV irradiation is a very fast process. When electrons overcome energetic bandgap in a semiconductor, it takes place the charge separation. There are at least three feasible pathways for further development of events. The first way is recombination of holes and electrons, the second is their trapping at defect sites in the bulk of the semiconductor and, the third, their diffusion to the surface and chemical reaction by electron transfer. Evidently, the first two ways are unproductive for photocatalysis.

The holes are oxidants that are able to oxidize water (forming OH radicals) or organic compounds if they are present in the reaction medium.

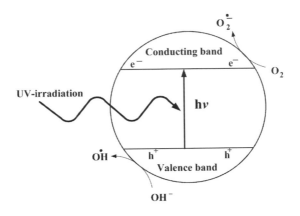

FIGURE 4.4 Schematic presentation of the photoinduced generation of radicals on the surface of semiconductor TiO_2 under UV irradiation ($\lambda < 390$ nm).

For a TiO_2, under UV irradiation, the generated electron may be trapped at defect sites on the surface or in the bulk, being localized on the d orbital of metal ion Ti^{4+}.

$$Ti^{4+} + e^- \rightarrow Ti^{3+}$$

Ti^{3+} metal ions may be identified by EPR spectroscopy. For instance, at least two types of Ti^{3+}, one of them axially symmetric ($g_\perp = 1.988$, $g_\parallel = 1.957$) and the other an octahedral structure (both were detectable only in the presence of hole scavengers)—were found on colloidal dispersed TiO_2, in acidic solution ($p^H = 2.2$).[33] From the results of the EPR investigations, it is known that there are at least two kinds of sites for generating superoxide ion-radicals from O_2 adsorption on the TiO_2. The first is five coordinated Ti^{3+} cations ($g = 2.023$), and the second is oxygen vacancy ($g = 2.020$).[34]

$$Ti^{3+} + O_2 \rightarrow \left[Ti^{4+} \ldots O_2^{\bullet -} \right]$$

Using the polycrystalline samples of TiO_2 Carter et al.[35] obtained the EPR spectra of surface $O_2^{\bullet -}$, which were assigned to "side-on" species in surface oxygen vacancies with $g_{xx} = 2.005$, $g_{yy} = 2.011$, and $g_{zz} = 2.019$. In turn, $O_2^{\bullet -}$ may enter into secondary reactions forming HO_2, H_2O_2, and OH.

The adsorbed oxygenated radicals, other than superoxide, may be formed in several ways. On the hydrated surface of TiO_2, the formation of HO_2 is feasible by the reaction

$$O_2^{\bullet -} + H^+ \rightarrow HO_2^{\bullet}$$

concurring with the decay of $O_2^{\bullet -}$ by capture of electrons.

$$O_2^{\bullet -} + e^- \rightarrow O_2^{2-}$$

O^- radical on the surface may appear as a result of the reaction of holes:

$$Ti^{4+} - O^{2-} + h^+ \rightarrow Ti^{4+} - O^-$$

There is EPR spectroscopic evidence for the ozonoid radicals O_3^-, formed in the adsorption of dioxygen on the lattice oxygen site of the surface TiO_2/SiO_2 ($g_1 = 2.0080$, $g_2 \approx 2.003$, $g_3 = 2.0026$).[36]

$$Ti^{4+} - O^- + O_2 \rightarrow Ti^{4+} - O_3^-$$

The reactive hydroxyl radical OH, playing an important role in oxidation of the organic compounds, may be formed as a result of the trapping of hydroxyl ions by holes.[34]

$$OH^- + h^+ \rightarrow OH^\bullet$$

Apparently, further concurrence of the reactions of OH by different pathways determine the chemical events in the reaction system. Recombination of OH on the surface gives H_2O_2, whose heterogeneous reactions on the solid surfaces (see Section 2.3.2.1), including TiO_2, are well known.

$$OH^\bullet + OH^\bullet \rightarrow H_2O_2$$

$$H_2O_2 + OH^\bullet \rightarrow HO_2 + H_2O$$

Another pathway to surface decay of $OH^\bullet$ is electron capture.

Singlet oxygen is not directly detectable on the surfaces by the EPR-method. However, there exist some investigations attempting to detect surface produced $^1O_2$ by spin-trapping methods.[34]

Note that TiO_2 is not only photosensitive material on the surface of which the generation of radicals was well proven experimentally. For example, evidence that radicals are formed directly on the surface of SiO_2 (rather than in the liquid phase) was obtained by the EPR method as described in the work of Macdonald et al.[37]

Note, too, another important peculiarity of the photocatalytic oxidation. Among major "actors" of the photocatalalytic oxidation, we have also mentioned the solvent. It appears that in photocatalytic oxidation the solvent may play the role of oxidant. In 2008, Yoshida et al.[38] showed that, in the absence of molecular oxygen, benzene and its alkyl derivatives may be oxidized by water on the surface of the photocatalyst (platinum-loaded titanium oxide) to phenol (correspondingly, also to derivatives of phenol) and hydrogen. Some authors believe that the role of oxidant plays electrophilic oxygen species (generated by photocatalysts under light irradiation), which selectively react with aromatic ring.[38] Indeed, the rate of reaction and distribution of products were different than in the presence of oxygen.[39] There are yet other examples of photocatalytic oxidation in the absence of dioxygen.[39 (p.657)] Methanol, ethanol, propanol, and 2-propanole in aqueous solution, on the Pt/TiO_2 photocatalyst, may be converted into formaldehyde, acetaldehyde and acetone with the formation of H_2, in the absence of oxygen or other oxidant. The overall photocatalytic reaction may be presented as

$$R\ R'CHOH \xrightarrow{Pt/TiO_2 + h\nu} R\ R'C{=}O + H_2$$

Oxidation occurs through the formation of holes and excited electrons of photocatalyst under irradiation. Holes play the role of oxidant of alcohol, and excited electrons reduce H^+ of water, finally giving H_2.[39]

As seen in the current chemical literature on photocatalysis, one can remark that the semiconductors and analogous materials, as effective photocatalysts, now have a predominant role in the chemistry of oxidation of the major classes of organic compounds, such as hydrocarbons, alcohols, and aldehydes.

According to Stephenson,[40] since the 1990s, many new radical reactions have been discovered, and they are being widely applied in chemical synthesis and in different areas of organic chemistry and biology. Stephenson labeled this development as the renaissance of radical organic chemistry. Analogously, the recent development of radical reactions on semiconductor surfaces in oxidation processes may also be considered to be part of the renaissance of radical mechanisms in the chemistry of heterogeneous catalytic reactions, which Volkenstein suggested more than half a century ago.[27, 65 in Ch. 1]

4.4 HETEROGENEOUS–HOMOGENEOUS MECHANISMS IN PHOTOCHEMICAL OXIDATION

4.4.1 OXIDATION OF SOME ORGANIC COMPOUNDS ON THE HETEROGENEOUS PHOTOCATALYSTS

The heterogeneous photocatalytic reactions of oxidation of the organic compounds with oxygen, briefly presented in this section, include selected examples of surface reactions occurring by formation of free radicals and their further transformations. Although not all details of the formation of radicals on the semiconductors, particularly, on a TiO_2 or TiO_2-based photocatalyst in oxidation are clear, the nature of key intermediates for the main classes of organic compounds are known. These examples clearly show the free radical character of the heterogeneous reactions on the photocatalysts. Nearly all suggestions about the mechanism of photoinitiated oxidation reactions with dioxygen relate to the generation of radicals, ion–radicals, and radical-like species on the surfaces of semiconductor materials, in rare cases, electronically excited species, as singlet oxygen under UV irradiation or sunlight. In reaction schemes, the main role is usually attributed to the reactions of OH radicals on the surfaces with the substrates. The detection and identification of radicals were presented in Section 4.4.3. The principal evidence for the heterogeneous-radical nature of processes in photocatalysis were obtained by the LIF and EPR investigations (often, by *in situ* measurements).[41] Indeed, not all intermediates of heterogeneous photocatalytic oxidation are made "visible" by these methods (for instance, the necessary conditions for the "visibility" or detectability of radicals by the EPR method were mentioned by Volodin et al. Ref.[1 in Ch. 2]). However, the combination of some methods with the spin trapping methods (spin scavengers) often permits obtaining valuable information about the nature of surface intermediates.[41]

For a long time, investigations on the thermal oxidation of methane to oxygenates[42–45] did not give the desired results. Direct partial oxidation of methane with molecular oxygen to oxygenated products, as methyl alcohol and formaldehyde, had been attractive to chemists since 1922, by the first work of Blair and Wheeler[42] in this area. Use of catalysts such as MoO_3/SiO and V_2O_5/SiO_2 (these were considered to be relatively effective in this reaction) does not provide high selectivity of the formation of formaldehyde in the conversion of methane. For example, the use of V_2O_5/SiO_2 gives only 3–4% selectivity in the formation of formaldehyde.[44] One of the limitations of selective oxidation of methane is that formaldehyde transforms into CO_2 and H_2O at temperatures much lower than those involved in the considerable conversion of methane (about 600°C). Unlike thermal oxidation, the photocatalytic oxidation of methane to formaldehyde may be carried out at relatively low temperatures. For example, nearly the same quantity of formaldehyde may be obtained at 180°C over

MoO_3/SiO_2 under UV irradiation and only at 570°C in the thermal oxidation over the same catalyst.[44] From this point of view, the photocatalytic pathway is more advantageous than thermal catalytic oxidation. In nearly all photochemical investigations on the oxidation of methane to formaldehyde on the metal oxide catalysts,[42] the formation of CH_3 radicals is considered a primary reaction.[44, 45]

$$CH_4 + (M^{n+})_x O_y \xrightarrow{h\nu} CH_3 + (M^{(n-1)+})_x O_{y-1}(OH)$$

This reaction is the heterogeneous dissociation of methane, after the adsorption on the surface of the catalyst. It is evidenced by the results of EPR spectroscopy.[44] According to the review of De Vekki and Marakaev[44] the mechanism of further transformations of the oxide catalysts relates to the reactions of lattice oxygen. For example, in the case of the heteropoly acid catalysts, it was suggested that the surface generated methoxy and hydroxyl species:

$$(M^{(n-1)+})_x O_{y-1}(OH) + O_2 \rightarrow (M^{n+})_x O_y + (OH)_s$$

$$CH_3 + (M^{n+})_x O_y \rightarrow (M^{(n-1)+})_x O_{y-1}(OCH_3)$$

$$(M^{(n-1)+})_x O_{y-1}(OCH_3) \rightarrow (M^{n+})_x O_y + (OCH_3)_s^- \searrow$$

diffusion into crystal

$$(OCH_3)_{cr}^- \rightarrow CH_2O + H_{cr}^-$$

$$(OCH_3)_{cr}^- + H_2O \rightarrow CH_3OH + OH_{cr}^-$$

where cr denotes species on/in the crystalline structure,

$$(M^{n+})_x O_y + H_{cr}^- \rightarrow (M^{(n-1)+})_x O_{y-1}(O^{\bullet-}) + \tfrac{1}{2} H_2$$

$$(M^{(n-1)+})_x O_{y-1}(O^{\bullet-}) + CH_4 \rightarrow (CH_3^{\bullet})_{ads} + (M^{(n-1)+})_x O_{y-1}(OH)$$

and the formation of formaldehyde may occur by the reaction

$$(CH_3O)_{ads} + O_{ads}(or\ O_s) \rightarrow CH_2O + (OH)_s$$

Finally,

$$(CH_3O)_{ads} + O_2 \rightarrow H_2O;\ CO;\ CO_2$$

$$OH_{cr} + OH_{cr} \rightarrow H_2O + O_{s,\ cr}$$

Photocatalytic oxidation of methane, in any case, occurs by the radical mechanism in the primary reaction. The mentioned behaviors, equally, concern other hydrocarbons. One of the first investigations of photocatalytic selective oxidation of C_2–C_8 alkanes, in the gas phase, on a TiO_2 was reported by N. Djeghri et al.[46] The main products of the oxidation, for both branched and nonbranched alkanes, under UV irradiation (210–390 nm) were aldehydes and ketones, accompanied by CO_2 and water. The selectivity of the oxidation into products (ketones and aldehydes) was 76% for n-alkanes, 80% for isoalkanes, and 57% for neoalkanes. The reactivity changed in the following order: $C_{tert} > C_{quat} > C_{sec} > C_{prim}$. Mechanism suggestions were based on the scheme of the consecutive formation of alkane → alcohol → aldehyde → ketone, without précising the nature of intermediates in each transformation.

Photocatalytic oxidation of ethylene on TiO_2 also occurs through the formation of radical intermediates.[47] Peroxy radicals were detected by the EPR method involving experiments with isotopically labeled species (C_2D_4, $^{17}O_2$, $^{13}C_2H_4$, etc.) in the photooxidation of ethylene with molecular oxygen on a TiO_2.[47] The following overall mechanism of the photocatalytic oxidation of ethylene on TiO_2 was proposed:

$$OH^- + \left(e^- + h^+\right) \rightarrow OH^\bullet + e^-$$

$$OH^\bullet + C_2H_4 \rightarrow (OH)CH_2CH_2^\bullet$$

$$(OH)CH_2CH_2^\bullet + O_2 \rightarrow (OH)CH_2CH_2OO^\bullet$$

The EPR characteristics of the adsorbed peroxy radicals were $g_1 = 2.035$, $g_2 = 2.008$, and $g_3 = 2.001$.

Oxidation of alcohols on a TiO_2 photocatalyst was investigated in a number of works mentioned in 3.5.3, using different methods, including EPR. It was shown that in both the thermal and photochemical oxidation of alcohols, the main intermediates were methoxy, as well as methylperoxy radicals. These results demonstrate the generation of surface radical species, the further reactions of which are responsible for formation of the final products of oxidation. In other words, the photoinduced heterogeneous catalytic reaction is primarily radical in nature.

The kinetic behaviors of photochemically induced reactions of oxidation with the oxygen of many organic compounds are similar to the thermal oxidation reaction. The thermal oxidation of acetaldehyde and propionic aldehyde with dioxygen was presented in Section 3.5.4.2. It was shown the heterogeneous–homogeneous nature of this reaction and the radical character of its heterogeneous constituent. The following examples of the photochemical oxidation of some aldehydes demonstrate the similarities between thermal and photocatalytic oxidation in this case. Already in 1928, Backstrom showed that through exposure of aldehydes in air or oxygen, under light, the quantum yields were $\Phi \gg 1$ (560–15000), demonstrating the chain-radical character of this reaction.[50] In photochemical oxidation of different aliphatic and aromatic aldehydes,[49, 50] like thermal oxidation, the main products were corresponding peroxyacids and acids. Thermal and photostimulated reactivity of the aliphatic and aromatic aldehydes adsorbed from the gas phase on the TiO_2 (rutile) were investigated by Jenkins and Murphy by the EPR method.[283 in Ch. 3] On the dehydroxylated surface of TiO_2, at low temperature (100 K), the mixture of the co-adsorbed aldehyde–oxygen under UV irradiation forms surface paramagnetic species identified as peroxyacyl species ($RCO_3^\bullet$), which are thermally stable until 250 K. The values of g-factors of the detected radicals—$g_1 = 2.0017$, $g_2 = 2.008$, $g_3 = 2.003$—were characteristic for peroxy radicals. According to the proposed mechanism, oxidation occurs as a consequence of the following reactions

$$RCHO_{ads} + h^+ \rightarrow RCO_{ads} + H\left(h^+\right)$$

$$RCHO_{ads} + TiO^- \rightarrow RCO_{ads} + TiOH^-$$

$$RCO_{ads} + O_{2(ads)} \rightarrow RCO_{3\,ads}$$

$$RCO_{3\,ads} + RCHO \rightarrow RCO_3H_{ads} + RCO_{ads}$$

It was shown that $RCO_{3\,ads}$ were leading radicals in the photocatalytic oxidation of aldehydes and that the concurrence of the rates of their decomposition and chain propagation reactions determines the overall rate of reaction and distribution of products. The heterogeneous-radical decomposition of RCO_3H_{ads} in reaction conditions was an additional source of radicals, accelerating the overall process. The authors of this investigation specially mention that the obtained data clearly demonstrate the heterogeneous propagation of chains in this reaction. Here, we obviously have a major analogy with the thermal oxidation of aldehydes. In particular, it fully confirms the conclusion made in the investigation of adsorbed radicals in the oxidation of aldehyde on the SiO_2 surface (see Section 3.5.4).

Thus, in certain conditions, the oxidation of aldehydes with oxygen, both in thermal and photochemical pathways, has a heterogeneous constituent, generating intermediates that are paramagnetic centers propagating chains.

Another example demonstrating the surface radical reaction mechanism is the photooxidation of acetone with molecular oxygen on $TiO_2(110)$.[51] It is also a good example of the heterogeneous–homogeneous reaction in photocatalysis. The main reaction is the transformation of acetone to acetate, occurring in two stages: (i) formation of the acetone–oxygen complex by thermal reaction with low activation barrier (10 kJ/mol), and (ii) photocatalytic decomposition of this complex, giving acetate and "ejecting" methyl radicals to the gas phase. The transfer of radicals to the gas phase was evidenced by the isotopically labeled experiments. The main surface intermediate species, forming in the photocatalytic oxidation of acetone, were RO_2 radicals, detected by the EPR method on both polymorphs of TiO_2 and, at the same time, confirmed by the labeled isotopic atoms experiments. Note that according to Attwood,[52] the EPR spectra RO_2 and RCO_3 are strongly different from O^-, O_2^-, or O_3^- spectra. By the detailed analyses of spectra obtained, it also showed the differences between the adsorbed radicals RCO_3 and RO_2 detected in the photocatalytic oxidations of aldehydes and acetone. The differences of two EPR spectra are shown in Figure 4.5.[52]

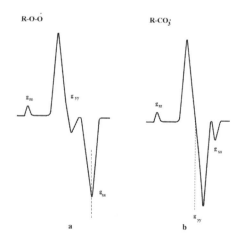

FIGURE 4.5 Schematic presentation of the EPR spectra on TiO_2: (a) obtained in co-adsorption of the acetone–oxygen mixture, analysis of the hyperfine tensor of RO_2 radical was based on g_{xx}; and (b) obtained in co-adsorption of an aldehyde–oxygen mixture, the analysis of hyperfine tensor of RCO_3 radical was based on g_{yy}. (Adapted from Ref.[52])

As shown above, the photochemical oxidation reactions in the presence of the solid catalyst, named heterogeneous photocatalytic reactions, in fact, are typically radical processes, occurring on the surface layer of catalysts. Like thermal oxidation reactions, the heterogeneous photocatalytic radical reaction may develop either as a heterogeneous–radical reaction or as a heterogeneous–homogeneous reaction. The heterogeneous–homogeneous occurrence of the photocatalytic oxidation will be given in the next section.

4.4.2 EXAMPLES OF HETEROGENEOUS PHOTOCATALYTIC REACTIONS OCCURRING BY THE HETEROGENEOUS–HOMOGENEOUS PATHWAY

For a long time, in the majority of investigations of the heterogeneous photooxidation processes of the organic compounds, it was considered that all reactions of the different types of surface species, generated under visible light or UV irradiation, occurred in the frameworks of the surface of photocatalysts and might be described by the Langmuir–Hinshelwood or Eley–Rideal mechanism.[53–57] In other words, these reactions, involving radicals or radical-like species, might have a heterogeneous but not a heterogeneous–homogeneous reaction mechanism. However, there is much experimental evidence about the heterogeneous–homogeneous nature of the photochemical oxidations of organic compounds on the surfaces, particularly on the photosensitive metal oxides and semiconductors. The following examples demonstrate the existence of bimodal reaction sequences in oxidation processes on the semiconductor photocatalysts (mainly, on TiO_2).

In general, the mechanism suggestions of heterogeneous photocatalytic oxidation are based on the predominant role of the photogenerated OH radicals.[58, 59, 83] Describing the behaviors of OH radicals on the surfaces of solid substances in thermal oxidation, such as metal oxides and salts, it was mentioned the existence of the different types of OH radicals and the limited surface mobility of part of them (see Section 3.4.2.3). Two different types of OH radicals, generated in the heterogeneous photocatalytic processes, were named free OH_f and surface-bonded OH_s hydroxyl radicals.[60 and ref. within] The generation of OH_s by the hole reaction with hydroxyl ion or water, happening in picoseconds, was presented in the previous Section 4.4.1.

The existence of two types of photogenerated OH radicals was suggested in rationalization of the results in a series of experiments[61] carried out in an aqueous suspension of anatase containing Fe(II) perchlorate (as a scavenger, potentially suppressing all chain reactions) and tertiary butanol or methanol as substrate. The quantum yields were 0.17 for butanol and 0.33 for methanol (in the presence of Fe(II) perchlorate). Previously, it was known that butanol reacted only with hydroxyl radicals and that methanol reacted not only with hydroxyl radicals but also with other radicals or oxidants. It appeared that the first half of the migrated to the surface holes produced hydroxyl radicals (strong oxidant) and that the second half gave another kind of oxidants (named "weaker oxidants"). According to these authors, on the anatase surface, about one-half of the hydroxyls was bonded to the single titanium-atom, while the other half was bonded by two titanium atoms. Correspondingly, the holes, migrating to the singly bonded hydroxyl, produced solution hydroxyl radicals, while the same holes that migrated to the doubly bonded hydroxyl produced weaker oxidizing radicals in solution. It was considered that here the different types of OH were nearly in equal quantities on the surface.[61 and ref. within]

Later, much experimental evidence was gathered indicating that a part of OH_f may desorb from the surface and diffuse to the volume of the reaction vessel.[60, 62–70] Note that these observations involve both liquid- and the gas-phase photocatalytic reactions.[60–62] The ratio of OH_s/OH_f may be changed, modifying the surface of TiO_2 by fluorides, phosphates, sulfates, etc.[62] Sun and Pignatello[63] observed a strong dependence of the reaction rate and composition of products on p^H in photochemical oxidation of 2,4-dichlorophenoacetic acid under UV irradiation on TiO_2. To explain the experimental data, they proposed a mechanism of the photooxidation, named dual-hole-radical mechanism, the peculiarity of which in photochemical oxidation was the following. It was assumed that there exist two competing reactions on the photocatalyst surface, the rates of which were dependent on the p^H of the reaction medium. One of them, hole oxidation of the organic compound by electron transfer reactions, occurs predominantly with carboxyl groups. The second is the reaction of the hole with hydroxyl groups of TiO_2, as the surface hydroxyl groups are well traps for holes and their reaction results in the generation of hydroxyl radicals. Finally, according to the authors, different types of OH radicals may be formed, as surface-attached OH_s, as well as surface-free OH_f radicals. The reactions of two types of OHs have different dependencies on the p^H of solution. Obviously, it is indirect evidence about the role of OH_f radicals in the bulk process, generated on the surface. Analogous conclusion was drawn by Minero et al.[64] in an investigation of the photochemical oxidation of phenol on TiO_2 and fluorinated-TiO_2 (F/TiO_2) in the presence of different alcohols (*tert*-butyl alcohol, 2-propanol, and furfuryl alcohol). They found that on a TiO_2, the "hole oxidation" is only 10% and that the main processes of oxidation occur via bonded OH radicals. Moreover, on a fluorinated-TiO_2 the reaction entirely occurs through the participation of the homogeneous hydroxyl radicals. Choi et al.[65] showed that the photogenerated OH radicals on a TiO_2 may be desorbed and diffused to the volume of the reaction mixture on the example of the photoinduced decomposition reaction of tetramethylammonium $(CH_3)_4N^+$ (originally used as $(CH_3)_4N^+OH^-$) in aqueous suspension of TiO_2 (P 25), using also FTIR-spectroscopy for the identification of surface intermediates. The rate of the kinetically first-order decomposition of $(CH_3)_4N^+OH^-$, as expected, was in strong dependence on the p^H of solution. The reactivity of the OH_f radicals, transferred to the liquid phase, was confirmed by their reaction with tert-butyl alcohol, occurring completely at $p^H = 3$ and partially at $p^H = 11$. Using the field emission scanning electron microscopy, it was investigated the decomposition of polyvinylchloride (PVC) on the composite material PVC/TiO_2 (PVC was embedded on TiO_2 in the form of a thin film) under UV irradiation, in air. It was found that the diameter of PVC cavities increased even in the absence of direct contact of PVC with TiO_2. The suggested explanation was that the active oxygen species, preferably OH_f, formed on a TiO_2, diffused across a few micrometers to the gas phase and reached to the surface of PVC. Recently, the photochemical oxidations of carbamazepine and ibuprofen,[66]

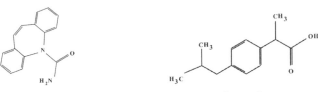

Carbamazepine   **Ibuprofen**

which are well-known pharmaceuticals, were investigated under visible light and UV irradiation on TiO_2 and ZnO photocatalysts, in the aqueous phase. Observing strong differences in the inhibiting effect of the rate of photooxidation in addition the effect of isopropanol (a scavenger of OH radicals) on the reaction mixtures containing carbamazepine and ibuprofen, it may be concluded that the photochemical reaction for two substances has different reaction mechanisms. The negligible inhibiting effect of the isopropanol addition to the reaction of ibuprofen was considered evidence of the predominant "hole mechanism," while the well-observed effect of its addition in the reaction of carbamazepine was seen as evidence of the "solution-phase mechanism." It is obvious that if the OH radicals are generated on the surface of semiconductors and they are main oxidant, the solution-phase reaction may take place only in the event of their transfer from the solid to the fluid phase.

Note also that OH-intermediated mechanisms are applicable both in the gas- and liquid-phase oxidation photocatalysis. Park and Choi[67] mentioned an interesting fact. The photogenerated radicals on TiO_2, further desorbed from the surface of TiO_2, may pass through air and reach an organic polymer membrane of 120 μm thickness, degrading dye substrates loaded within its pores.[67] Apparently, other intermediates, oxygen-active species as superoxide anion-radical or as hydroperoxide radical, play less important roles than hydroxyl radicals do. However, in some cases, when HO_2 radicals, or H_2O_2 are formed in the photooxidation,[68 and ref. within] the probability of the heterogeneous–homogeneous pathway of overall reaction may be enhanced, as the transfer of heterogeneously formed HO_2 radicals into fluid phases, including in photochemical conditions, is a well-established experimental fact. (See discussion of the decomposition of hydrogen peroxide on metal oxides and references in Chapter 2.)

Many other experimental facts, evidencing the heterogeneous–homogeneous mechanisms in photochemical oxidation reactions by the participation of radicals, were given in the recent valuable review of Choi and co-authors.[68] Predominantly, the OH-intermediated oxidation mechanisms were highlighted in these and many other works in spite of the existing opinion that, in some cases, the role of OHs is overestimated.[69] A powerful oxidant, OH radical, can abstract H-atoms from organic compounds, carry out additional reaction to the double bond of unsaturated compounds, and, in general, react with the surface-bonded organic compounds. As a result, the new radicals formed are mainly carbon-centered radical-like species, which in the presence of dioxygen easily form organic peroxy radicals. In dependence on the reaction conditions, they may continue the further oxidation to other oxygenated products on the surface or be transferred to the fluid phases. Like all mentioned examples of heterogeneous–homogeneous reactions in thermal oxidation, the possibility of bimodal mechanisms in photocatalytic reactions depends first of all on the nature of the photocatalyst and the state of its surface. For instance, the above-described experimental results, evidencing the reactivity of oxygen intermediates on the surface and in the fluid phases, in the case of their transfer into the fluid phases were obtained on the anatase TiO_2 samples. According to some opinions,[68] rutile and anatase polymorphs are not very different in their reductive conversion reactions, where OH radicals do not play a determining role, whereas in oxidative conversion, anatase exhibits higher activity than rutile. Other authors observe that in oxidation of the organic compound on rutile, the reaction zone is confined to the surface of a sample, whereas on anatase the reaction zone has relatively greater radius and extends far from the surface.[70] Obviously, these differences relate only to the surface nature of two polymorphs of TiO_2.

The role of OH radicals remains an important subject of many discussions in the chemical literature. It has repeatedly been mentioned that, according to some researchers, its role may be overestimated[69] in both thermal and photocatalytic oxidations. According to Nosaka et al., the radical OH is not the main oxidant, whereas "the surface trapped holes play the role of oxidant."[71]

Although the above-mentioned works did not use the term *heterogeneous–homogeneous* to characterize photocatalysis, the experimental observations were described exactly as chemical reactions occurring with the surface generation of radicals, part of which desorbs from the surface and then diffuses to the reaction mixture and enters into the different reactions with the components of the mixture. Thus, in these contexts, all are proven examples of heterogeneous–homogeneous processes, as photoinitiated reactions.

4.4.3 DETECTION OF RADICALS IN PHOTOCATALYTIC OXIDATION

Heterogeneous photocatalysis, presently, use a great number of modern and "classic" methods of investigation,[71] including methods of the surface science and neighboring areas.[72] Among them, the different applications of the EPR method play a basic role in identification of the major intermediates, produced in photocatalysis.[69] Some examples of the application of the EPR-measurements were given in previous Section 4.4.1. Detailed description of other methods for detection of the reactive oxygen species (ROS) can be found in special chemical literature.[74–76] One of the methods widely applied in investigations of photocatalysis is LIF.[74, 75] The application of the so-called single-molecule detection spectroscopic technique (the LIF experiments of Tachikawa and Majima) was covered in Section 2.3.2.1, as it relates to the heterogeneous-radical reactions of decomposition. Our aim here is to cite a few more experimental methods that reveal the heterogeneous–homogeneous nature of photocatalytic reactions as affected by the participation of radicals. Although the semiconductors compose a wide class of photocatalysts, mainly the reactions on bare TiO_2 samples are discussed here, as it is the most investigated photocatalyst.

One of the clearest confirmations of the heterogeneous–homogeneous character of photocatalytic oxidation of organic compounds on TiO_2 was obtained by Nosaka et al.[74] in experiments using the LIF method. The simplified scheme of the experimental

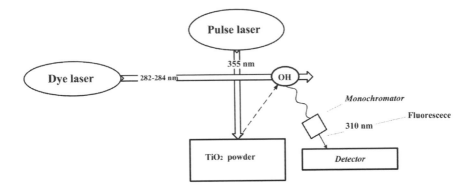

FIGURE 4.6 Scheme of Nosaka's experiment.[3] The detection of OH radicals desorbed from the surface of TiO$_2$ powder and diffused to the volume. (Altered and adapted from the work of Nosaka.[74])

setup for detecting OH radicals is represented in Figure 4.6. For samples of TiO$_2$ powder, 355 or 266 nm UV irradiation was used. Dye laser irradiation of 282–284 nm was used for excitation of OH, generated on the surface of TiO$_2$. The resulting fluorescence emission of 310 nm was detected and assigned to OH radicals. The heterogeneous–homogeneous occurrence of the photocatalytic reaction was confirmed in the "classical example" of the photoinduced splitting of water on TiO$_2$.[74] In parallel experiments, surface species were detected also by EPR (trapped holes and electrons, at low temperature) and proton NMR spectroscopies. Note that the LIF method stands out by its very high sensitivity in detecting OH radicals (10^6 molecule/cm^3).

Measurements of Nosaka by the application of the LIF method showed that OH radicals formed on the surface of TiO$_2$ diffuse about 5 mm away from the surface (in 100 µs time). Nosaka's experiment proved that the detected species is the OH radical and that it forms on the surface of TiO$_2$.

Using the same experimental technique, Zang and Nosaka[77] demonstrated the differences in the catalytic actions of anatase and rutile TiO$_2$ in photooxidation of organic compounds. For instance, in coumarin and coumarin-3-carboxylic acid aqueous solutions, OH radicals were detected in the reactions of both anatase and rutile. Apparently, the generation of OH radicals is a result of water splitting. Zang and Nosaka observed that the concentration of OH radicals was higher in the near-surface region and that it decreased in the bulk of the solution. On the other hand, the amount of radicals desorbed from the anatase surface and diffused to the bulk of solution was larger than that for the rutile surface. Adding H$_2$O$_2$ to these solutions, they observed that in the case of anatase, the concentration of OH radicals decreased but increased in the case of rutile. The differences in the generation of OH radicals were explained by the structural features of two TiO$_2$ polymorphs. According to the proposed mechanism, the trapped hole on the anatase surface produces OH radicals, while the formation of radicals on the rutile surface occurs through combination of two trapped holes, forming the Ti-peroxo (Ti-OO-Ti) structure. The catalytic events, according to Zang and Nosaka,[77] demonstrating the differences of the generation of radicals on the anatase and rutile TiO$_2$ surfaces, schematically are represented in Figure 4.7.

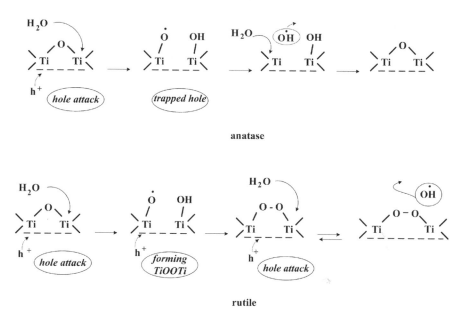

anatase

rutile

FIGURE 4.7   Scheme of the generation of OH radicals on the surfaces of anatase and rutile TiO_2 according to the presentation of Zang and Nosaka. (Ref.[77])

In another series of works, Nosaka et al.[74] investigated the mechanism of the generation of active intermediate species from adsorbed water on anatase and rutile TiO_2 surfaces, surveying the quantities not only of OH radicals, but also $^{\bullet}O_2^-$, as well as hydrogen peroxide. The mechanism suggestions were based on so-called ATR-IR spectroscopic data (ATR, or attenuated total reflection, is a sampling technique used in infrared spectroscopy). It was observed that the formation rate of $^{\bullet}O_2^-$ for rutile was higher than that for anatase. For OH radicals, as already mentioned, the amounts were in inverse relation for two polymorphs of TiO_2. Differences in the mechanisms of the generation of active intermediate species on two surfaces are represented in Figure 4.8.[78]

Desorption of radicals in the photocatalytic oxidation of organic compounds was observed experimentally on the surface of the monocrystalline $TiO_2(110)$, under UV irradiation.[77, 80] Photoinitiated desorption of methyl radicals from the surface of $TiO_2(110)$ in oxidation of the adsorbed ketones ($R(CO)CH_3$, where R is H, CH_3, C_2H_5, C_6H_5), under UV irradiation, was investigated using the pump-probe laser ionization technique.[79] In the presence of the surface-adsorbed oxygen species, as a result of the charge transfer, ketones form photoactive η^2-bonded diolate species:

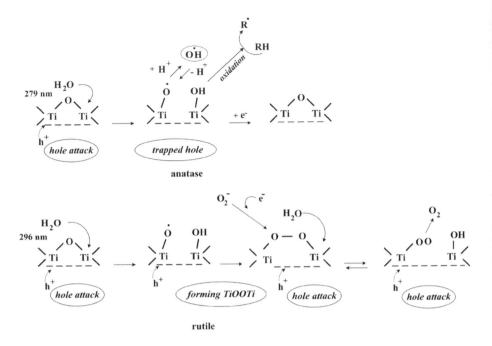

FIGURE 4.8 Scheme of the generation of OH and O_2^- radicals on the surfaces of anatase and rutile TiO_2 according to the presentation of Nosaka et al.[78]

It has been measured the translational energy distribution of methyl radicals in laser UV-induced cleavage of α-C-C(H_3) bonds. Apparently, in this case, the R(CO)O fragment rests on the surface. In analogous experiments of the photon-stimulated decomposition of acetone and butanone, in the presence of co-adsorbed oxygen, on a TiO_2, it was observed desorption of methyl and ethyl radicals, from surface diolate species, leaving acetate and propionate species on the surface.[80] In this work, it was observed also other interesting details. For example, in the case of butanone, only ethyl radical desorbs. Unlike methyl radical, in the case of acetone, ethyl radicals also react on the surface, at 100 K, in the presence of co-adsorbed oxygen, preferentially producing ethylene by dehydrogenation reaction.

Investigating the photocatalytic oxidation of ethanol and propanol on a $TiO_2(110)$ using the same technique it was shown the desorption of methyl radicals from the surface.[81] Emission of radicals from the surface was observed, followed by the dehydrogenation of acetaldehyde and acetone which formed on the surface as intermediates of alcohol oxidation.[81]

The formation of singlet oxygen from radicals is well-known (recombination of RO_2; $2H + 2O_2^-$; $OH + O_2^-$).[5 in Ch. 3] Although O_2 on the surface of TiO_2 produces singlet oxygen, under UV-irradiation (355 nm), detectable by near-infrared phosphorescence, at 1270 nm,[82] the small quantum yield (0.12) indicates that its contribution in overall oxidation is very small. However, it may not be neglected in the mechanism suggestions.[82]

The formation of radicals in photoassisted reactions was shown by the EPR method using the example of dye (ethyl ester of fluorescein (FLEt)), under 532 nm irradiation, on TiO_2, in both aerobic and nonaerobic oxidation of alcohols.[83, 84] The following reactions were suggested:

$$Dye + h\nu \ (visible) \rightarrow Dye^*$$

$$Dye^* + TiO_2 \rightarrow {}^{\bullet}Dye^+ + TiO_2\left({}^-\right)$$

${}^{\bullet}Dye^+$ was detected by the EPR method (g = 2.0047). In the presence of dioxygen, the formation of peroxy radical from the dye was observed. Apparently, further oxidation of alcohols also occurs by the radical mechanism.

Heterogeneous–homogeneous reaction in photochemical condition (UV irradiation, 375 nm), in a gaseous mixture composed from H_2O_2 and O_2 on TiO_2 films was investigated by the so-called cavity ring down spectroscopy (CRDS, a highly sensitive method of the absorption spectroscopy) in the flow reactor. HO_2 radicals were detected by this method in the gas phase.[85] It was shown that the detected radicals were produced in the heterogeneous decomposition of H_2O_2 but not in the reduction reaction of O_2 on the TiO_2. Different types of TiO_2 were tested, and it was concluded that they exhibited various activities in the generation and further diffusion of HO_2 to the gas phase in decomposition reactions.[85]

These examples again clearly demonstrate that the majority of photooxidation reactions of organic compounds by participation of photosensitive semiconductor materials are heterogeneous–homogeneous processes. The existence both the heterogeneous and homogeneous constituents was established by the direct detection of reaction intermediates. However, this important circumstance is not always taken into consideration in the photocatalysis of new materials.

4.5 CONCLUDING REMARKS ABOUT THE HETEROGENEOUS– HOMOGENEOUS MECHANISMS IN OXIDATION ON THE HETEROGENEOUS PHOTOCATALYSTS

In the book of Albini and Fagnoni,[86] devoted to the photochemically generated intermediates, light, metaphorically named the "ideal green reagent," makes it possible to transform the chemical compounds but "leaves no residue." Moreover, the possibility of the selective oxidation of organic compounds with an oxidant, such as molecular oxygen, in "mild" conditions, makes heterogeneous photocatalysis environmentally friendly and an easily available chemical process.

Different combinations of light and catalytic action may take place in heterogeneous photocatalysis. Depending on the conditions and chemical nature of photocatalysis, the chemical processes on the solid surface may involve photoinitiated catalytic conversion of the adsorbed compound; noncatalytic heterogeneous reaction, under permanent irradiation; and interaction of the homogeneous and heterogeneous reactions by the photogenerated intermediates.

In the interest of wide applications in both water and air treatment, investigations of heterogeneous photocatalysis were for a long time focused on oxidative degradation or decomposition reactions of the organic pollutants.[87, 88] Heterogeneous photocatalysis is also a very attractive domain for selective oxidation to target products. Another major domain of investigations in photocatalysis relates to fundamental understanding of atmospheric heterogeneous processes, playing a key role in depletion of the ozone layer and other phenomena in atmospheric chemistry.

The experimental results represented in this chapter demonstrate some important peculiarities of heterogeneous photocatalytic oxidation of organic and inorganic compounds, in comparison with thermal oxidation reactions. Particularly, on the photosensitive semiconductor materials, the action of light creates electrons and holes, trapping of which by surface species results in the formation of free radicals on the surface. The secondary reactions of the radicals may be of two types: (i) heterogeneous, when the adsorbed active intermediate species, mainly radicals or radical-like species, react with the initial organic compounds, and (ii) homogeneous, when these species continue to react in the fluid phases, followed by their "escape" from the surface of photocatalysts. In turn, the heterogeneous reaction may be a radical-chain or non-chain process. A proven example of the heterogeneous radical-chain reaction is the oxidation of aliphatic and aromatic aldehydes on TiO_2 powder, as evidenced experimentally by direct, *in situ* EPR detection of the peroxyacyl radicals, propagating surface chains. Another example of the photoinduced heterogeneous radical-chain reaction is hydrosilylation of unsaturated hydrocarbons (polymerization, see Section 2.4.2). An example of the heterogeneous non-chain-radical process is photo-Kolbe reaction, presented in Section 2.4.1.

Thus, like thermal oxidation with molecular oxygen, heterogeneous photocatalytic reactions have complex multistage and often multiphase mechanisms. However, the mechanism of the photocatalytic oxidation is very different from that of thermal oxidation. In the majority of heterogeneous photocatalytic oxidation reactions, in an initial stage, the energy of visible light or UV irradiation may be used for generation of the active intermediate species via electron transfer or interactions with holes. Although the nature of the photoinitiated intermediates may vary, the excited species, then, adsorbed radicals, including ion–radicals and radical-like species, are main intermediate species in heterogeneous photocatalytic reactions. They include photogenerated oxygen-active species, as well as organic radicals and peroxides formed in secondary reactions. Part of them may be detected by different *in situ* experimental methods (EPR, LIF, IR, Raman spectroscopies). The existence of others is postulated in different schemes of the mechanisms proposed for the photocatalytic reactions. Evidently, part of light energy dissipates in the form of heat in the solid substances and in the surrounding environment. In general, the light energy becomes an initiator of radicals, mainly via secondary reactions. Further transformation of intermediates may occur nearly by a "scenario" similar to thermal oxidation reactions. Photocatalytic oxidation, which occurs directly by singlet oxygen or other excited-state intermediates, transferred to fluid phases, may also be attributed to this class of reactions.

Here, we must make note of the enhanced probability of the photostimulated desorption of surface intermediates, radicals, or others to the fluid phases, where their further reactions are inevitable. That is why, seemingly, heterogeneous–homogeneous photocatalytic processes are more widespread than thermal bimodal reactions in comparable conditions. These very brief notes on heterogeneous photocatalytic reactions and presented examples are further evidence that the opinion of two to three decades ago about a non-wide spreading of heterogeneous–homogeneous mechanisms in the oxidation of organic compounds is not validated.

REFERENCES

4.1

1. Arora A. Text Book of Inorganic Chemistry. Discovery Publishing Pvt. Ltd., 2005, New Delhi, 802 pages, p. 740.
2. Albini A. Photochemistry: Past, Present and Future. 2016, Springer, Heidelberg, 302 pages.
3. Albini A., Dichiarante V. The "belle époque" of photochemistry. Photochem. Photobiol. Sci., 2009, v. 8, p. 248–254. DOI: 10.1039/B806756B
4. Ciamician G. The photochemistry of the future. Science, 1912, v. 36, p. 385–394.
5. Schroll P. Early pioneers of organic photochemistry. In: König B. (ed.). Chemical Photocatalysis. 2013, Walter and Gruyter, Berlin, Boston, p. 7–11.
6. Kaji M., Kragh H., Pallo G. Early Responses to the Periodic System. Oxford UP, 2015, USA, 344 pages, p. 271.
7. Nebbia G., Kauffman G. B. Prophet of solar energy: A retrospective view of Giacomo Luigi Ciamician (1857–1922), the founder of green chemistry, on the 150th Anniversary of his birth. Chem. Educator, 2007, v. 12, p. 362–369.

4.2

8. Wardle B. Principles and Applications of Photochemistry, John Wiley & Sons, UK, 2009, 267 pages.
9. Parmon V., Emeline A. V., Serpone N. Glossary of terms in photocatalysis and radiocatalysis (A Preliminary Version of IUPAC's Project 2001-036-1), International Journal of Photochemistry, v. 4, 2002, p. 91–131.
10. Hennig H., Billing R., Knoll H. Photocatalysis: Definitions and classifications, p. 51–69. In: Kalyanasundaram K., Grätzel M. (eds). Photosensitization and Photocatalysis Using Inorganic and Organometallic Compounds. Catalysis by Metal Complexes, v. 14, Springer Netherlands, 1993.
11. Tappeiner H. *Ueber die Wirkung fluorescierenden Stoffe auf Infusioren nach Versuchen von O. Raab. Munch. Med. Wochenschr.* 1900, v. 47, p. 5–7.
12. Raab O. *Ueber die Wirkung fluorescierenden Stoffe auf Infusorien. Zeitschrift fur Biologie,* 1900, v. 39, p. 524–526.
13. Yogo T., Urano Y., Ishitsuka Y., Maniwa F., Nagano T. Highly efficient and photostable photosensitizer based on BODIPY chromophore. J. Am. Chem. Soc., 2005, v. 127, v. 35, p. 12162–12163. DOI: 10.1021/ja0528533.
14. Blossey E. C., Neckers D. C., Thayer A. L., Schaap A. P. Polymer-based sensitizers for photooxidations. Journal of the American Chemical Society 1, 1973, v. 95, 17, p. 5820–5822.
15. Roberts J. D., Caserio M. C. Basic principles of organic chemistry: Study guide. Chapter 28, Photochemistry. p. 1371, W. A. Benjamin Inc., London, 1977, 1596 page.
16. Albini A., Fagnoni M. Photochemically-Generated Intermediates in Synthesis, John Wiley and Sons, 2013, 380 pages.

4.2.1

17. Chanon M. (ed). Homogeneous Photocatalysis (Wiley Series in Photoscience and Photoengineering), 1997, John Wiley, New York, 426 pages.
18. König B. (ed). Chemical Photocatalysis. 2013, Walter and Gruyter, Berlin, Boston.
19. Bauer R., Fallmann H. The photo-Fenton oxidation - a cheap and efficient wastewater treatment method. Research on Chemical Intermediates. 1997, v. 23, 4, p. 341–354.
20. Cornils B., Herrmann W. A., Beller M., Paciello R. (eds). Applied Homogeneous Catalysis with Organometallic Compounds: A Comprehensive Handbook in Four Volumes, 2018, Wiley-VCH, Weinheim (Germany), 1737 pages.

4.2.2

21. Colmenares J. C., Xu Y-J (eds). Heterogeneous Photocatalysis: From Fundamentals to Green Applications. 2016, Springer-Verlag, Berlin, 415 pages.
22. Coronado J. M. A historical introduction to photocatalysis. p. 1–4. In: Coronado J., Fresno F., Hernández-Alonso M. D., Portela R. (eds). Design of Advanced Photocatalytic Materials for Energy and Environmental Applications. 2013. Springer-Verlage, London. Green Energy and Technology.
23. Gaya I. U. Heterogeneous Photocatalysis Using Inorganic Semiconductor Solids 2013, Springer, Netherlands, 212 pages.
24. Ibhadon A. O., Fitzpatrick P. Heterogeneous photocatalysis: Recent advances and applications (Review). Catalysts, 2013, v. 3, p. 189–218; doi:10.3390/catal3010189. www.mdpi.com/journal/catalysts
25. Suib S. L. (ed.). New and Future Developments in Catalysis: Solar Photocatalysis, 2013, Elsevier, 492 pages, p. 2.
26. Fujishima A., Honda K. Electrochemical photolysis of water at a semiconductor electrode. Nature. 1972, v. 238 (5358), p. 37–38.
27. Vol'kenshtein F. F. Experiment and the electronic theory of catalysis. Russ. Chem. Rev., 1966, v. 35, 7, p. 537–546. DOI: 10.1070/RC1966v035n07ABEH001494
28. Strasser P. Fuel cells, Chapter 3. In: Schlögl R. (ed). Chemical Energy Storage. 2013, Walter de Gruyter, Berlin, Boston.
29. Ohji T., Singh M. Engineered Ceramics: Current Status and Future Prospects. Wiley, 2016, 536 pages, p. 439.
30. Dette C., Pérez-Osorio M. A., Kley C. S., Punke P., Patrick C. E., Jacobson P., Giustino F., Jung S. J., Kern K. TiO_2 anatase with a bandgap in the visible region. Nano Lett., 2014, v. 14, 11, p. 6533–6538.

4.3

31. Weiss J. *Über das Auftreten eines metastabilen, aktiven Sauerstoffmoleküls bei sensibilisierten Photo-Oxydationen, Naturwissenschaften.* 1935, v. 23, 35, p. 610.
32. Franck J., Livingston R. Remarks on the fluorescence, phosphorescence and photochemistry of dyestuffs. J. Chem. Phys., 1941, v. 9, p. 184–190.
33. Howe R. F., Gratzel M. l. EPR observation of trapped electrons in colloidal titanium dioxide. J. Phys. Chem., 1985, v. 89, 21, p. 4495–4499.
34. Wang Z., Ma W., Chen C., Ji H., Zhao J. Probing paramagnetic species in titania-based heterogeneous photocatalysis by electron spin resonance (ESR) spectroscopy (A mini review). Chemical Engineering Journal, 2011, v. 170, p. 353–362.
35. Carter E., Carley A. F., Murphy D. M. Evidence for O_2^- radical stabilization at surface oxygen vacancies on polycrystalline TiO_2. J. Phys. Chem. C, 2007, v. 111, p. 10630–10638.

36. Murata C., Hattori T., Yoshida H. Electrophilic property of O_3^- photoformed on isolated Ti species in silica promoting alkene epoxidation. J. Catal., 2005, v. 231, p. 292–299.
37. Macdonald R., Rhydderch S., Holt E., Grant N., Storey J. M. D., Howe R. F. EPR studies of electron and hole trapping in titania photocatalysts. Catalysis Today, 2012, v. 182, p. 39–45.
38. Yoshida H., Yuzawa H., Aoki M., Otake K., Itoh H., Hattori T. Photocatalytic hydroxylation of aromatic ring by using water as an oxidant. Chem. Commun., 2008, p. 4634–4636. DOI: 10.1039/B811555A
39. Yoshida H. Photocatalytic organic syntheses. Chapter 27, p. 647–669. In: Anpo M., Kamat P. V. (eds). Environmentally Benign Photocatalysts: Applications of Titanium Oxide-based Materials Springer, 2010, p. 650 and 657.
40. Stephenson C. R. J., Curran D. P. The renaissance of organic radical chemistry – deja vu all over again. Beilstein J. Org. Chem., 2013, v. 9, p. 2778–2780.

4.4

4.4.1

41. Augugliaro V., Bellardita M., Loddo V., Palmisano G., Palmisano L., Yurdakal S. Overview on oxidation mechanisms of organic compounds by TiO_2 in heterogeneous photocatalysis (review). Journal of Photochemistry and Photobiology C: Photochemistry Reviews, 2012, v. 13, 3, p. 224–245.
42. Blair E. W., Wheeler T. S. The oxidation of hydrocarbons, with special reference to the production of formaldehyde. J. Soc. Chem. Ind., 1922, v. 41, p. 303–310T.
43. Wada K., Yoshida K., Watanabe Y., Suzuki T. J. The selective photooxidation of methane and ethane with oxygen over zinc oxide and molybdena-loaded zinc oxide catalysts. Chem. Soc., Chem. Commun., 1991, p. 726–727.
44. De Vekki A. V., Marakaev S. T. Catalytic partial oxidation of methane to formaldehyde. Russian Journal of Applied Chemistry, 2009, v. 82, 4, p. 521–536.
45. Kolmakov K. A., Pak V. N. Photocatalytic oxidation of methane to formaldehyde on tungsten oxide surface. Zhurnal Prikladnoy Khimii, 2000, v. 73, 1, p. 159–161.
46. Djeghri N., Formenti M., Juillet F., Teichner S. J. Photointeraction on the surface of titanium dioxide between oxygen and alkanes. Faraday Discussions of the Chemical Society, 1974, v. 58, p. 185–193.
47. Gonzalez-Elipe A. R., Che M. ESR study of the radicals involved in the photo-oxidation of ethylene on TiO_2. J. Chem. Phys., 1982, v. 79, p. 355.
48. Murthy D. M., Giamello E. EPR of paramagnetic centers of solid surfaces. Chapter 7, p. 183, 190. In: Gilbert B. C., Davies M. J., Murphy D. M. (eds). Electronic Paramagnetic Resonance, v. 18, 2002, Royal Society of Chemistry, Cambridge.
49. Murthy D. M., Giamello E. EPR of paramagnetic centers of solid surfaces. Chapter 7, p. 183, 190. In: Gilbert B. C., Davies M. J., Murphy D. M. (eds). Electronic Paramagnetic Resonance, v. 18, 2002, Royal Society of Chemistry, Cambridge.
50. McNesby J. R., Heller C. A. Jr. Oxidation of liquid aldehydes by molecular oxygen. Chem. Rev., 1954, v. 54, 2, p. 325–346. DOI: 10.1021/cr60168a004
51. Henderson M. A. Photooxidation of acetone on $TiO_2(110)$: conversion to acetate via methyl radical ejection. J. Phys. Chem. B, 2005, v. 109, 24, p. 12062–12070.
52. Attwood A. L. Characterization and Reactivity of Thermal and Photo-Generated Surface Defects and Radical Centers over TiO_2; An EPR Investigation. Ph-Doctoral Thesis. Cardiff University, 2004, p. 214.

4.4.2

53. Gaurina-Medjimurec N. (ed). Handbook of Research on Advancements in Environmental Engineering, 2015.
54. Chatterjee D., Dasgupta S. Visible light induced photocatalytic degradation of organic pollutants. J. Photochem. Photobiol. C., 2005, v. 6, p. 186–205.
55. Augugliaro V., Loddo V., Palmisano G., Palmisano L. Clean by Light Irradiation: Practical Applications of Supported TiO_2. RSC Publishing, Cambridge, 2010, p. 9.
56. Khezrianjoo S., Revanasiddappa H. D. Langmuir-Hinshelwood kinetic expression for the photocatalysis. Degradation of metanil yellow aqueous solution by ZnO catalyst. Chem. Sci. Journal, 2012, v. 3, p. 1–9.
57. Matsuoka M., Saito M., Anpo M. Photoluminescence spectroscopy. Chapter 4, p. 149–184. In: Che M., Vedrine J. C. (eds). Characterization of Solid Materials and Heterogeneous Catalysts: From Structure to Surface Reactivity, v. 1–2, 2012, Wiley-VCH, 1284 pages.
58. Nosaka Y., Nosaka A. Y. Kinetic processes in the presence of photogenerated charge carriers, Chapter 8, p. 163–184; p. 174. In: Schneider J., Bahnemann D., Ye J., Puma G. L., Dionysiou D. D. (eds). Photocatalysis: Fundamentals and Perspectives, RSC Publishing, 2016, 436 pages.
59. Dvoranová D., Barbieriková Z., Brezová V. Radical Intermediates in Photoinduced Reactions on TiO_2 (An EPR Spin Trapping Study). Molecules, 2014, v. 19, p. 17279–17304. Doi:10.3390/molecules191117279
60. Park H., Kim H-i., Moon G-h., Choi W. Photoinduced charge transfer processes in solar photocatalysis based on modified TiO_2. Energy Environ. Sci., 2016, v. 9, p. 411–433.
61. Oliver B. G., Carey J. H. Photodegradation of wastes and pollutants in aquatic environment (a chapter), p. 629–650. In: Pelizzetti E., Serpone N. (eds). Homogeneous and Heterogeneous Photocatalysis. v. 174, NATO ASI Series, 1986.
62. Murakami Y., Kenji E., Nosaka A. Y., Nosaka Y. Direct detection of OH radicals diffused to the gas phase from the UV-irradiated photocatalytic TiO_2 surfaces by means of laser-induced fluorescence spectroscopy. J. Phys. Chem. B, 2006, v. 110, 34, p. 16808–16811. DOI: 10.1021/jp063293c
63. Sun Y., Pignatello J. J. Evidence for a surface dual hole-radical mechanism in the titanium dioxide photocatalytic oxidation of 2,4-D. Environmental Science and Technology, 1995, v. 29, p. 2065–2072. http://dx.doi.org/10.1021/es00008a028
64. Minero C., Mariella G., Maurino V., Vione D., Pelizzetti E. Photocatalytic transformation of organic compounds in the presence of inorganic ions. 2. Competitive reactions of phenol and alcohols on a titanium dioxide–fluoride system. Langmuir, 2000, v. 16, 23, p. 8964–8972. DOI: 10.1021/la0005863
65. Choi W., Kim S., Cho S., Yoo H-I., Kim M-H. Photocatalytic reactivity and diffusing OH radicals in the reaction medium containing TiO_2 particles. Korean Journal of Chemical Engineering. 2001, v. 18, 6, p. 898–902.
66. Georgaki I., Vasilaki E., Katsarakis, N. Study on the degradation of Carbamazepine and Ibuprofen by TiO_2 and ZnO photocatalysis upon UV/visible-light irradiation. American Journal of Analytical Chemistry, 2014, v. 5, p. 518–534. http://dx.doi.org/10.4236/ajac.2014.58060
67. Park J. S., Choi W. Remote photocatalytic oxidation mediated by active oxygen species penetrating and diffusing through polymer membrane over surface fluorinated TiO_2. Chemistry Letters, 2005, v. 34, 12, p. 1630–1631.
68. Park H., Kim H., Moon G., Choi W. Photoinduced charge transfer processes in solar photocatalysis based on modified TiO_2. Energy Environ. Sci., 2016, v. 9, p. 411–433. DOI: 10.1039/C5EE02575C

69. Wang Z., Maa W., Chena C., Ji H., Zhaoa J. Chemical engineering probing paramagnetic species in titania-based heterogeneous photocatalysis by electron spin resonance (ESR) spectroscopy. A mini review. Chemical Engineering Journal, 2011, v. 170, p. 353–362.
70. Kim W., Tachikawa T., Moon G., Majima T., Choi W. Molecular-level understanding of the photocatalytic activity difference between anatase and rutile nanoparticles. Angew. Chem., Int. Ed., 2014, v. 53, p. 14036–14041.
71. Nosaka Y., Nishikawa M., Nosaka A. Y. Spectroscopic Investigation of the Mechanism of Photocatalysis Review. Molecules, 2014, v. 19, p. 18248–18267; doi:10.3390/molecules191118248.

4.4.3

72. Albini A., Fagnoni M. (eds). Handbook of Synthetic Photochemistry. Wiley-VCH, 2010, 484 pages.
73. Henderson M. A. A surface science perspective on TiO_2 photocatalysis. Surface Science Reports, 2011, v. 66, 6–7, 185–297.
74. Nosaka Y. Surface Chemistry of TiO_2 Photocatalysis and LIF Detection of OH Radicals (Chapter 8). p. 205. In: M. Anpo, P. V. Kamat (eds). Environmentally Benign Photocatalysts: Nanostructure Science and Technology, Springer, Science +Business Media, 2010.
75. Murakami Y., Kenji E., Nosaka A. Y., Nosaka Y. Direct detection of OH radicals diffused to the gas phase from the UV-irradiated photocatalytic TiO_2 surfaces by means of laser-induced fluorescence spectroscopy. J. Phys. Chem. B, 2006, v. 110, 34, p. 16808–16811. DOI: 10.1021/jp063293c
76. Zielonka J., Lambeth J. D., Kalyanaraman B. On the use of L-012, a luminol-based chemiluminescent probe, for detecting superoxide and identifying inhibitors of NADPH oxidase: a reevaluation. Free Radic. Biol. Med., 2013, v. 65, p. 1310–1314. Doi: 10.1016/j.freeradbiomed.2013.09.017. Epub 2013 Sep 27.
77. Zhang J., Nosaka Y. Mechanism of the OH radical generation in photocatalysis with TiO_2 of different crystalline types. J. Phys. Chem. C, 2014, v. 118, 20, p. 10824–10832. DOI: 10.1021/jp501214m
78. Kakuma Y., Nosaka A. Y., Nosaka Y. Difference in TiO_2 photocatalytic mechanism between rutile and anatase studied by the detection of active oxygen and surface species in water. Phys. Chem. Chem. Phys., 2015, v. 17, p. 18691–18698. DOI:10.1039/C5CP02004B
79. Wilson D. P., Sporleder D., White M. G. Final state distributions of methyl radical desorption from ketone photooxidation on $TiO_2(110)$. Phys. Chem. Chem. Phys., 2012, v. 14, p. 13630–13637. DOI: 10.1039/C2CP42628E
80. Henderson M. A. Ethyl radical ejection during photodecomposition of butanone on $TiO_2(110)$. Surface Science, 2008, v. 602, 20, p. 3188–3193. http://dx.doi.org/10.1016/j.susc.2007.06.079by
81. Kershis M. D., White M. G. Photooxidation of ethanol and 2-propanol on $TiO_2(110)$: evidence for methyl radical ejection. Phys. Chem. Chem. Phys., 2013, v. 15, 41, p. 17976–17982.
82. Daimon T., Nosaka Y. Formation and Behavior of Singlet Molecular oxygen in TiO_2 photocatalysis studied by detection of near-infrared phosphorescence. J. Phys. Chem. C, 2007, v. 111, 11, p. 4420–4424.
83. Zhao J. C., Chen C. C., Ma W. H. Photocatalytic degradation of organic pollutants under visible light irradiation. Top. Catal., 2005, v. 35, p. 269–278.
84. Zhang M., Chen C. C., Ma W. H., Zhao J. C. Visible-light-induced aerobic oxidation of alcohols in a coupled photocatalytic system of dye sensitized TiO_2 and TEMPO. Angew. Chem. Int. Ed., 2008, v. 47, p. 9730–9733.

85. Yi J., Bahrini C., Schoemaecker C., Fittschen C., Choi W. Photocatalytic decomposition of H_2O_2 on different TiO_2 surfaces along with the concurrent generation of HO_2 radicals monitored using cavity ring down spectroscopy. J. Phys. Chem. C, 2012, v. 116, 18, p. 10090–10097. DOI: 10.1021/jp301405e

4.5

86. Albini A., Fagnoni M. Photochemically-Generated Intermediates in Synthesis. John Wiley and Sons, 2013, 380 pages.
87. Gaurina-Medjimurec N. (ed). Handbook of Research on Advancements in Environmental Engineering, 2015, IGI Global Hershey (USA) 605 pages, p. 60.
88. Chatterjee D., Dasgupta S. Visible light induced photocatalytic degradation of organic pollutants. J. Photochem. Photobiol. C. 2005, v. 6, p. 186–205.

5 Heterogeneous–Homogeneous and Homogeneous–Heterogeneous Processes in Atmospheric Chemistry

Before the 1970s, a number of heterogeneous and heterogeneous–homogeneous reactions through the participation of some free radicals, playing a key role in atmospheric processes, were mainly investigated in the context of the problems of physics and chemistry of areas other than atmospheric chemistry. At about that time, the important role of atmospheric aerosols, dusts, salts, liquid clouds and reactions on their surfaces in atmospheric processes was revealed. The systematic investigations in this field showed that the ozone cycle in the atmosphere, as well as other atmospheric processes, is closely related to problems involving the heterogeneous reactions of pollutants on condensed-phase particles. The heterogeneous–homogeneous and homogeneous–heterogeneous processes have specific expression in atmospheric processes. As is true of every scientific discipline, atmospheric science also, in general, uses its own terminology; this domain of investigations is known as multiphase atmospheric chemistry. It includes heterogeneous reactions of gas-phase atmospheric components, photogenerated radicals, and electronically excited species on/in solid- and liquid-phase particles (aerosols and suspended liquid droplets).

This chapter is not intended to be a review of multiphase and heterogeneous reactions in atmospheric processes. Only some selected examples of these reactions will be represented to demonstrate the existence and peculiarities of the homogeneous–heterogeneous or heterogeneous–homogeneous reaction sequences in natural processes that occur in the atmosphere.

5.1 PECULIARITIES OF CHEMICAL REACTIONS IN THE ATMOSPHERE

Reaction conditions in the atmosphere are completely different from those for chemical reactions in the laboratory and industry or in natural conditions on Earth's surface.[1-5] The specific dependence of the temperature and pressure of the atmosphere on the altitude from Earth's surface determines the main parameters of the occurrence of the atmospheric reactions, including the heterogeneous or heterogeneous–homogeneous reactions of atmospheric species. Light and cosmic irradiation have their contribution in chemical transformations in the atmospheric processes. As the atmosphere is a dynamic medium, other main factors affecting chemical transformations are permanent input and output of energy and mass. As will be shown, photochemical processes have their own peculiarities in atmospheric conditions. Atmospheric processes occur in "wall-less" and thermodynamically open reaction systems. Although the reactor walls do not exist in the atmosphere, an effect resembling "wall effects" in thermal oxidation reactions in reactors does exist. The solid or liquid particles in the atmosphere, such as aerosols, soot, mineral dusts, organic compounds, and inorganic salts (e.g., NaCl, NH_4HSO_4, $(NH_4)_2SO_4$), clouds, liquid droplets, and others, in interaction with the components of the atmosphere, play a role close to the "effects of reactor walls."[3] The role of "walls effects" is exhibited by the specific reactions of solid or liquid particles with gas-phase radicals and other components of the atmosphere. The lifetime and other characteristic parameters of radicals, ions, electronically excited molecules, and other transient active species are changed in atmospheric conditions. All the above-mentioned factors indicate that atmospheric reactions have specific behaviors.

The heterogeneous chemistry of atmosphere is one of the subdisciplines of the atmospheric science, investigating the interactions of atmospheric gas components with solid and liquid particles. There is a great number of fundamental works devoted to the problems of atmospheric chemistry, especially the problems of heterogeneous reactions in the atmosphere. We would not like, here, to provide a long list of literature references on atmospheric heterogeneous reactions. The majority of the references of the books and reviews on this subject, the reader may find within the given bibliography. Here, the reactions on the surface of atmospheric particles, such as aerosols and soot, will be discussed only from the point of view of the processes exhibiting the peculiarities of bimodal reaction sequences.

5.2 HETEROGENEOUS REACTIONS OF FREE RADICALS ON THE CONDENSED PHASE PARTICLES IN ATMOSPHERIC PROCESSES. UPTAKE COEFFICIENTS OF ACTIVE INTERMEDIATES

Heterogeneous reactions on the surface of atmospheric particles with the participation of free radicals, electronically excited molecules, and other active intermediates have an important role in many atmospheric processes in the troposphere and low stratosphere, which finally determine the quality of air, state of the environment, and climate on Earth's surface.[6, 7] The term *aerosol*, signifying colloidal solution in the air, was proposed as a term analogous to *hydrosol* (a colloidal solution in water).[8] Aerosol particles in the atmosphere have diameters between about 0.002 μm and 100 μm. Although the average concentration of atmospheric condensed-phase particles is low (10^2–10^5 cm^{-3}),[3, p.423; 7] their role is significant. Heterogeneous reactions on the aerosols have specific peculiarities related not only to the chemical nature of the condensed phase particles, but also to specific reaction conditions on their surfaces.

Aerosols may be divided into different main groups: solid and liquid (i), on the one hand, and primary and secondary (ii), on the other hand. Natural dusts, volcanic eruptions, sea salts, water droplets, smog, and particles of anthropogenic origin, released from Earth's surface, constitute the separate groups of primary aerosols. Primary aerosols often become centers for nucleation and condensation of different pollutants, involving organic compounds and forming secondary aerosols. There exist an enormous number of individual compounds (about 10^5),[9] in measurable quantities, composing secondary organic and inorganic aerosols. In atmospheric conditions, aerosol particulates permanently change their composition, sizes, and properties under the influence of different dynamic factors (light, irradiation, biogenic and anthropogenic emissions from the Earth's surface, temperature and pressure gradients, chemical interactions between the gaseous components of atmosphere, etc.).

The significant role of heterogeneous reactions in atmospheric processes, for the first time, was revealed in the 1970s, in relation to the first investigations of acid rains containing H_2SO_4 and HNO_3.[10–12] Dutch atmospheric chemist P. Crutzen's[13] discovery of the role of the reactions of NO_x, leading to the depletion of ozone, became an important stimulus for large investigations of the atmospheric catalytic cycles and the role of heterogeneous reactions on the aerosol particles in atmosphere. For his works on tropospheric ozone depletion, in 1995 he received Nobel Prize in Chemistry with M. Molina and F. S. Rowland.[13]

Nearly all studies on the heterogeneous reactions of atmospheric radicals directly or indirectly relate to problems involving the depletion of ozone.[11, 12] In the 1990s, Calvert revealed the crucial role of aerosol particles (polar stratospheric clouds, PSCs) in the Antarctic's ozone depletion at the end of winters and in extremely cold conditions.[10] The results of the investigations of the heterogeneous chemistry on the polar stratospheric clouds were successfully applied in modeling of ozone depletion problems at low stratosphere, which led to important conclusions in atmospheric chemistry.

In early investigations exemplifying the significant differences of the reactivity of aerosol particles toward atmospheric radicals, all the condensed-phase atmospheric particles were conditionally divided into three groups:[14] "natural" (e.g., NH_4NO_3, $(NH_4)_2SO_4$, NH_4Cl, Na_2SO_4, $NaNO_3$); "human made" (e.g., ash-fly and emissions of automobiles, FeO_x-$FeSO_4$, $Zn(NO_3)_2$, $Pb(NO_3)_2$, K_2CO_3); and organic aerosols (e.g., malonic acid, glycine, potassium propionate). Early laboratory studies, carried out by UV resonance fluorescence (to detect OH) and chemiluminescence (to detect H, O, HO_2), showed observable differences in the reactivity of radicals on different aerosols. The OH radicals were more reactive towards nearly all investigated aerosols than HO_2, H, and O. It was found that the rate of loss of OH radicals was higher, especially, toward aerosols of the second group than for other groups.[14, 15] On the other hand, it was also observed that the surface of model aerosols might be completely changed (as in the case of the reaction OH + FeO_x-$FeSO_4$ or O + NH_4Cl) or unchanged (as in the case of the recombination of radicals on the solid surfaces) in these interactions. It was also found that the so-called loss coefficient (γ) of OH in interaction with malonic acid (organic aerosol) particles is more than unity ($\gamma > 1$). Later, it was found that the atmospheric aerosols in the oxidation of organic pollutants might exhibit either catalytic or noncatalytic behaviors.[6, 14, 16] The majority of the condensed-phase particles does not exhibit catalytic behaviors in heterogeneous oxidation or decomposition reactions of organic and inorganic pollutants with the participation of atmospheric radicals. However, certain classes of condensed-phase particles, such as soot, dusts, some metal oxides, and transition metal salts, have catalytic properties toward a number of VOCs (volatile organic compounds, PAHs (polycyclic aromatic hydrocarbons), and inorganic compounds (ozone, azote oxides, and others), either in oxidation or decomposition reactions, under sunlight irradiation, or even, in the dark.[14–16] Under UV or sunlight irradiation, aerosol particles often play a role of specific heterogeneous photocatalysts in atmospheric conditions. The mechanism suggestions in this case are not much different from those discussed in Chapter 4 for photocatalytic oxidation of some organic compounds. A number of examples of heterogeneous photocatalytic reactions of atmospheric interest via radical intermediates are given in the review article of George et al.[15]

The radicals in the atmospheric medium are very varied by their composition (oxygen, azote, halogen, carbon, sulfur and other elements in different combinations) and potential role in the homogeneous and heterogeneous atmospheric processes. Among them, the class of oxygen centered radicals (some azote oxides, OH, HO_2, RO_2, ClO, BrO, etc.) have a pivotal role in atmospheric processes.

In the following section, as an example of bimodal (heterogeneous–homogeneous or homogeneous–heterogeneous) interactions with the atmospheric condensed-phase particles, among the different oxygen-centered radicals, the role of OH radicals will be briefly examined.

Atmospheric models are a combination of mass and energy transport processes with overall chemical reaction models. No chemical system may be fully described without knowledge of the experimental kinetic data and rate constants of all or at least the most important reactions. Unfortunately, in atmospheric science, the complete kinetic data of reactions in the reacting system are only rarely known. To describe a complex interaction in the gas-liquid (on droplets or clouds) or gas-solid (on aerosols, soot, or dust particles) systems, atmospheric chemists or physicists use parameters such as adsorption and reaction coefficients. Sticking coefficient, uptake coefficient, accommodation coefficient, loss coefficient, or condensation coefficient are close parameters used in describing the interactions of molecules of the fluid phases with solid or liquid particulates.[8 in Ch. 1; 17; 18] They are useful parameters, permitting to estimate the contribution of heterogeneous reactions to overall atmospheric processes. In general, these parameters show a fraction of molecules interacting with the surfaces. For instance, the uptake coefficient refers to a number of collisions, which lead to chemical transformation, divided by the total number of collisions. According to the IUPAC definition,[8 in Ch. 1] the sticking coefficient is "the ratio of the rate of adsorption to the rate at which the adsorptive strikes the total surface, i.e., covered and uncovered." Obviously, it is a function of the surface coverage, at the given temperature, and surface structure of the adsorbent. Analogously, the accommodation coefficient is a measure of the efficiency of the capture of molecules or atoms that collide with aerosol particles, cloud droplets, etc. It is a fraction of the collisions that lead to the capture of molecules (atoms, radicals, etc.) and is not reflected from the surface. The accommodation coefficient,[17] usually denoted as ε or γ, is a unitless parameter

$$\gamma = \frac{net\ loss\ of\ trace\ gas\ molecules\ removed\ from\ the\ gas\ phase}{total\ number\ of\ trace\ gas\ collisions\ with\ the\ surface}$$

Experimentally determined uptake, accommodation, or other analogous coefficients are well documented in atmospheric chemistry.[17–19] In general, the values of the uptake coefficients, according to the definitions, will be changed in the interval $0 < \gamma < 1$. For instance: the uptake coefficients of HO_2 and RO_2 radicals were obtained on aerosol particles NaCl, KCl, NH_4NO_3, $(NH_4)_2SO_4$.[18] The data show that for RO_2 on NaCl, the uptake coefficient is about $6 \cdot 10^{-4}$, which is five times lower than that for HO_2 ($\gamma_{HO2} = 3 \cdot 10^{-3}$).[18] However, there are a number of cases for which the experimental values of γ are more than unity ($\gamma > 1$) in interaction of the components of atmosphere with the solid- or liquid-phase aerosols. Some examples and suggested explanations of these observations in atmospheric chemistry are given in Section 5.3.

The above-mentioned coefficients are close in their physical meaning but are not equivalent. For example, the nonequivalence of the uptake coefficient and sticking or accommodation coefficient was shown in the interaction of OH with the alkane aerosols (for instance, on the example of the "oxidative "aging" of the aerosol").[19]

In heterogeneous and heterogeneous–homogeneous reactions proceeding via the formation of active intermediates, the coefficient of accommodation (ε) is an important parameter for determining the rate constants. An active intermediate or other reactant reaches the surface from the gas phase via the diffusion. In the case of the fast diffusion of active species, at the given temperature, the rate of heterogeneous reaction depends on the number of stocks to the surface. The following differential equation is known from the molecular kinetic theory of gases.[51 and 39 in Ch. 1]

$$V\,(dn/dt) = -\varepsilon\,S\,(n\,v_{av}/4)$$

where n is the number of chemical species, v_{av} is the average thermal velocity of them, and V and S are, respectively, volume and surface of the reaction vessel. In the integral form, this equation will be

$$n = n_0 exp\left[(-S/V)(\varepsilon V_{av}/4)\right]$$

where, n_0 is the initial concentration of species.

$$k = (S/V)(\varepsilon v_{av}/4)$$

k is the rate constant of the heterogeneous reaction on the surface,

$$\varepsilon = k, \quad if \quad (S/4V)/v_{av} = 1$$

It is known[5] that in the kinetic regime of reaction

$$1/k = 1/k_d + a/\varepsilon$$

where k_d is the diffusion constant of accommodation (a is another constant). If $k_d \gg k$, the reaction will occur in the kinetic region. Therefore, the value of ε in the heterogeneous reaction, also k, may be determined experimentally by different methods, some of which were described by Krilov and Shub.[51 in Ch. 1]

5.3 HOMOGENEOUS–HETEROGENEOUS REACTIONS OF OH RADICALS IN THE ATMOSPHERE

The role of the most powerful oxidant in atmospheric processes is attributed to the radicals OH.[20–22] Their role is important in the formation of smog, in depletion of the ozone layer, and in other atmospheric processes. There are also other oxidant agents in atmosphere, such as O_3, singlet oxygen, and NO_x and the reactions of OH radicals are often involved in the atmospheric catalytic cycles of these oxidants.

In atmosphere, the concentration of OH radicals is very low. Its average values are in the range of $1 \cdot 10^5$ to $2 \cdot 10^7$ molecules/cm^3 and are dependent on the altitude from the Earth's surface.[23] With increase in altitude, they decrease. Simultaneously, the decrease of the concentration of OHs relates to the decrease of the concentration of water in the atmosphere, which also decreases with the increase in altitude. This dependence becomes explicable when we consider that the primary generation of OH radicals occurs by the reaction of photoexcited oxygen-atom with water, followed by the photolysis of ozone under UV irradiation.

$$O_3 + h\nu(\lambda < 320 \text{ nm}) \rightarrow O(^1D) + O_2(^1\Delta)$$

$$O(^1D) + H_2O \rightarrow 2OH$$

The nighttime concentration of OHs of the primary formation, practically, may be zero. A fraction of $O(^1D)$, formed in photolysis of ozone, produces atomic oxygen in the ground state. Atmospheric N_2 or O_2 molecules play the role of a third particle in this process.

$$O(^1D) + N_2 \rightarrow O(^3P) + N_2$$

$$O(^1D) + O_2 \rightarrow O(^3P) + O_2$$

Secondary sources of OH radicals involve a number of chain-radical reactions by photolysis of HONO, H_2O_2, CH_3OOH, and other pollutants, as well as some atmospheric reactions of NO_x.[21] According to Lelieveld et al.,[20] the concentration of OH in the atmosphere, formed from secondary sources, is about two times larger than that in the primary formation.

Depending on sunlight irradiation and water concentration in the atmosphere, the concentration of OH changes in different geographic zones. Mainly, it is low at the atmosphere of the pole regions of the Earth. However, the global average concentration of OH is relatively stable in clean air.[20] In 2011, systematic observations and measurements of OH at the atmosphere, above the tropical Pacific Ocean (South West),[24] showed that it was low in this region in comparison with the global average concentration of OH (10^6 molecules/cm^3).[25] Surprisingly, measurements taken in 2014 showed the absence of OH in the mentioned region. These observations were considered to be discovery of hydroxyl radical holes in the atmosphere.[26] Because the OH radicals serve as the "detergent" of the atmosphere, their absence is related to atmospheric pollution and the formation of ozone holes.

The atmospheric reactions of OH radicals are conditionally divided into two groups: homogeneous and heterogeneous reactions. But as will be shown later, they are interrelated atmospheric processes and often occur as unique, bimodal processes.

Apparently, the main loss of OH radicals on a global scale takes place via the homogeneous reactions with CO (40%), organic substances (VOCs, 30%), CH_4 (15%), as well as ozone, hydrogen, azote oxides, and other trace gasses (15%) in the atmosphere.[21, 22] The reaction of OH with CO produces H-atoms and, consequently, with dioxygen, HO_2 radicals.[24]

$$CO + OH \rightarrow CO_2 + H$$

$$H + O_2 \rightarrow HO_2$$

The reaction of OH with organic substances, including methane, occurs chiefly by the abstraction of H-atoms and the formation of radicals (R) that readily transform into organic peroxy radicals RO_2. The peroxy radicals (HO_2 and RO_2) in turn participate in ozone depletion and many other atmospheric reactions.[25]

Other significant loss of OH is caused by the reaction with NO_2 giving HNO_3[22, 23, 25]

$$OH + NO_2 \rightarrow HNO_3$$

On the other hand, NO_2 may also be an important source of OH, via decomposition of HONO

$$NO_2 + H_2O \rightarrow HONO + HNO_3$$

In general, NO_x has an important role in local air pollution in urban areas.[25]

Thus, part of the OH radicals, generated as a result of atmospheric photochemical processes in homogeneous conditions, initiates radical reactions of oxidation with components of the atmosphere. Other parts of OH radicals react through the heterogeneous pathway. Although the concentration of OH in the troposphere is low, because of its very short lifetime (< 1s) and high reactivity in homogeneous reactions, and only a small part of OH radicals reaches the solid surfaces by diffusion,[27] the heterogeneous reactions of OH on the aerosol particles have a very important role in atmospheric processes.[2] The functionality of the OH radicals has been revealed in two types of heterogeneous atmospheric processes.[2, 22, 27] On the one hand, the interactions of OH with the aerosols initiate heterogeneous oxidation of organic particulates, and, on the other hand, they become responsible for the heterogeneous activation of halogens on the surfaces of the clouds in marine and polar troposphere, which is a major contributor to ozone depletion.

In the first case, the heterogeneous interactions of OH radicals with aerosols involve different types of oxidation reactions:

OH + adsorbed compound on the aerosol → oxidation products

as well as reactions of the so-called "aging" of aerosols:

OH + SOA (secondary organic aerosol) → oxidation products

Reactions of both types may occur under sunlight and cosmic irradiation or in the dark. The role of OH radicals in the heterogeneous generation of halogens, related to the problem of ozone depletion in the atmosphere, will be discussed in the next section.

Heterogeneous oxidation reactions of the atmospheric OH radicals with adsorbed compounds have been investigated on a number of aerosol particulates, including mineral dusts, soot, metal oxides, and salts.[27–29] Among them, the reactions on the mineral dust aerosols occupy an important place, as much as one-third to one-half of the aerosols emitted into the atmosphere are dust particles. According to data presented by Bian et al.,[28] the global mean decrease of OH concentration due to dust pollution is 11.1% (calculated as a ratio of the heterogeneous loss of OH radicals to their summary loss by homogeneous-photochemical and heterogeneous pathways). However, in the atmosphere over regions excessively polluted by dust emissions from the surface of the Earth, it can be much higher, as much as 60%.[28] The reaction of OH + HCHO on the surface of $(Si(OH)_4)_n$ may serve as a model reaction of the interaction of OH radicals with the adsorbed organic compound on the aerosol particulates, as it is the basic building block of silicates, clays, and phyllosilicates, simulating the dust particles. This system was studied by quantum computational methods,[30] which showed

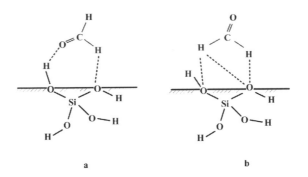

a   b

FIGURE 5.1 Optimized structures of formaldehyde adsorbed on the surface of Si(OH)$_4$ (a) through oxygen, interacting with hydroxyl of the surface, (b) through two hydrogen atoms of formaldehyde with the surface oxygen atoms. (Altered and adapted from Ref.[30])

that the adsorbed formaldehyde can react in the presence of gas-phase OH radicals, producing surface-bound formyl radicals and water.[30, 31]

$$HCHO + {}^{\bullet}OH \rightleftarrows HCO^{\bullet} + H_2O$$

The second pathway may be the OH addition reaction to formaldehyde.[30, 31]

$$HCHO + {}^{\bullet}OH \rightleftarrows {}^{\bullet}CH_2O(OH)$$

The results confirm the feasibility of the formation of [formaldehyde … OH] complex with a very small activation barrier, assumed in experimental works.[Ref. within 30, 31] Formaldehyde may be adsorbed on the surface of Si(OH)$_4$ in two forms (Figure 5.1): **(a)** through oxygen, interacting with hydrogen of the surface hydroxyl, and one hydrogen atom interacting with oxygen of the surface, and **(b)** through two hydrogen atoms of formaldehyde with the surface oxygen atoms. As the spectroscopic data of the adsorption of formaldehyde on the phyllosilicates indicate a predominant bonding by hydrogen bonds of **(a)** type, the H-abstraction by OH radicals was considered to be a main reaction channel (Figure 5.2).[29–31 and ref. within]

The computational results of the reaction rate constant were compared with the rate constant of the analogous gas-phase reaction of formaldehyde and OH. Surprisingly, it appears that the heterogeneous rate constant of the reaction is at least one order of magnitude lower than that in the gas phase. Therefore, the contribution

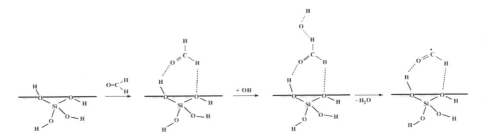

FIGURE 5.2 Formation of the adsorbed formyl radical on Si(OH)$_4$ by H-abstraction reaction of adsorbed formaldehyde with OH radical, according to the quantum-mechanical calculations and descriptions in the works.[30–31]

of this heterogeneous reaction to the loss of OH radicals on the dust will be lower, correspondingly, in comparison with the gas phase. In other words, in this case, the homogeneous reactions of OH will be predominant over the heterogeneous ones.[30, 31]

In analogy with adsorbed formaldehyde, the reactions of C_2-C_5 aldehydes with OH, on the mineral aerosols, occur mostly by the abstraction of H atom from the CHO group of adsorbed aldehyde.[32] The H-abstraction from aldehyde in interaction of NO_2 with aldehydes on aerosol particulates (SiO_2), forming nitrous acid and acyl radicals, also was considered "an exclusive pathway" of this heterogeneous reaction.[33]

Li et al.[29] investigated the heterogeneous reactions of some volatile organic compounds, such as acetaldehyde, propionaldehyde, and acetone, adsorbed on the oxide particles (SiO_2, Al_2O_3, Fe_2O_3, TiO_2, and CaO) of atmospheric interest. Applying FTIR and UV/visible spectroscopic measurements, they obtained values for the uptake coefficients (10^{-4} to 10^{-6} at 298 K) for these pollutants. Based on these data, they found that the heterogeneous loss of these compounds is comparable with that, due to photolysis and reactions with OH radicals in the gas phase, in the troposphere.

The reactions OH radicals, with a great number of adsorbed polycyclic aromatic hydrocarbons (PAHs) or their different derivatives on the dust or carbonaceous particles, were investigated.[34–36] All are evidence of the reactivity of adsorbed compounds with OH radicals, diffusing from the gas phase to the surface of aerosols.

These and many other examples show that homogeneously photoinitiated radicals OH, in turn, initiate heterogeneous oxidation of pollutants on the surfaces of certain types of aerosol particulates. However, as mentioned, the majority of the atmospheric aerosols either did not exhibit catalytic properties or did not work as classical heterogeneous catalysts. Often, the products of the heterogeneous reactions on the atmospheric particles rest on the surface rather than desorb to the gas phase. That is why in atmospheric chemistry this kind of reactions were named particle-to-phase reactions.[2]

The problem regarding the nature of the heterogeneous interaction of radicals with the aerosols remains unresolved. Apparently, the heterogeneous reaction sometimes occurs not only as a radical process, but also as a chain-radical process.[16 in Ch. 4]

Another interesting and more abundant aspect of research of the heterogeneous reactions of OH that is of atmospheric interest, is their interactions with the aerosols, which are in the form of tin films[37–38] or submicron organic particles,[2, 39] termed secondary chemical processes in atmospheric studies.[37–50] The experimental investigations in this area were carried out either in a stationary flow tub or in continuous-flow stirred tank reactors, described in a number of works.[2 and ref within] As an analytical technique, the mass-spectrometric method was applied.[2] Experimental data permit to determine or estimate the uptake coefficients of OH radicals with organic pollutants on/in the condensed-phase aerosols. The values of uptake coefficients may be taken as an indication about the nature of the heterogeneous process. If the value of the uptake coefficient is more than unity, it means that in a heterogeneous process, in each collision of OH radicals, more than one molecule of the condensed-phase molecules were involved. This may be interpreted as evidence about the "secondary chain chemistry" by the participation of the condensed-phase molecules.[2, 39, 47] Note that the exactitude of the measurements of the uptake coefficients and their assignation to the heterogeneous reactions, in this case, play a decisive role in coming to further conclusions. Unfortunately, many uncertainties in determination of the uptake coefficients may not be excluded; therefore, according to George et al.,[40] the data on uptake coefficients are not strong indicators of the "secondary chemistry" on aerosols. However, the direct interaction of OH radicals with the surface molecules on the condensed-phase particles, giving products or chemically modifying aerosol particles (named "aging"), is obvious.[2, 39, 40] Here, the detailed mechanism of the heterogeneous reaction is unclear. Is it a heterogeneous chain, or is it a heterogeneous non-chain-radical process?

The general scheme shown in Figure 5.3 for the interactions of OH radicals and the aerosols, including the radical reactions of different kind, was given by Russell[44] in early 1957 for liquid-phase processes. An analogous mechanism also was suggested for OH oxidation of condensed-phase hydrocarbons[45] (e.g., OH + dioctyl sebacate aerosol[45]). Obviously, the interaction of OH begins at the heterogeneous stage, producing radicals R by H-atom abstraction from the organic molecule, which, consequently, may form RO_2 and HO_2 radicals. In the further reactions, the radicals either form molecular products or participate in the reactions of chain propagation.

Let us briefly examine some examples of the "particle-phase secondary chain chemistry" obtained in certain works. The uptake coefficients for OH on dioctyl sebacate particles were found to be $\gamma = 2.0$[42] or 1.3.[40] Lambe et al.,[43] investigating the

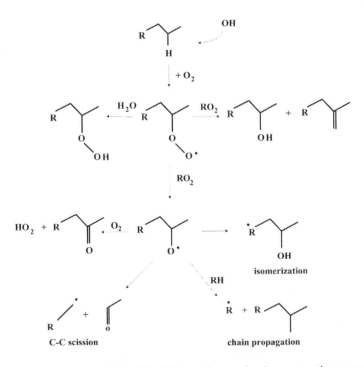

FIGURE 5.3 Scheme of the OH + RH reaction mechanism suggestion according to Russell.[44]

reaction of OH with n-alkanes, hopanes, and steranes, have found that $\gamma = 1$–40 for overall homogeneous and heterogeneous reactions. They concluded that the homogeneous constituent of the process prevails over the heterogeneous one. Obviously, the fact of the existence of the parallel channels for OH consumption is evidence about the homogeneous–heterogeneous character of the overall process. Kinetic investigations of the reaction of OH-radicals with oleic acid, linoleic acid, and linolenic acid secondary organic aerosols (in the presence of 20.7 ppm H_2O_2 in a 10% mixture of O_2 in N_2), showed the effective uptake coefficients 1.72, 3.75, and 5.7, respectively.[46] In a series of analogous experimental measurements with squalene ($C_{30}H_{50}$, containing six C=C bonds) aerosols, it was shown that the effective uptake coefficient of OH was equal to 2.34.[47] Liu et al.,[48] investigating the heterogeneous reaction of chlorine atoms with submicron squalane droplets in the flow tube reactor, in which chlorine atoms were produced by the photolysis of Cl_2 under irradiation of 365 nm, have found that the effective uptake coefficient was about 3 and concluded the presence of the secondary chain chemistry. Other analogous examples may be found in the review articles about the heterogeneous interactions of aerosols with atmospheric oxidants.[2]

In general, one of the main conclusions from all the above experimental and quantum-mechanical calculation works[38–50] was disclosure of the so-called particle-phase secondary chain chemistry, which occurred in the condensed phase. In other words, the interaction of OH or other radicals (for instance, Cl atom) initiate chain-radical reactions in the condensed phase. Investigating the detailed mechanism of the interaction of OH with alkane and alkene-thiol monolayers deposited on gold, using the molecular beam scattering technique, accompanied by spectroscopic observations, D'Andrea et al.[49] proposed the radical-induced surface polymerization mechanism in these reactions. The values of the uptake coefficients ($\gamma > 1$) obtained may be explained by polymerization processes in the particle phase. It should be noted that the polymerization in the interaction of OH with organic compounds on the condensed-phase particles is one but not the only pathway of reactions (as it is represented in the review[23]). The OH radical-induced reaction also may be a chain process of another type.[24] Recently, an analogous experimental investigation of the OH radical-induced reaction with SOA (squalene) was carried out in the presence of NO_x ($NO + NO_2$).[50] Kinetic measurements and mass-spectroscopic analyses showed that the oxidation on/in SOA, in this case, was a chain-radical reaction, propagated by alkoxy radicals (formed inside the aerosol).

$$RO_2 + NO \rightarrow RO + NO_2$$

then,

$$RH + RO \rightarrow ROH + R$$
$$R + O_2 \rightarrow RO_2$$

In the suggested mechanism, the initiation reaction in/on aerosols occurs by OH-radicals:

$$OH + RH \rightarrow R + H_2O$$

The chain reactions, according to Richards-Henderson et al., "travel" in aerosol particles, an indication of which is the effective value of uptake coefficient $\gamma = 9$.[50]

All presented examples again approve general conclusion drawn above that in atmospheric processes, the radicals OH, generated mainly by a photochemical pathway in the gas phase, react in a parallel manner with the components of the gas phase by homogeneous reactions and with the particle-phase components by heterogeneous reactions. This process is similar to the homogeneous chain-radical reaction in thermal oxidation, when the intermediate radicals of the homogeneous reaction interact with the walls of the reactor or surfaces of the solid substances present in the reaction medium. Here, the "wall effects" exist in a reacting system "without walls." In other words, the overall reaction is a photoinitiated homogeneous–heterogeneous process occurring in atmospheric conditions.

5.4 HOMOGENEOUS–HETEROGENEOUS AND HETEROGENEOUS–HOMOGENEOUS PROCESSES IN THE DECOMPOSITION OF ATMOSPHERIC OZONE

Heterogeneous interactions of ozone in the atmosphere may be conditionally divided into the reactions of the stratospheric ozone and those of the tropospheric ozone.[51] Ozone plays a different role in the processes at low atmosphere and in the altitude around 15–20 km from the Earth's surface, where it exists in the form of about a 3-mm-thick layer. This layer has an exceptional protective function from the UV and other cosmic irradiations for the life of our planet. Unlike stratospheric ozone, in the troposphere, close to Earth's surface, beginning from certain concentrations (> 0.1 mg$\cdot$m^{-3}), ozone is a contaminant of the air, being highly toxic for human and other living organisms.[51] In scientific-popular literature, taking into consideration the controversial effect of ozone on life, the tropospheric ozone is called "bad ozone" and stratospheric ozone "good ozone".

Approximately 85–90% of the ozone in the atmosphere lies in the stratosphere and the rest in the troposphere. Therefore, it seems that the reactions of ozone decomposition, including the direct heterogeneous decomposition on the condensed-phase particles in the low troposphere, will play a minor role in the overall balance of ozone and the stability of the ozone layer in the stratosphere. However, tropospheric ozone is a primary source of OH radicals, the major oxidant of the atmosphere. In this regard, its heterogeneous reactions on the different mineral aerosols, SOAs, dust, soot, metal oxide, and other pollutants are also important factors in the atmospheric chemistry of ozone. Note that the heterogeneous catalytic and photocatalytic reactions of the decomposition of ozone are of interest not only in atmospheric science, but also in water treatment technologies, chemical syntheses, etc.[51]

The catalytic role of metals (Pt, Pd, Ru, Ag, W) in the direct, heterogeneous decomposition of ozone was known more than a century ago.[51, 52] Early investigations showed that the relative activity in decomposition of ozone was higher for Ag than for other metals. The metal oxides usually exhibit more activity than corresponding metals.[52] In general, transition metal oxides are good catalysts in the decomposition of ozone.

According to one of the reviews on the heterogeneous decomposition of ozone,[51] the uptake coefficient of ozone changes in a wide range (10^{-11}–10^{-5}), depending on the nature of atmospheric condensed-phase particles (aerosols, dusts, soot, etc.).

For instance, on relatively inert surfaces such as quartz, silica, and glass, it is in the order of 10^{-11}; for some salts, such as NH_4HSO_4 or $(NH_4)_2SO_4$ 10^{-7}–10^{-6}; but for the so-called Saharan dust 10^{-5}. Heterogeneous catalytic decomposition of ozone on metals and metal oxides is well documented in some important reviews.[2, 11, 51] To understand the recent developments in the chemistry of heterogeneous reactions of ozone with atmospheric organic condensed-phase particles, the reader is referred to Chapleski et al.'s review.[2]

In both the catalytic and noncatalytic decomposition of ozone, the only major final product is molecular oxygen, and the first kinetic order of decomposition is characteristic for decomposition on the majority of solid substances.[12] The mechanism of the heterogeneous decomposition of ozone is disputable in the chemical literature.[51] On the surfaces of the majority of transition metal oxides, the decomposition of ozone under UV irradiation occurs by the formation of O_3^- radicals via electron transfer from the metal cation:

$$O_3 + e \rightarrow O^- + O_2$$
$$O_2 + O^- \rightarrow O_3^-$$

The EPR spectrum of O_3^- is well known in the chemical literature (see Chapter 4, Section 4.3, and Chapter 3, Section 3.4.2.1). For instance, on NiO, the following values of the g-factor of the EPR spectrum, $g_1 = 2.0147$, $g_2 = 2.0120$ and $g_3 = 2.0018$, were reported.[51] However, the radical pathway of reaction by the formation of O_3^- may be a minor channel in decomposition; for instance, in the case of NiO, it is only about 5%. It was considered that the main channel of the decomposition had nonradical mechanism.[51]

The general mechanism of the heterogeneous decomposition of ozone[within 51, 15] in a simplified form may be represented as follows:

$$O_3 + S \rightarrow O_3\text{-}S \rightarrow O_2 + O\text{-}S$$
$$O_3 + O\text{-}S \rightarrow O_2 + O_2\text{-}S$$
$$O_2\text{-}S \rightarrow O_2 + S$$
$$2O\text{-}S \rightarrow O_2 + S$$

where S is surface. On some metal oxide aerosols containing Al_2O_3, SiO_2, an analogous mechanism of ozone decomposition may be suggested.[12]

Taking into account the aims of this work, we will not discuss the details of the heterogeneous catalytic reactions of ozone decomposition on the surfaces of different solid substances, modeling aerosol particles. In general, this reactions have no direct and revealed relation to heterogeneous–homogeneous, or homogeneous–heterogeneous reactions in atmospheric chemistry. However, some more complex atmospheric processes involving ozone exhibit heterogeneous–homogeneous behaviors. The examples given below demonstrate the existence of some interacting homogeneous and heterogeneous processes in the atmospheric chemistry of ozone.

One of the main problems in atmospheric chemistry revolves around gaining a detailed information about the mechanism of ozone depletion, known as the formation of ozone holes in stratosphere. This is a global environmental problem and is of exceptional importance for all of humanity.[53]

About 20% of ozone depletion may be predicted by calculations using the Chapman mechanism (see Section 3.4.2.1), which involves the catalytic cycles of HO_x, ClO_x and NO_x).[54] Presently, it must also be completed by the reaction cycle of BrO_x, having made a significant (about 30%) contribution to ozone depletion.[53] Already in the 1970s, Crutzen[13] had concluded that the anthropogenic factor (related to the formation NO and NO_2 from N_2O) in ozone depletion was crucial. The appearance of chlorine in the troposphere was related mostly to pollution caused by chlorofluorocarbons, as well as carbon-tetrachloride, methylchloroform, methylchloride, and other chlorinated compounds. Among them, only methylchloride is produced by microorganisms in the oceans (about 20% being among all sources of Cl); other sources have an anthropogenic origin.[53]

Cl atoms may be produced by photolysis from the mentioned compounds

$$XCl + h\nu(\lambda < 215 \text{ nm}) \rightarrow X + Cl$$

where X is an organic radical.

The Cl-atom initiates a chain reaction in the depletion of ozone, catalyzed by ClO:

$$Cl + O_3 \rightarrow ClO + O_2$$

$$ClO + O \rightarrow Cl + O_2$$

Systematic investigations of the ozone column (ozone concentration in other units of measurement), starting in 1985, showed a dramatic decrease in the Antarctic region at the end of the polar winter. These observations were not possible to explain by the existing models of the atmospheric processes, which at that time involved only homogeneous photochemical reactions. In 1986, Solomon et al.[55] suggested the halogen emission to the atmosphere, followed by the heterogeneous reaction of HCl and $ClONO_2$ on polar stratospheric clouds (PSCs) over the Antarctic region in the stratosphere. Later, the exceptional role of heterogeneous reactions in the Antarctic region ozone depletion was confirmed by a great number of investigations.[53 and refs. within] Although the ozone depletion in this region has a seasonal character, however, its effects are dramatic for climate. First of all, a correlation was established between the destruction of the ozone layer and the presence of chlorine, bromine compounds, and azote oxides in the atmosphere, which was considered an important peculiarity of this atmospheric system.[53] It appeared that the heterogeneous reactions on the polar stratospheric clouds had a pivotal role for the explanation of Antarctica's ozone depletion, also taking into account the specific meteorological factors during the polar winter's extreme conditions. The stable polar vortex isolates the Antarctic stratosphere and cools it down to $-78°C$ at the end of the Antarctic winter.[51, 53, 55]

The polar stratospheric clouds and aerosols have a complex chemical composition.[54] The principal components of these clouds and aerosols are water, sulfate, nitrate, ammonium salts, organic compounds, and black carbon soot, also including soil dust (calcium carbonate, silicates, and metal oxides).[51] Clouds are formations of water and primary aerosols, in a supercooled liquid state and are composed from droplets of different compositions. In the polar vortex region of the Antarctic stratosphere, at least two types of polar stratospheric clouds are formed under extremely cold conditions in winter, and they are responsible for ozone depletion.[56] The first type consists of the solution of nitric acid, sulfuric acid, and water ice (HNO_3-H_2SO_4-H_2O), as well as frozen nitric acid trihydrates; and the second type is water ice particles that form when the temperature decreases to below 188 K.[54]

During the above-mentioned homogeneous catalytic cycles, part of NO_2 and ClO form $ClONO_2$.

$$NO_2 + ClO \rightarrow ClONO_2$$

$ClONO_2$ molecules are unreactive to ozone molecules. Thus, a part of NO_x and also, temporarily, a part of ClO_x may be removed from ozone deletion processes.

The chlorine-containing species HCl and $ClONO_2$ present in the atmosphere convert to Cl_2 and HNO_3 only by the heterogeneous reaction

$$ClONO_2 + HCl \rightarrow Cl_2 + HNO_3$$

This heterogeneous reaction is crucial for understanding the formation of ozone holes.[53, 54, 56] It is noteworthy that the absence of this reaction in the gas phase has been proven experimentally.[56] The heterogeneous nature of the following five reactions on the polar stratospheric clouds has also been firmly established.[53, 54]

$$ClONO_2 + H_2O \rightarrow HOCl + HNO_3$$

$$N_2O_5 + HCl \rightarrow ClNO_2 + HNO_3$$

$$N_2O_5 + H_2O \rightarrow 2HNO_3$$

$$HOCl + HCl \rightarrow Cl_2 + H_2O$$

$$BrONO_2 + H_2O \rightarrow HOBr + HNO_3$$

Cl_2 is released from the surface of clouds, producing the ozone-depleting chlorine atoms under UV irradiation, while the molecules of HNO_3 remain on the surface (Figure 5.4). In fact, this process (also called "denoxification") removes HNO_3 from the stratosphere, where the product of its photolysis NO_2 can easily react with ClO. Moreover, $ClONO_2$ is much less active in the reaction with ozone than ClO.

According to some authors,[56] the overall mechanism of this reaction is not clear, but it is not a radical reaction and may occur like a typical hydrolysis reaction in solution. Following volcanic eruptions, the role of heterogeneous reactions involving species such as $ClONO_2$ and HOCl dramatically increases, as the concentration of sulfate aerosols also increases.

To understand ozone depletion, it is important to also consider two halogen cycles[53] involving halogen containing intermediates such as catalysts (ClO and BrO):

$$2\,(Cl + O_3 \rightarrow ClO + O_2)$$
$$2ClO + M \rightarrow Cl_2O_2 + M$$
$$Cl_2O_2 + h\nu \rightarrow Cl + ClO_2$$
$$\underline{ClO_2 + M \rightarrow Cl + O_2 + M}$$
$$2O_3 \rightarrow 3O_2$$

$$Cl + O_3 \rightarrow ClO + O_2$$
$$ClO + BrO \rightarrow Cl + Br + O_2$$
$$\underline{Br + O_3 \rightarrow BrO + O_2}$$
$$2O_3 \rightarrow 3O_2$$

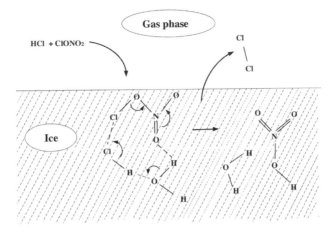

FIGURE 5.4   Scheme of the heterogeneous reaction of ClONO$_2$ with HCl on the ice surface. (Altered and adapted from Ref.[56])

Thus, as a result of the overall heterogeneous processes, the Cl$_2$ molecules are released to the gas phase, and they fill the atmospheric reservoir. Depletion of ozone increases, Cl$_2$, as well as HOCl, under UV irradiation, rapidly dissociates into Cl atoms and OH participating in the depletion of ozone.

Another example exhibiting the role of the complex homogeneous–heterogeneous-processes is reaction of OH radicals on sea-salt aerosols in the sulfur cycle of the marine boundary layer.[57–60] Knipping et al.,[58] based on a series of laboratory experiments and kinetic modeling data for a simplified system of pure NaCl particles, proposed the following net reaction

$$OH_g + Cl^- \rightarrow 0.5Cl_{2\,g} + OH^-$$

Although the impact of this reaction to sulfur cycle is not fully accepted,[59] the release of Cl$_2$ seems to be real. The reaction sequence of the overall process is analogous to the above-described phenomenon of ozone depletion via heterogeneous reactions. The overall reaction here is also a result of the combination of homogeneous–heterogeneous processes (homogeneous generation of OH and heterogeneous interaction on NaCl), and heterogeneous–homogeneous processes (heterogeneous formation of Cl$_2$ and its homogeneous dissociation, followed by desorption and formation of halogen atom-radicals, which participate in the depletion of ozone in the gas-phase). Although there are not investigations quantitatively describing the significance of this reaction in ozone depletion, it is evident that it may not be neglected in schemes of the atmospheric modelling.

We may ask, what is the relationship between these heterogeneous processes and the subject of the present work? Let us once again mentally follow the overall process of high ozone depletion in the Antarctic region and in the above-described conditions. The appearance of N$_2$O and halogenated hydrocarbons as pollutants in the atmosphere produce HCl and ClONO$_2$ by homogeneous reactions. Then, reaching on the solid surface, they produce Cl$_2$ molecules by the heterogeneous reactions, which are active in ozone depletion. They "escape" from the surface to the homogeneous medium, where they dissociate into two Cl-atoms under UV or sunlight irradiation. From the point of view of the overall process of ozone depletion, the atoms of Cl, as well as ClO, are intermediates. In this context, the overall process of ozone depletion may be considered to be a reaction sequence of the homogeneous–heterogeneous and, then, heterogeneous–homogeneous processes. Each stage of the overall process involves different reactions, but they are all linked by the intermediates, transferred from one phase to the other. The only difference from other examples of the heterogeneous–homogeneous reactions is that some of the intermediates are not radical species, or radical-like species, but they do transform into them under UV or sunlight irradiation. In this regard, this is one of the specific peculiarities of the exhibition of bimodal reaction sequences in atmospheric conditions. Also, it is obvious that the overall process of ozone depletion reflects the complex, multicomponent, multiphase, and multistage nature of the reaction in this chemical system.

Taking into account all of these peculiarities, one should note again that the overall process of ozone depletion by the schemes mentioned in this chapter may be regarded as a complex combination of heterogeneous–homogeneous and homogeneous–heterogeneous reactions in the atmospheric conditions.

5.5 CONCLUDING REMARKS ON THE EMERGENCE OF BIMODAL REACTION SEQUENCES IN ATMOSPHERIC PROCESSES

The main chemical processes in the atmosphere are linked to oxidation reactions by the participation of molecular oxygen or other oxidants under specific conditions (light, irradiation, variable composition of the reaction system, specific changes in temperature and pressure, presence of aerosols). Here it is important to again note the major role of the emissions of pollutants from the Earth's surface to the atmosphere, producing negative consequences for life on our planet.[61–63] The present atmospheric science clearly shows that the imbalance in atmospheric processes, the greenhouse effect, climate changes, and many other global phenomena are the consequences of the anthropogenic pollution of the atmosphere. More than 100 000 compounds of natural and anthropogenic origin are known in the Earth's atmosphere in measurable quantities. The majority of them undergo different chemical transformations, mainly oxidation, with the participation of molecular oxygen in atmospheric conditions. These reactions involve photochemical processes, producing active intermediates, radicals, and electronically excited species that react not only with components of the atmosphere in the gas phase, but also on/in the condensed-phase particles. As a rule, the photogenerated radicals in the gas phase initiate chain-radical reactions. Interactions of the atmospheric radicals with aerosols and other atmospheric particles are an important constituent of the overall atmospheric oxidation processes, opening the heterogeneous pathway of oxidation and producing products other than those in the gas-phase reaction.

The bimodal sequence of reactions in atmospheric processes has some specific peculiarities. The primary reaction of the generation of radicals or electronically excited species is a photochemical process, occurring under sunlight and cosmic irradiation. Although the possibility of heterogeneous generation (nonphotochemical) of active atmospheric species is not excluded, as shown by the above examples, the primary photogeneration of active intermediate species in the gas phase either was the only ("exclusive") pathway or, at least, prevailed over the heterogeneous pathways. Apparently, this is due mainly to the noncatalytic properties of most condensed-phase particles in oxidation reactions, in atmospheric conditions. Subsequently, part of the primary intermediates in the gas phase produces secondary species or products. Other parts of them diffuse to the surface of aerosol or droplet particles, reacting with the adsorbed species or components of condensed-phase atmospheric particles.

Thus, the overall atmospheric process, from the viewpoint of bimodal reaction sequences, consists of the parallel heterogeneous and homogeneous reactions, which often have the same intermediate, for example, OH radicals. Although there are similarities between the chain-radical thermal oxidation and atmospheric reactions, there are many important differences too. In the case of thermal oxidation reactions, the generation of radicals at low temperatures occurs on the surface of the solid substances rather than in the gas phase. In contrast, in atmospheric oxidation processes, the primary radicals are almost always photogenerated in the gas phase. On the other hand, essential differences exist between the heterogeneous constituents of reactions in gas-phase thermal oxidation and in atmospheric processes. In the majority of cases, the condensed-phase particles in the atmosphere absorb the gas-phase radicals rather than only adsorb them. The further reactions on/in condensed phases are stoichiometric rather than catalytic reactions. In the case of thermal as well as photochemical oxidations in the gas phase, the solid surface, participating in the reaction, unlike atmospheric processes, does not undergo transformations in the bulk of the solid. In any event, the described examples are particular cases of homogeneous–heterogeneous processes (example: OH + SOA, etc.), exhibiting some specificity in the atmospheric processes.

The atmospheric homogeneous–heterogeneous processes may also be the reactions of the specific types. According to Devise et al.,[64] the so-called gas-to-particle conversion is a special class of heterogeneous reactions in the atmosphere, when the species transfer from the gas to the condensed phase or form new condensed-phase particles. These authors mentioned three types of this kind of process: (i) homogeneous–homomolecular (gas-phase molecules condense on the particle composed from the same molecule, (ii) homogeneous–heteromolecular (different gas molecules compose new particles), and (iii) heterogeneous–heteromolecular ("growth of preexisting particles due to deposition of molecules from the gas phase,"[64] apparently, other "particles").

Another peculiarity of oxidation in the atmosphere is that the activation of molecular oxygen or other oxidants is involved in the catalytic cycles of NO_x, ClO_x, HO_x, and others. The reactions of RO_2 and HO_2 radicals, which form in the secondary reactions, also have an essential role in the gradual oxidative degradation or, conversely, in "assembling" or polymerization reactions of organic pollutants. The above peculiarities in oxidation processes are common for nearly all classes of organic pollutants and are closely linked with the consequences of pollution in the atmosphere.

One pivotal problem of atmospheric chemistry, depletion of the ozone layer, is also an interesting example of the complex combination of homogeneous–heterogeneous and heterogeneous–homogeneous processes. The seasonal ozone depletion in the stratosphere of the Antarctic polar region is a specific multiphase process in complex meteorological conditions. The heterogeneous–homogeneous nature of this process is well proven in the case of the organic pollutants, such as halogenated hydrocarbons in the atmosphere. Although the heterogeneous process on the PSC does not have the radical mechanism, the overall ozone depletion process is a consecutive, multiphase, homogeneous–heterogeneous, and heterogeneous–homogeneous process, by the participation of radical species.

Let us again return to the problems of the scientific terminology related to the atmospheric processes. When describing the homogeneous–heterogeneous or heterogeneous–homogeneous processes, the scientists working in atmospheric chemistry use the terms *multiphase and heterogeneous chemistry*.[61] Interestingly, the terms *multiphase* and *heterogeneous* have a specific meaning in atmospheric processes. Poschl[62] maintains that "the term 'heterogeneous reaction,' generally, refers to reactions of gases at the atmospheric particle surface, whereas the term 'multiphase reaction' refers to reactions in the particle bulk, involving species from the gas phase." However, these nuances do not essentially change the homogeneous–heterogeneous (OH + SOA-squalane aerosol) or heterogeneous–homogeneous (depletion of ozone by halogens emitted from the surface of the PSC, as a result of the heterogeneous reactions) character of the complex process.

What is the importance of this knowledge? In modeling studies, we must view the atmospheric chemical processes, having bimodal sequences of occurrence, in the context of the overall peculiarities of these kinds of reactions. This requires improving the simulation investigations of the atmospheric processes involving more realistic schemes and different parameters of heterogeneous processes. Only good models may predict spatial and temporal distribution of the important chemical species, and, therefore, also the real processes in the atmosphere.[63]

The heterogeneous–homogeneous and homogeneous–heterogeneous complex processes in the atmosphere again demonstrate the widespread existence of this chemical phenomenon not only in non-natural chemical systems (as reactions in the laboratory or industrial reactors), but also in a natural chemical system (as the Earth's atmosphere). Finally, this knowledge marks a step on the pathway to controllable oxidation processes in nature.

REFERENCES

5.1

1. Akimoto H. Atmospheric Reaction Chemistry, Chapter 6, Heterogeneous reactions in atmosphere and uptake coefficients. Springer, Atmospheric Sciences, Tokyo, 2016, 432 pages.
2. Chapleski R. C. Jr., Zhang Y., Troya D., Morris J. R. Heterogeneous chemistry and reaction dynamics of the atmospheric oxidants, O_3, NO_3, and OH, on organic surfaces. Chem. Soc. Rev., 2016, v. 13, p. 3731–3746. DOI: 10.1039/C5CS00375J

3. Purmal A. P. Interaction of radicals with atmospheric aerosols, p. 423–435. In: Minisci F. (ed.). Free Radicals in Biology and Environment. NATO ASI Series, v. 27, 1997, Springer-Science, Kluwer Academic Publishers, Bardolino, Italy, 500 pages.
4. Schryer D. R. Heterogeneous Atmospheric Chemistry, v. 28, 1982, American Geophysical Union, 273 pages.
5. Davis D., Niki H., Mohnen V., Liu S. Homogeneous and heterogeneous transformations. In: Global Tropospheric Chemistry: A Plan for Action. National Academic Press. Washington, 1984, 78 pages, p. 194.

5.2

6. Prinn R. G., Huang J., Weiss R. F., Cunnold D. M., Fraser P. J., Simmonds P. G., McCulloch A., Harth C., Salameh P., O'Doherty S., Wang R. H. J., Porter L., Miller B. R. Evidence for substantial variations of atmospheric hydroxyl radicals in the past two decades. Science, 2001, v. 292, pp. 1882–1888.
7. Jacob D. J. Heterogeneous chemistry and tropospheric ozone, Atmospheric Environment, 2000, v. 34, p. 2131–2159.
8. Spurny K. R. Aerosol Chemical Processes in the Environment, CRC Press, Lewis Publishers, Washington, 2000, 617 pages, p. 3.
9. Goldstein A. H., Galbally I. E. Known and unknown organic constituents in the Earth's atmosphere. Environ Sci. Technol. 2007, v. 41, 5, p. 1514–1521.
10. Calvert J. G. SO_2, NO and NO_2 oxidation mechanisms: Atmospheric Considerations, 1984, Butterworth, Boston.
11. Gillespie A. Climate Change, Ozone Depletion and Air Pollution: Legal Commentaries with Policy and Science Considerations. Leiden, The Netherlands, Martinus Nijhoff Publishers, 2006, 405 pages.
12. Calvert J. G., Orlando J. J., Stockwell W. R., Wallington T. J. The Mechanisms of Reactions Influencing Atmospheric Ozone, 2015, Oxford University Press. N.Y., 589 pages.
13. Swedin E. G. Science in the Contemporary World: An Encyclopedia. 2005, ASC-CLIO Santa Barbara, USA, 381 pages, p. 218.
14. Schryer D. R., Jech D. D., Easley P. G., Krieger B. B. Kinetics of reactions between free radicals and surfaces (aerosols) applicable to atmospheric chemistry. p. 107–120. In: Schryer D. R. Heterogeneous Atmospheric Chemistry, v. 28, American Geophysical Union, 1982, 273 pages. DOI: 10.1029/GM026p0107
15. George C., Ammann M., D'Anna B., Donaldson D. J., Nizkorodov S. A. Heterogeneous photochemistry in the atmosphere. Chem. Rev., 2015, v. 115, 10, p. 4218–4258. DOI: 10.1021/cr500648z
16. Wallington T. J., Nielsen O. J. Atmospheric degradation of anthropogenic molecules, Chapter 3, p. 68: In: Boule P. (ed.). Environmental Photochemistry, Part 1, 1999, Springer-Verlag, Berlin, 365 pages.
17. Shen X., Zhao Y., Chen Z., Huang D. Heterogeneous reactions of volatile organic compounds in the atmosphere. Atmospheric Environment, 2013, v. 68, p. 297–314.
18. Gershenzon Y. M., Grigoreva V. M., Ivanov A. V., Reorov R. G. O_3 and OH sensitivity to heterogeneous sinks of HO_x and CH_3O_2 radicals on aerosol particles. Faraday Discussions, 1995, v. 100, p. 83–100.
19. Houle F. A., Hinsberg W. D., Wilson K. R. Oxidation of a model alkane aerosol by OH radical: The emergent nature of reactive uptake. Phys. Chem. Chem. Phys., 2015, v. 17, p. 4412–4423.

5.3

20. Lelieveld J., Gromov S., Pozzer A., Taraborrelli D. Global tropospheric hydroxyl distribution, budget and reactivity. Atmos. Chem. Phys., 2016, v. 16, p. 12477–12493.
21. Finlayson-Pitts B. J., Pitts J. N. Jr. Chemistry of the Upper and Lower Atmosphere: Theory, Experiments, and Applications, 2000, Academic, San Diego, Calif., 993 pages.
22. Oxidation in the Atmosphere ESPERE Climate Encyclopaedia – www.espere.net - Lower Atmosphere More – Part 1: Oxidation and OH radicals page 4.
23. Molina M. J., Ivanov A. V., Trakhtenberg S., Molina L. T. Atmospheric evolution of organic aerosol, Geophys. Res. Lett., 2004, v. 31, L22104. DOI:10.1029/2004GL020910
24. Rex M., Wohltmann I., Ridder T., Lehmann R., Rosenlof K., Wennberg P., Weisenstein D., Notholt J., Krüger K., Mohr V., Tegtmeier S. A tropical West Pacific OH minimum and implications for stratospheric composition. Atmos. Chem. Phys., 2014, v. 14, p. 4827–4841. DOI:10.5194/acp-14-4827-2014
25. Monks P. S. Gas-phase radical chemistry in the troposphere. Chem. Soc. Rev., 2005, v. 34, p. 376–395.
26. Thompson A. Huge Hole in Earth's 'Detergent' Layer Found Over Pacific. April, 2014, Website: Climate Central; http://www.climatecentral.org/news/huge-hole-in-earths-detergent-layer-found-over-pacific-17302
27. Liu Y., Ivanov A. V., Molina M. J. Temperature dependence of OH diffusion in air and He. Geophysical Research Letters, 2009, v. 36, L03816, p. 1–4. DOI: 10.1029/2008GL036170
28. Bian H., Zender C. S. Mineral dust and global tropospheric chemistry: Relative roles of photolysis and heterogeneous uptake, J. Geophys. Res., 2003, v. 108, D21, p. 4672. DOI:10.1029/2002JD003143
29. Li P., Perreau K. A., Covington E., Song C. H., Carmichael G. R., Grassian V. H. Heterogeneous reactions of volatile organic compounds on oxide particles of the most abundant crustal elements: Surface reactions of acetaldehyde, acetone, and propionaldehyde on SiO_2, Al_2O_3, Fe_2O_3, TiO_2, and CaO. Journal of Geophysical Research, 2001, v. 106, D6, p. 5517–5529.
30. Iuga C., Olea R. E., Vivier-Bunge A. Mechanism and kinetics of the $OH^{\cdot}$ radical reaction with formaldehyde bound to an $Si(OH)_4$ monomer. J. Mex. Chem. Soc. 2008, v. 51, 4, p. 36–46.
31. Iuga C., Vivier-Bunge A., Hernández-Laguna A., Sainz-Díaz I. C. Quantum chemistry and computational kinetics of the reaction between OH Radicals and formaldehyde adsorbed on small silica aerosol models. J. Phys. Chem. C, 2008, v. 112, 12, p. 4590–4600. DOI: 10.1021/jp077557m
32. Iuga C., Sainz-Dıaz C. I., Vivier-Bungea A. On the OH initiated oxidation of C_2–C_5 aliphatic aldehydes in the presence of mineral aerosols. Geochimica et Cosmochimica Acta, 2010, v. 74, p. 3587–3597.
33. Ji Y., Wang H., Chen J., Li G., An T., Zhao X. Can silica particles reduce air pollution by facilitating the reactions of aliphatic aldehyde and NO_2? J. Phys. Chem. A, 2015, 119, p. 11376–11383.
34. Esteve W., Budzinski H., Villenave E. Heterogeneous reactivity of OH radicals with phenanthrene. Journal of Polycyclic Aromatic Compounds, 2003, v. 23, 5, p. 446–451.
35. Miet K., Budzinski H., Villenave E. Heterogeneous reactions of OH radicals with particulate-pyrene and 1-nitropyrene of atmospheric interest. Journal of Polycyclic Aromatic Compounds, 2009, v. 29, 5, p. 267–281.

36. Jariyasopit N., Zimmermann K., Schrlau J., Arey J., Atkinson R., Yu T. W., Dashwood R. H., Tao S., Simonich S. L. Heterogeneous reactions of particulate matter-bound PAHs and NPAHs with NO_3/N_2O_5, OH radicals, and O_3 under simulated long-range atmospheric transport conditions: reactivity and mutagenicity. Environ. Sci. Technol. 2014, v. 2, 48, 7, p. 10155–10164. DOI: 10.1021/es5015407

37. Molina M. J., Ivanov A. V., Trakhtenberg S., Molina L. T. Atmospheric evolution of organic aerosol, Geophys. Res. Lett., 2004, v. 31, L22104. DOI: 10.1029/2004GL020910

38. Vlasenko A., George I. J., Abbatt J. P. D. Formation of volatile organic compounds in the heterogeneous oxidation of condensed-phase organic films by gas-phase OH, J. Phys. Chem. A, 2008, v. 112, p. 1552–1560.

39. McNeill V. F., Yatavelli R. L. N., Thornton J. A., Stipe, C. B., Landgrebe O. The heterogeneous OH oxidation of palmitic acid in single component and internally mixed aerosol particles: vaporization, secondary chemistry, and the role of particle phase, Atmos. Chem. Phys., 2008, v. 8, p. 5465–5476.

40. George I. J., Vlasenko A., Slowik J. G., Broekhuizen K., Abbatt J. P. D. Heterogeneous oxidation of saturated organic aerosols by hydroxyl radicals: uptake kinetics, condensed-phase products, and particle size change, Atmos. Chem. Phys., 2007, v. 7, p. 4187–4201.

41. Smith J. D., Kroll J. H., Cappa C. D., Che D. L., Liu C. L., Ahmed M., Leone S. R., Worsnop D. R., Wilson K. R. The heterogeneous reaction of hydroxyl radicals with sub-micron squalane particles: a model system for understanding the oxidative aging of ambient aerosols Atmos. Chem. Phys., 2009, v. 9, p. 3209–3222.

42. Hearn J. D., Lovett A. J., Smith G. D. Ozonolysis of oleic acid particles: evidence for a surface reaction and secondary reactions involving Criegee intermediates, Phys. Chem. Chem. Phys., 2005, v. 7, p. 501–511.

43. Lambe A. T., Miracolo M. A., Hennigan C. J., Robinson A. L., Donahue N. M. Effective rate constants and uptake coefficients for the reactions of organic molecular markers (n-alkanes, hopanes, and steranes) in motor oil and diesel primary organic aerosols with hydroxyl radicals. Environ. Sci. Technol., 2009, v. 43, 23, p. 8794–8800. DOI: 10.1021/es901745h

44. Russell G. A. Solvent effects in the reactions of free radicals and atoms, J. Am. Chem. Soc., 1957, 79, p. 3871–3877.

45. Hearn J. D., Renbaum L. H., Wang X., Smith G. D. Kinetics and products from reaction of Cl radicals with dioctyl sebacate (DOS) particles in O_2: A model for radical-initiated oxidation of organic aerosols. Physical Chemistry Chemical Physics, 2007, v. 9, p. 4803–4813.

46. Nah T., Kessler S. H., Daumit K. E., Kroll J. H., Leone S. R., Wilson K. R. OH-initiated oxidation of sub-micron unsaturated fatty acid particles. Phys. Chem. Chem. Phys., 2013, v. 15, p. 18649–18663. DOI: 10.1039/C3CP52655K

47. Nah T, Kessler S. H., Daumit K. E., Kroll J. H., Leone S. R., Wilson K. R. Influence of molecular structure and chemical functionality on the heterogeneous OH-initiated oxidation of unsaturated organic particles. J. Phys. Chem. A. 2014, 12, 118 (23), p. 4106–4119. DOI: 31.10.1021/jp502666g. Epub 2014 Jun 2.

48. Liu C-L., Smith J. D., Che D. L., Ahmed M., Leone S. R., Wilson K. R. The direct observation of secondary radical chain chemistry in the heterogeneous reaction of chlorine atoms with submicron squalane droplets. Phys. Chem. Chem. Phys., 2011, v. 13, p. 8993–9007.

49. D'Andrea T. M., Zhang X., Jochnowitz E. B., Lindeman T. G., Simpson C., David D. E., Curtiss T. J., Morris J. R., Ellison G. B. Oxidation of organic films by beams of hydroxyl radicals. J. Phys. Chem. B, 2008, v. 112, p. 535–544.

50. Richards-Henderson N. K., Goldstein A. H., Wilson K. R. Large enhancement in the heterogeneous oxidation rate of organic aerosols by hydroxyl radicals in the presence of nitric oxide. Journal of Physical Chemistry Letters, 2015, v. 6, 22, p. 4451–4455.

5.4

51. Batakliev T., Georgiev V., Anachkov M., Rakovsky S., Zaikov G. E. Ozone decomposition. Interdiscip. Toxicol., 2014, v. 7, 2, p. 47–59.
52. Monchot W., Kampschulte W. *Über die Einwirkung von Ozon auf metallisches Silber und Quecksilber. Berichte der deutschen chemischen Gesellschaft.* 1907, v. 40, 3, p. 2891.
53. Chipperfield M. P. Global Atmosphere – The Antarctic ozone hole, p. 1–33. In: Hester R. E., Harrison R. M. (eds.). Still Only One Earth: Progress in the 40 Years since the First UN Conference on the Environment, 2015. Issues on Environmental Science and Technology, RSC, Cambridge, 285 pages.
54. Lagzi I., Mészáros R., Gelybó G., Leelőssy Á. Atmospheric Chemistry, 2013. Website: http://www.renderx.com/ Atmospheric Chemistry
55. Solomon S., Garcia R. R., Rowland F. S., Wuebbles D. J. On the depletion of Antarctic ozone. Nature, 1986, v. 321, p. 755–758.
56. Zhang M-T. Leu L. F. Keyser heterogeneous reactions of $ClONO_2$, HCl, and HOCl on liquid sulfuric acid surfaces. J. Phys. Chem. 1994, v. 98, p. 13563–13574.
57. Von Glasow R. Importance of the surface reaction OH +Cl⁻ on sea salt aerosol for the chemistry of the marine boundary layer – a model study. Atmos. Chem. Phys., 2006, v. 6, p. 3571–3581.
58. Knipping E. M., Lakin M. J., Foster K. L., Jungwirth P., Tobias D. J., Gerber R. B., Dabdub D., Finlayson-Pitts B. J. Experiments and molecular/kinetics simulations of ion-enhanced interfacial chemistry on aqueous NaCl aerosols. Science, 2000, 288, p. 301–306.
59. Laskin A., Gaspar D. J., Wang W., Hunt S. W., Cowin J. P., Colson S. D., Finlayson-Pitts B. J. Reactions at interfaces as a source of sulfate formation in sea-salt particles. Science. 2003, 301(5631), p. 340–344.
60. Behnke W., Zetzsch C. Heterogeneous photochemical formation of Cl atoms from NaCl aerosol, NO_x and ozone. Journal of Aerosol Science, 1990, v. 21, Suppl. 1, p. S229–S232.

5.5

61. Cwiertny D. M., Young M. A., Grassian V. H. Chemistry and Photochemistry of Mineral Dust Aerosol. Annual Review of Physical Chemistry, 2008, v. 59, p. 27–51. DOI:10.1146/annurev.physchem.59.032607.093630
62. Poschl U. Atmospheric aerosols: Composition, transformation, climate and health effects. Angew. Chem. Int. Ed. 2005, v. 44, p. 7520–7540.
63. Sahagian D. A User's Guide for Planet Earth: Fundamentals of Environmental Science. 2013, Cognella Academic Publishing, USA, Science, Math, and Engineering Titles, 220 pages.
64. Devise D., Niki H., Mohmen V., Liu S. Homogeneous and heterogeneous transformation, p. 84. In: Global Tropospheric Chemistry: A Plan for Action, part 2, Assessment of current understanding. National Academic Press, Washington, USA, 1984, 195 pages.

6 Bimodal Reaction Sequences as Nonequilibrium Processes

The Nobel Prize winner in Chemistry in 1977, Ilya Prigogine, wrote: "A chemical reaction is a typically irreversible process, driving a system towards equilibrium: there is no possible reduction of chemistry to quantum mechanics."[1] Nonequilibrium processes play a key role in heterogeneous catalysis, as well as in heterogeneous–homogeneous and homogeneous–heterogeneous reactions. The spontaneous evolution of any real system, involving chemical reactions, to the minimum of G (Gibbs energy) may not be fully described by the thermodynamics of equilibrium processes, where the time parameter is absent. Here it is pertinent to mention that the "thermodynamics of nonequilibrium processes relates to every chemical transformation."[102 in Ch.1] The mechanism of the majority of oxidation reactions relates to the dynamics of the systems moving from a nonequilibrium to an equilibrium state.[2] The word "dynamics" as used here concerns both chemical reaction and energy transfer processes. In this advancement of the chemical system, the different phenomena (multiplicity of the steady states, critical phenomena, oscillations of the concentrations and reaction rates, appearance of the spatial and temporal patterns, etc.) may be observed. The majority of them are exhibitions of a more common phenomenon, self-organization in the chemical systems, operating far from the equilibrium state.[3] In this regard, the theoretical explanation of many phenomena in heterogeneous–homogeneous reactions and revelation of their detailed mechanisms are closely related to the self-organization phenomenon and formation of dissipative structures.[2]

6.1 SOME TURNING POINTS IN THE HISTORY OF HETEROGENEOUS CATALYSIS

Knowledge about bimodal (heterogeneous–homogeneous and homogeneous–heterogeneous) reactions relates to the theoretical perceptions in heterogeneous catalysis, on the one hand, and to the reaction dynamics in the gas or liquid phases, on the other hand. Historically, they were developed separately. However, the modern natural sciences, particularly chemistry and physics, can examine many phenomena from the single point of view. Such an opportunity makes possible the theory of kinetics and thermodynamics of nonequilibrium processes and a more common concept, self-organization, integrating the understanding of the different phenomena.[1–3]

Before presenting some examples related to bimodal reactions and this concept, let us review some historical stages in the development of the theory of heterogeneous catalysis. It may be useful to understand why modern heterogeneous catalysis has entered a new stage of development related to the concept of self-organization.

Since the discovery of one of the first clear evidences about the participation of the solid surface in the reaction between coal gas and air on Pt, by Sir Humphry Davy in 1817, a great number of theories have been posited to explain this phenomenon. However, real comprehension of heterogeneous catalysis, from the modern viewpoint, began with the results of the investigations of Langmuir, clearly demonstrating that catalysis by the solid substances is a chemical process on their surfaces.[4] It begins by the adsorption of initial reactants and terminates by the desorption of final products, after a certain surface transformation of initial reactants has taken place.

According to Masel,[121 in Ch. 3] one of the main ideas that changed the understanding of heterogeneous catalysis was recognition of the role of the formation and stabilization of intermediates on the surfaces of catalysts. For a successful catalysis, the stabilization of intermediates must be optimal between the strong and weakly chemical bonding,[2] the first qualitative perceptions of which were given by Sabatier's principle.[121 in Ch. 3; 5; 6] Masel also paid attention to the data showing the partitioning of free energy of reaction over the elementary steps in the steady-state heterogeneous catalytic reaction.[121 in Ch. 3] In his book, he illustrated this concept by the hydrogen oxidation reaction on Pt. Based on the free energy diagram calculated on the basis of the mechanism suggestions given by Anton and Cadogan,[7, 8] one can see how the decrease of free energy was partitioned between the elementary steps (Figure 6.1). The mechanism of the hydrogen + oxygen reaction on Pt, suggested by these authors and represented in Masel's book, involves the following elementary steps:

$$O_2 + 2S \rightarrow 2O_{ads} \qquad (1)$$

$$O_{ads} + H_2 \rightarrow OH_{ads} + H_{ads} \qquad (2)$$

$$2O_{ads} + H_2 \rightarrow 2OH_{ads} \qquad (3)$$

$$2OH_{ads} \rightarrow H_2O + O_{ads} \qquad (4) \qquad\qquad (I)$$

$$O_{ads} + H_{ads} \rightarrow OH_{ads} \qquad (5)$$

$$H_{ads} + OH_{ads} \rightarrow H_2O \qquad (6)$$

$$H_2 + 2S \rightleftarrows 2H_{ads} \qquad (7)$$

In general, the complex reaction is more favored when it occurs as a stepwise process. By this example, it was shown that if the values of ΔG_i (free energy decreases) were partitioned nearly equal or close to equality for elementary steps, the rate of overall reaction might seek the maximum.[5] If the intrinsic activation barriers are higher in one or some steps than in others, the rate of reaction may be reduced in comparison with the pathway by nearly equal partitions of the free energy decrease between steps (stages). Note, however, that this is not a general rule and that there are a number of exceptions, some of which Masel described.[121 in Ch. 3] The problem of the stepwise and simultaneous occurrence of catalytic reactions was examined in detail by Boreskov.[68 in Ch.1]

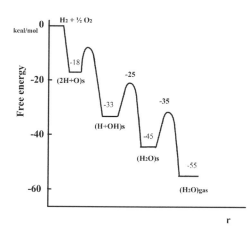

FIGURE 6.1 Free energy diagram for the catalytic cycle of the formation of water (involving reaction (4) of scheme (I) dominating in excess of oxygen) in the oxidation of hydrogen with oxygen on Pt, calculated on the basis of data Anton and Cadogan,[8] r is reaction coordinate. (Adapted from Masel.[121 in Ch. 3])

In the past, development of theoretical perceptions in the heterogeneous catalysis was related to two important theories. One of them was the theory of the so-called "assembles" of Balandin that was very briefly presented in Section 3.5.2.1. The other was the theory of active sites on the surface, developed by Taylor.[9, 10] He showed that a small fraction of the surface was responsible for the catalytic action. The existence of active sites (one or more types) leads to the consideration of the role of surface atoms by unusual coordination ("varying degree of saturation") and their specific geometry.[11] Later, these ideas were developed by Thompson, who showed the role of different defects on the surfaces of solid catalysts (or inhibiting surfaces), including dislocations, which often play a role of active sites. Further experimental data indicated that the activation energies on crystal defect sites and step edges were lower than on the homogeneous surface sites, on the terraces of crystals.[121 in Ch. 3] Developing Taylor's theory, his follower, Boudart in the 1960s, divided all heterogeneous catalytic reactions into "structure-sensitive" and "structure-insensitive" reactions, taking into consideration rate dependence on the size of solid particles of the catalyst, for instance, for metallic catalysts in the oxidation of hydrocarbons (Section 3.5.2). This phenomenon was related mostly to changes in the number of crystallographic faces and the sizes of particles of the catalyst. On the other hand, the existence of centers of unusual coordination were revealed. Presently, these and many other fundamental ideas are essential for the catalyst design, including modern nanocatalysts, by the creation of more open, undercoordinated sites on nanoscale catalysts. Note also that, in all cases, geometric factors play an important, often pivotal, role in successful catalytic action.

In the 1960–1970s, it became evident that existing theoretical perceptions were not sufficient to explain the mechanisms of many complex heterogeneous catalytic reactions and newly discovered phenomena at that time (self-oscillations, critical phenomena, etc.).[53 in Ch.1] Application of the thermodynamics of nonequilibrium processes in heterogeneous catalysis was a new stage, permitting the creation of a new approach in the theoretical interpretation of existing experimental data. The next two sections are devoted to the role of some important revelations in this development. At that time, UHV (ultra-high vacuum) experiments on monocrystals were being intensively developed in surface science, giving valuable information about the elementary chemical acts and collective effects (relaxation, reconstruction, and reorganization of surfaces) in heterogeneous catalysis.[134 in Ch. 1] It became evident that the catalyst surface might undergo both reversible and irreversible transformations during reactions, related to the creation of new states (nonsteady-state concentrations of active centers) in the reacting system, which led to different expressions of self-organization phenomena.

Since the 1990–2000s, experimental data have led to the overall acceptance of these conceptions, the evidence of which was the recognition of Ertl's works with the Nobel Prize in Chemistry in 2007[34 in Ch. 3] in the field of heterogeneous catalysis.

In the 1990–2000s, Ertl and his co-workers in Germany's Fritz Haber Institute applied surface science methods (especially low-energy electron diffraction (LEED), ultraviolet photoelectron spectroscopy (UPS), scanning tunneling microscopy (STM), and many others) in the detailed kinetic studies of the catalytic ammonia synthesis over iron and oxidation of carbon monoxide over platinum. Ertl also analyzed the phenomena of the kinetic oscillations, surface dissipative structures, and developed nonlinear kinetics in heterogeneous catalysis.

The modern theory of heterogeneous catalysis is based on several concepts, although a number of perceptions may be considered common or general for them:

a. In many cases, active sites on the surface arise during catalytic reaction, if they do not exist or if they were mechanically not created before reaction.

b. More than one type of active centers may exist or may be formed in the catalytic reaction.

c. Relaxation, reconstruction, and reorganization of the solid surface take place in catalytic reactions, although they can be experimentally unobservable when their rates are very great.

d. Nonequilibrium processes in catalytic reactions are responsible for the creation of the nonsteady-state concentrations of active centers, and they become the cause of the appearance of autocatalysis, multiplicity of steady states, critical phenomena, oscillations of rates, emergence of patterns, etc.

e. Exhibitions of all of the above-mentioned phenomena permit us to characterize catalytic action as a cyclic (multistage and multicomponent), self-organizing chemical process, where the system itself regenerates active centers through reorganization (reconstruction) of the surface.

f. Chain-radical and heterogeneous catalytic reactions have different mechanisms, but they are analogous in the creation of cyclic processes, controlling the regeneration of the reaction active centers.

g. In many complex and multistep catalytic reactions, the spatial (therefore, also the temporal) separation of steps takes place, one particular case being the phase transfer of intermediates (phase separation) as an important stage of overall heterogeneous and heterogeneous–homogeneous reactions.

h. Many experimental observations in the heterogeneous catalysis and heterogeneous–homogeneous reactions (oscillations, chemical waves, and critical phenomena) may be explained as exhibitions of the more common phenomena of self-organization.

6.2 ANALOGIES BETWEEN HETEROGENEOUS CATALYTIC AND RADICAL CHAIN REACTIONS

Close similarities between homogeneous reactions in the fluid phases and heterogeneous catalytic processes in oxidation were not revealed until the early 1950s. Previously, as mentioned above, it was known that both the heterogeneous catalytic and homogeneous chain-radical reactions of oxidation occurred via the formation and destruction of active intermediates. However, theoretical notions about these types of reactions were developed separately.

Christiansen[12] was the first one who noticed close similarities between radical-chain and catalytic reactions in homogeneous and heterogeneous processes. He found that the two types of reactions were similar in their multistep (stepwise) and cyclic character via the formation of reaction intermediates. Moreover, he indicated that these reactions themselves create the conditions for repeated cycles.

Subsequently, the wide kinetic investigations and development of the general theory of chemical reactions in both areas not only confirmed the existence of similarities, but also revealed new, more profound relationships, even analogies between them.[36 in Ch. 3; 13] Note again that the concepts mentioned concern only complex, multi stage, and multicomponent reactions, but not concerted reactions, which occur without formation of intermediates and are only one-step chemical acts.

In order to demonstrate the similarities between these two types of reactions, let us examine the following two examples of oxidation reactions with dioxygen, with the same initial reactants, but carried out in the presence and absence of the heterogeneous catalyst. Correspondingly, one of them may be a typically radical chain reaction in the gas phase, in the absence of the heterogeneous catalyst, and the other a "purely" heterogeneous reaction occurring on the surface of a catalyst. Obviously, the reaction conditions or parameters (temperature, pressure, concentrations of reactants, etc.) are different for homogeneous and heterogeneous reactions.

The first example is the reaction $H_2 + O_2$. The mainly accepted mechanism suggestions of this reaction in the gas phase were given in Chapter 1 (Section 1.3) and on the Pt(111) monocrystal surface in Chapter 3 (Section 3.5.1.2).

A simple comparison of the schemes for the mechanism suggestions represented in Sections 3.5.1.2 (Ertl et al.) and 6.1 (Anton and Cadogan) shows that they are not identical, but they are similar. The combination of steps (2) and (3) in the scheme (1) presented in Section 6.1 results in the formation of O_{ads} species and the combination of steps 3–5 in Section 3.5.1.2 in the formation of OH_{ads} species. Both O_{ads} and OH_{ads} are intermediates that play analogous roles in the catalytic cycles of two schemes. Other steps in two schemes are identical.

Let us also compare the schemes, based on the mechanism suggestions for the gas-phase oxidation of hydrogen (in the absence of catalysts) presented in Section 1.3 with a brief scheme of the mechanism suggested by Hinshelwood that is represented below (II).[121 in Ch. 3] They are analogous if we remove the heterogeneous stages from the scheme represented in Section 1.3.

$$X + O_2 \rightarrow 2O + X \qquad (1)$$
$$O + H_2 \rightarrow OH + H \qquad (2)$$
$$H_2 + OH \rightarrow H_2O + H \qquad (3) \qquad (II)$$
$$H + O_2 \rightarrow OH + O \qquad (4)$$
$$X + 2H \rightleftarrows H_2 + X \qquad (5)$$

Now, let us compare the schemes of the suggested mechanisms for gas-phase and heterogeneous catalytic reactions. Common intermediates in both cases are O, H, OH (in some schemes, also involving HO_2 and H_2O_2). Although not all elementary reactions coincide exactly, the main stages of the chain reaction, including the branching of the chain are, in certain considerations, analogous in both reactions. Indeed, an important difference is that the intermediates H, O, OH, HO_2 in heterogeneous catalytic reactions are adsorbed species on the catalyst but are not free radicals. Note that here consideration of the exact nature of the reaction intermediates (radicals, radical-like, or adsorbed species) are less important than consideration of the overall peculiarities of the development of two processes. From this point of view, these two reactions are analogous as processes having similar intermediates, as well as, in this case, even the same final product. Also, the cyclic character of two processes is obvious.

The second example, taken from the work of Masel,[121 in Ch. 3, p. 449] is the oxidation reaction of alcohols with dioxygen: deuterated methanol oxidation to formaldehyde on Ag(110) catalyst with oxygen, investigated by Barteau and Madix in 1982.[14] The proposed mechanism of reaction was represented by the following scheme:

$$O_2 \rightleftarrows O_{2(ads)} \qquad (1)$$
$$CH_3OD \rightleftarrows CH_3OD_{(ads)} \qquad (2)$$
$$O_{2(ads)} \rightarrow 2O_{(ads)} \qquad (3)$$
$$CH_3OD_{(ads)} + O_{(ads)} \rightarrow CH_3O_{(ads)} + OD_{(ads)} \qquad (4)$$
$$CH_3OD_{(ads)} + OD_{(ads)} \rightarrow CH_3O_{(ads)} + D_2O_{(ads)} \qquad (5) \qquad (III)$$
$$CH_3O_{(ads)} \rightarrow CH_2O_{(ads)} + H_{(ads)} \qquad (6)$$
$$CH_2O_{(ads)} \rightarrow CH_2O_{(g)} \qquad (7)$$
$$D_2O_{(ads)} \rightarrow D_2O_{(g)} \qquad (8)$$
$$2H_{(ads)} \rightarrow H_{2(g)} \qquad (9)$$

Masel compared the above-proposed mechanism of reaction (1–9) (III) with the mechanism of the gas-phase oxidation of methanol, known as a typically chain-radical reaction.[121 in Ch. 3] In the above scheme, step (3) is analogous to the initiation of radical chains (considering that two $O_{(ads)}$ are radicals or radical-like species on the surface). Steps (4) and (5) correspond to the propagation of chains. The main analogy appears in (3)–(6) and (9) steps, which also exist in the mechanism of the gas-phase oxidation of methanol. The mechanism of gas-phase oxidation was completed by the following two reactions

$$OH + CH_3OH \rightarrow CH_2O + H_2O$$
$$O + CH_3OH \rightarrow CH_2OH + OH$$

which are hindered on the surface, caused by the steric limitations in transformations of the intermediates in their adsorbed state. The overall conclusion was that there exists a general analogy between the gas phase and the heterogeneous catalytic reaction of the oxidation of methanol, as cyclic processes.

A number of similar examples exhibiting analogies between the heterogeneous catalytic and chain-radical reactions were given by many authors at different times (Krilov,[15] Boudart,[16] Masel[121 in Ch. 3] and others). Analogies between the homogeneous chain and heterogeneous catalytic reactions are exhibited not in a precise replication of elementary reactions, but in general trends of the evolution of chemical systems as cyclic processes.

Analogies between radical-chain and heterogeneous catalytic reactions are more obvious when not only the gas-phase reaction, but also the heterogeneous reaction occur according to chain-radical mechanisms, some examples of which were presented in the earlier sections of this book: low-temperature thermal oxidation of alcohols and aldehydes in Chapter 3; photochemical oxidation of aldehydes, alcohols, and ketones on the surfaces of catalysts in Chapter 4; some atmospheric reactions on aerosols, such as squalene or squalane on the SOA in Chapter 5; and others.

Thus, even, a simple and formal comparison of two types of reactions shows the existence of at least certain analogies between them:

a. Both the radical chain and heterogeneous reactions are complex, multi-component, and multistage, often also being multiphase chemical reactions occurring by the formation of active intermediates;

b. They are cyclic and stepwise processes involving the generation of active intermediate species and their reactions, including the regeneration, termination, or deactivation of the intermediates;

c. There are analogies between the formal kinetic equations describing two types of reactions.

Moreover, in both chemical systems, there may arise:

d. Feasibility of the thermodynamic and kinetic coupling (conjugation) of the stages (steps);

e. Accumulation of the super steady-state (nonequilibrium) concentrations of intermediates;

f. Multiplicity of the stationary states of the system;

g. Critical phenomena (oscillation, spatiotemporal pattern, hysteresis, etc.);

and

h. Formation of dissipative structures and different, revealed and non-fully revealed, exhibitions of the self-organization phenomenon.

Although this list is probably not complete, it involves some deep analogies in the nature of these reaction systems.

Formal kinetic schemes of the heterogeneous catalytic reactions, demonstrating some analogies with the radical chain reactions, were discussed by Krilov.[15] One of these reactions was a simple reaction scheme (known as the two centers scheme) of the catalytic reaction

$$A \rightarrow B$$

on the heterogeneous catalyst

$$A + Z \rightarrow A' + Z^* \qquad k_1$$

$$A + Z^* \rightarrow B + Z^* \qquad k_2$$

$$Z^* \rightleftarrows Z \qquad\qquad k_3 \text{ and } k_{-3}$$

where Z is the active center and Z^* denotes the "excited" state of that center. This scheme involves three basic stages in the terminology of radical-chain reactions: initiation, propagation, and termination.

Let us follow Krilov's consideration in his work.[15] The condition

$$k_1[A] \gg k_{-3}$$

corresponds to the case, when the activation by chemical reaction prevails over the thermal activation of the reaction centers.

Consider that the concentration of A is n

$$dn/dt = -k_2 n[Z^*]$$

$$d[Z^*]/dt = k_1 n[Z] - k_3[Z^*]$$

and, if N_0 is the total number of active centers

$$d[Z^*]/dt = k_1 N_0 n - (k_1 + k_3)[Z^*]$$

The equation corresponds to the quasi-stationary state of $[Z^*]$. Hence, it permits us to find the characteristic time (t_1) of the quasi-stationarity state:

$$t_1 \sim 1/(k_1 n + k_3)$$

As the experimental results show, in the general case, $[Z^*]_{qs} \gg [Z^*]_{eq}$, where $[Z^*]_{qs}$ represents quasi-stationary concentrations and $[Z^*]_{eq}$ equilibrium concentrations of the active centers. The latter is determined as

$$[Z^*]_{eq} = [Z^*] (k_{-3}/k_3)$$

Then, the stationary state concentration may be written as

$$[Z^*]_{st} = k_1 N_0 n/(k_1 n + k_3) \gg [Z^*]_{eq}$$

If W_{st} is the stationary rate of reaction,

$$W_{st} = k_1 k_2 n^2 N_0 /(k_1 n + k_3)$$

From the point of view of formal kinetics (mathematics), the recent equation is similar to Langmuir's famous equation in heterogeneous catalysis. If $k_1 n \ll k_3$,

$$W_{st} = (k_1 k_2 /k_3) n^2 N_0$$

It is obvious that the activation barrier of reaction is $E = E_1 + E_2 - E_3$. It will take place in the case of the ordinary, nonbranched chain-radical reaction. Note, again, that the physical meanings of the constants are different, though there is the same formal kinetic description of both the radical-chain and heterogeneous catalytic processes.

Krilov[15] also discussed the formal kinetic analogies between the heterogeneous and radical chain reactions in the case, when the active centers (sites) arise during the reaction and they do not exist before reaction. The reaction scheme in this case may be presented as

$$A + Z \rightarrow B + Z \qquad (1 - \alpha)k_1$$

$$A + Z \rightarrow B + Z^* \qquad \alpha k_2$$

$$A + Z^* \rightarrow B + Z^* \qquad k_2$$

$$Z^* \rightleftarrows Z \qquad\qquad k_3 \text{ and } k_{-3}$$

where α is the probability of the excitation of the active center by the reaction $Z \to Z^*$ (it is considered a slow reaction).

The analysis of this scheme, when $k_2 \gg k_1$ and $\alpha k_1 n \gg k_3$, gives the kinetic equations

$$dn/dt = k_1 N_0 n - k_2 n[Z^*]$$

$$d[Z^*]/dt = \alpha k_1 N_0 n - (\alpha k_1 n + k_3)[Z^*]$$

and the following expressions for t_1 and $[Z^*]_{st}$

$$t_1 \sim 1/(\alpha k_1 n + k_3)$$

$$[Z^*]_{st} = \alpha k_1 N_0 n/(\alpha k_1 n + k_3) \gg [Z^*]_{eq}$$

Here, it is interesting to analyze some limiting conditions, exhibiting analogies between heterogeneous reactions and chain reactions.

If $k_2/k_1 \gg k_3/k_{-3}$, and $t \ll t_1$, the kinetic equation shows the first-order reaction

$$dn/dt = -(k_2 k_1/k_3) N_0 n = -K_0 n$$

where $K_0 = (k_2 k_1/k_3) N_0$. If $\alpha k_1 n \ll k_3$,

$$dn/dt = -\alpha(k_2 k_1/k_3) N_0 n^2 = -K_{st} n^2$$

$K_{st} = \alpha (k_2 k_1/k_3) N_0$, the reaction becomes second order. If, however, $\alpha k_1 n \gg k_3$ (the activation of the reaction center is fast and the "relaxation" is a slow process), $[Z^*]_{st} \approx N_0$ and, correspondingly

$$dn/dt = -k_2 N_0 n = -K_{st} n$$

This is an equation of the first-order reaction demonstrating that the regime of the reaction is changed.

$$K_0/K_{st} = k_3/k_{-3} \gg 1$$

The condition of the overall slow reaction is $k_2/k_1 \ll k_3/k_{-3}$. In this case, when $t \ll t_1$

$$dn/dt = -k_1 N_0 n = -K_0 n$$

If, in addition, $\alpha k_1 n \ll k_3$

$$[Z^*]_{st} = \alpha(k_1/k_3) N_0 \ll N_0$$

Returning to the conclusion of the previous case, we see that the rate of reaction will be equal.

$$dn/dt = -k_1 N_0 n - \alpha(k_1 k_2/k_3) N_0 n^2$$

Here, the total order of reaction may be changed from 1 to 2, depending on the constants and α.

Finally, let us follow with an interesting case of kinetic analysis given by Krilov[15] that is related to the following scheme in heterogeneous catalysis. Initially, the reaction occurs on the excited center Z^*; then, during the reaction, a feasibility of the generation of one more Z^* center with probability α arises. The reaction scheme is the following:

$$A + Z^* \to B + Z^* \qquad (1-\alpha)k_1$$
$$A + Z^* \to B + 2Z^* \qquad \alpha k_1$$
$$Z^* \rightleftarrows Z \qquad k_2 \gg k_{-2}$$

According to Krilov, this scheme of a heterogeneous catalytic reaction is an analog of the branched chain reaction by linear termination of chains.

$$dn/dt = -k_1[Z^*]n$$

$$d[Z^*]/dt = \alpha k_1 n[Z^*] - k_2[Z^*] + k_{-2}[Z] = (\alpha k_1 n - k_2)[Z^*] + k_{-2}N_0$$

$$[Z^*] + [Z] = N_0$$

If $\alpha k_1 n > k_2$, a "chain explosion" takes place by a time parameter τ

$$\tau = 1/(\alpha k_1 n - k_2)$$

In analogy with the chain reactions, the condition $\alpha k_1 n = k_2$ indicates a possibility of the existence of the "up limit" (see p-T diagrams in the gas-phase reaction in Section 1.3, Figure 1.1), when the reaction may occur with an explosion. The difference is that the "up limit" in the gas-phase reaction is the result of the quadratic reaction of the termination of chains, while in this case, the "up limit" is limited by the number of active centers on the surface, as $[Z^*]_{max} = N_0$.

The active centers may exist not only on the surface, but also in the bulk of the solid, as it was considered by Krilov.[15] On the other hand, an additional number of active centers may also exist in the gas phase. The latter case may take place in heterogeneous–homogeneous or homogeneous–heterogeneous reactions. This case was not analyzed in Krilov's work.[15] However, it is obvious that, in a heterogeneous–homogeneous reaction, the number of active centers in the overall reaction will be

$$N_0 = [Z^*]_{max} + [Z^*]_{hom}$$

where $[Z^*]_{hom}$ will be the number of active centers in the gas phase. Not all other formal kinetic behaviours of the system, in this case, *a priori*, will be notably changed. Thus, the formal kinetic analogies between the heterogeneous catalytic reaction and chain-radical reactions, from our point of view, may be extended for cases of the complex heterogeneous–homogeneous reactions, if the active centers also exist in/on both phases.

As an example of the heterogeneous–homogeneous reaction, where the similar active intermediates, leading active centers, exist both in the gas phase and on the surface of the catalyst may be served the oxidation of propionic aldehyde in the reaction vessel with the developed surface (KCl-coated surface of glass), described in Section 3.5.4.2. The experimental data of this heterogeneous–homogeneous reaction indicate that RCO_3 radicals in the gas phase are leading active centers. It was established by direct detection of the radicals in the gas phase using the matrix isolation EPR method and analyzing the overall kinetic data (*ibid.*, 3.5.4.2). On the other hand, in the same conditions, the heterogeneous consumption of aldehyde with the selective formation of final products (peroxypropionic and propionic acids), as well as the common kinetic behaviors of the system, indirectly indicate the existence of the parallel reaction on the surface. The radical nature of the leading active centers, which, apparently, are adsorbed RCO_3 radicals, also has been evidenced by experimental data obtained by the EPR method (Section 3.4.5). Furthermore, there are experimental evidences of the heterogeneous-chain character of the surface reaction: such as the possibility of initiation by the radicals of the "purely" heterogeneous reaction, in conditions excluding the homogeneous propagation of chains, and direct reactions of radicals with the adsorbed aldehyde, indicating the role of adsorbed radicals on the surface process. Other evidences about the role of adsorbed radicals, formed in the reaction, were established in the photochemical oxidation reaction of acetaldehyde, also by direct detection of the same radicals adsorbed on the surfaces of the solid substances (for instance, on TiO_2[283 in Ch. 3]). Thus, the radicals (mainly RCO_3) apparently produce the same products, in different proportions, on the solid surface and in the gas phase. However, like other adsorbed species, the adsorbed RCO_3 and the same radical in the gas are different species. In this particular case, the important observation was that the active centers, propagating the reaction, existed both on the surface and in the gas phase. Similar observations were made in the case of the reaction $H_2 + O_2$ on Pt, in conditions of heterogeneous–homogeneous reaction, at high-temperature catalytic combustion: similar parallel reactions on the surface and in the gas phase give the same product by different pathways.[74 in Ch. 1; 121 in Ch. 3]

All the mentioned schemes and their analyses show that both types of reactions may be described by similar formal kinetic laws. Note again that the physical meaning of the rate constants, as well as the nature of active centers in two phases, are different; however, in some conditions, the common kinetic behaviors of the system may be described by analogous kinetic equations. This kinetic analysis and discussed cases also show that, from the point of view of formal kinetic behaviors, in some aspects, heterogeneous–homogeneous reactions may be considered a particular case of heterogeneous catalytic reactions, when active centers or leading active intermediates may exist not only on the surface, but also elsewhere in the system, on the other solid phase of the catalyst and even in the gas phase. In addition, note also that these conclusions concern cases when both constituents of the overall reaction have stepwise and cyclic mechanisms that are similar to each other.

6.3 ABOUT SPATIAL AND TEMPORAL SEPARATION OF THE STAGES IN COMPLEX HETEROGENEOUS–HOMOGENEOUS REACTIONS

In this and the next sections, we will attempt to approach the other fundamental problem, namely: "Why, in certain conditions, may the heterogeneous–homogeneous pathway of reaction become more accessible for a complex reaction system than only a "purely" homogeneous or a "purely" heterogeneous pathway?" This question is approached in the context of the kinetics and thermodynamics of the nonequilibrium process.

To understand the causes of the appearance of the heterogeneous–homogeneous reaction, occurring through the formation of radicals or other active intermediate species by their further transfer to the fluid phases, we must examine the phenomenon that is named spatial separation of reaction steps (stages). It is well known in heterogeneous catalysis.[15; 34, 51 in Ch. 1] This means that two or more single steps of the complex heterogeneous reaction may take place on different active sites. In certain cases, the surface active sites may even be on different solid phases. There are a sufficient number of examples clearly demonstrating that the different steps or stages of the catalytic reaction may occur on the sites of the different phases of the multiphase and multicomponent solid catalyst. As an example, consider the oxidation of propylene to acrolein on $Bi_2(MoO_4)_3$ catalysts, discussed in Section 3.5.2.3. According to Krilov,[15] if this catalyst also contains Fe and Co elements, composing $FeMoO_4$, $Fe_2(MoO_4)_3$ and $CoMoO_4$ separate phases, primary activation of propylene and formation of the allylic complex occur by the reaction presented in Section 3.5.2.3, on the $Bi_2(MoO_4)_3$ phase, and adsorption of dioxygen takes place on $FeMoO_4$ ($O_2 + 4e \rightarrow 2O^{2-}$). Then, O^{2-} migrates to the phase $Fe_2(MoO_4)_3$ and reacts in the boundary between this phase and the $Bi_2(MoO_4)_3$ phase with the C_3H_5 radical, producing acrolein. The $CoMoO_4$ phase plays a stabilizing role for the β-FeMoO phase (reduced from $Fe_2(MoO_4)_3$).

The spatial separation of the stages or steps may arise spontaneously, as the affinities and temperature dependencies of the elementary reactions toward different active sites are not the same in complex heterogeneous processes. This is also one of the main causes for the appearance of gradients of the different parameters in a reaction system. Thereby, the mass diffusion may take place from one active site to another, from one phase to another, or from one zone of the reaction to another.[15]

In heterogeneous catalysis, the spatial separation of the steps or stages in a complex multistep reaction favors shifting the reaction to the right. How can it happen? Here, let us again follow Krilov's scheme,[15] written for heterogeneous catalytic reactions, occurring through the formation of intermediates in stationary or quasi-stationary, but not in equilibrium, concentrations

$$A \rightleftarrows B \rightarrow C \qquad (a)$$

having the rate constants k_1 for direct reaction $A \rightarrow B$, k_{-1} for reverse reaction $B \rightarrow A$, and k_2 for reaction $B \rightarrow C$. Consider a case when $K = k_1/k_{-1} \ll 1$ and the reaction $A \rightarrow C$ occurs via implementation of two consecutive steps: one through active centers X and the second on active center Y. The scheme can be written as

$$A \rightleftarrows B_x \rightarrow B_y \rightarrow C \qquad (b)$$

If the catalyst is a mixture or a composition of different metal oxides, depending on the chosen conditions, the first reaction $A \rightleftarrows B_x$ may occur on X centers of metal oxide and the reaction $B_x \rightarrow B_y$ on the other Y centers of the second metal oxide. It is obvious that the active centers are spatially separated. The overall reaction may be accelerated due to the permanent outcome of an intermediate to the other site or other zone of the reaction. In other words, the first intermediate passes from one surface site to another site or other phase, separated spatially each from another.[15] Apparently, some advantages of the application of the multiphase complex catalysts in certain catalytic reactions also relate to the overcoming of the thermodynamic restrictions via the separation of stages or steps.[15]

Formally, the above general schemes (a) and (b) correspond also to the case of the kinetic description of heterogeneous–homogeneous reactions. Considering that B is an intermediate, produced by X center and then desorbed to the gas phase, we may represent the above scheme as

$$A \rightleftarrows B_x \rightarrow B_{y(gas)} \rightarrow C \qquad (c)$$

where $B_x \rightarrow B_{y(gas)}$ is the desorption step of the intermediate to the gas phase. All the above-mentioned considerations taken into account in elucidations of the occurrence of the heterogeneous catalytic reactions are also valid in the case of heterogeneous–homogeneous reactions. The simplest case of that is scheme (c).

As examples, there are a number of decomposition reactions by the further transfer of active intermediates to the gas phase, presented in Section 2.3. Here, the situation is analogous to that of the spatial separation of active sites in heterogeneous catalysis. The only difference is that $B_{y(gas)}$ is not an adsorbed species, but rather a radical or other active intermediate species that appears in the gas phase. The final product C may be the same or different depending on different reaction pathways. In a number of cases, the final product(s) are the same, for example, in decomposition of hydrogen peroxide (H_2O, O_2) and ozone (O_2); oxidation of hydrogen (H_2O) and aldehydes (carboxylic acids), etc. In these cases, a general explanation for the existence of heterogeneous–homogeneous reactions may be given from the viewpoint of the phase separation of the steps. The slow step or stage $A \rightarrow B_x$ may be accelerated by the outcome of B_x from the reaction zone, in the given case, by the desorption of radicals from the surface. The desorption, in general, is an endothermic process. To overcome this stage, it is necessary to remove the product (intermediate or final), using the energy of coupled steps. In this case, the overall process may become energetically favorable. As a result, the $A \rightleftarrows B$ (or $A \rightleftarrows B_x \rightarrow B_y$) will be shifted to right, making possible the acceleration of the overall heterogeneous–homogeneous reaction.

The existence of the phase separation of steps (stages) was well demonstrated by Krilov and Shub in some examples taken from the literature of heterogeneous catalysis some decades ago.[51 in Ch. 1] As it has been mentioned above, the heterogeneous–homogeneous reactions are a particular case of the spatial separation of steps or stages by the diffusion of active intermediates not to other solid phases, but to the gas phase. In this regard, there are profound analogies between "purely" heterogeneous and heterogeneous–homogeneous reactions.

From the practical point of view, the effect of the phase separation in complex heterogeneous reactions often serves to control selective oxidation. In this aim, the different stages of the heterogeneous catalytic reaction may be separated both spatially and temporally. Spatiotemporal separation of the stages are widely applied for the redox type of heterogeneous catalytic reactions. It may be represented by the following simplified scheme[17] consisting of two stages of the complex process. The first of them is an oxidation reaction of any substrate by oxidant that may also be named pseudo-catalyst (metal oxides, complexes containing metal ions of variable valence, and oxo or peroxo atoms, etc):

$$Substrate + Pseudocatalyst(oxidant) \rightarrow Oxygenated \; substrate +$$

$$Pseudocatalyst(reduced \; form) \qquad (I)$$

The second stage, spatially and temporally separated from the primary, is the reoxidation (regeneration) of the reduced substance used in primary reaction by oxidant agents as dioxygen, hydrogen peroxide, etc.

$$Pseudocatalyst \; in \; reduced \; form + O_2 \rightarrow Pseudocatalyst(oxidant) \qquad (II)$$

The two above reactions (I)–(II) formally compose a catalytic cycle. This approach was successfully applied for selective oxidation with dioxygen in many industrial catalytic processes.[17–19] Sometimes it is one way to increase the selective formation of desired products.[17, 18–21] For example, the oxidation of butane to maleic anhydride using the $(VO)_2P_2O_7$ catalyst, which is known as the DuPont process in the chemical industry,[18, 19] is based on the spatial and temporal separation of the oxidation of butane and reoxidation of catalysts.[20]

Let's take another example. Environmentally, DDT (dichlorodiphenyltrichlorethane) is one of the most dangerous compounds and is known as a persistent pollutant.[22]

The synthesis and wide application of DDT was related to the names of two famous chemists, both of them Nobel Prize winners. DDT was synthesized by Austrian chemist O. Zeidler (1850–1911) in 1874, under the supervision of German chemist Adolf Von Baeyer (1835–1917), who was the founder of the famous pharmaceutical company, named after him. Baeyer was the 1905 recipient of the Nobel Prize in Chemistry. In 1948, "for the discovery of the high efficiency of DDT as a contact poison against several arthropods," the Swiss chemist P. Müller (1899–1965) was awarded the Nobel Prize in Physiology or Medicine, despite the fact that he was a chemist and not a medical researcher." For a long time, DDT was used to prevent malaria by "high efficiency."[22] However, according to the World Health Organization (WHO), since 2009, DDT has been restricted in most countries as a persistent organic pollutant strongly affecting human heath, as well as destroying animal life, especially birds. In 2015, WHO affirmed that use of DDT was linked to cancer.

In ordinary conditions, DDT is unreactive with air or oxygen and even with some strong oxidants, such as chrome oxide in glacial acidic acid or nitric acid.[23] However, the catalytic oxidation of DDT with a simpler oxidant, such as dioxygen, becomes possible in "mild" conditions, by the temporal separation of the catalytic oxidation stage, according to the above scheme *(I)-(II)*, using a TiO_2 supported dioxo-Mo-bipyridine complex (dioxo-Mo^{VI}-dichloro[4,4′-dicarboxylato-2,2′-bipyridine]) – as a catalyst. In the first stage, the main product of the interaction between DDT and complex is dicofol (alcohol). This stage may be accelerated photochemically, for instance, by UV irradiation. Stopping the reaction in a certain moment (stopping also UV irradiation), the reduced complex may be easily reoxidized under dioxygen, in the second stage, separated temporarily from the first one. The effects of the separation of the stages in the given example and in a complex heterogeneous–homogeneous reaction are almost similar (though not analogous). However, in a bimodal reaction, the separation occurs without external intervention or additional procedure. It arises spontaneously in the system and corresponds to the overall energetic balance of the process.

6.4 GENERATION AND DESORPTION STAGES OF ACTIVE INTERMEDIATES IN NONEQUILIBRIUM PROCESSES

Let us return to the desorption of active intermediates now, in light of the perceptions of nonequilibrium processes, as it is a pivotal stage for determining the bimodal sequence of the complex reaction.

Nearly everywhere adsorption is an exothermic process, or ΔH has a negative value.

$$\Delta H = \Delta G + T\Delta S$$

TΔS in the Gibbs equation has a negative sign, as at constant temperature and pressure, the adsorbed molecules or atoms lose a part of their transitional or rotational entropy. As the change in the Gibbs potential (ΔG) is negative for a spontaneous process, taking into account the entropic factor, the adsorption will be favorable at low temperatures and, conversely, unfavorable at high temperatures. The desorption, in general, is an endothermic process, for it is a reverse process of adsorption.

Therefore, at high temperatures, the desorption will be a thermodynamically favorable process. In the simplest case, for a reversible adsorption in an equilibrium state, when the adsorbent–adsorbate interactions are weak (10–50 kJ/mol), the activation barrier of desorption will be

$$E_d = (-\Delta H_a) + E_a$$

where ΔH_a is the thermal effect of the reversible adsorption and E_d and E_a are activation barriers of adsorption and desorption. Even if $E_a = 0$ for the adsorption, the desorption remains an activated process.[24] For example, the adsorption of methyl radical on a diamond(100) monocrystalline sample is not an activated (or nearly nonactivated) process, but the desorption has an activation barrier equal to 286.3 kJ /mol.[25]

According to the classical models of the heterogeneous catalytic reactions, the desorption of products is the last main stage of the multistage process. Obviously, in a heterogeneous–homogeneous reaction, the desorption of species from the surface is an intermediate stage in the overall heterogeneous–homogeneous process. It takes place in competition with the other surface reactions.[26] Each of these two main stages (surface reaction and desorption) has its own temperature dependencies. Therefore, the competition of the rates of the surface reaction and desorption determines what kind of reaction, either heterogeneous or heterogeneous–homogeneous, prevails in the overall process. A simple comparison may be done if the rates of desorption and chemical reaction, as well as their dependencies on the temperature on the surface, are known. For example, in the case of adsorbed methanol on the surface of Ni(100), using experimental data,[27] one may calculate the rates of its desorption (r_{des}) to the gas phase and the rate of its decomposition (r_{dec}) on the surface. It has been shown[4] that at about 250 K and above, $r_{des} \geq r_{dec}$.

desorption $\qquad r_{des} = 10^{14} sec^{-1} \qquad E = 59$ kJ/mol

decomposition $\qquad r_{dec} = 2 \cdot 10^9 sec^{-1} \qquad E = 38$ kJ/mol

Obviously, with increase in temperature, the desorption becomes a more prevalent process. At low temperatures (T < 250 K), the main process is a heterogeneous decomposition reaction. Analogously, in the case of adsorbed radicals on the surface, the temperature may be predicted when the desorption of radicals to the fluid phases prevails over the heterogeneous reaction, if the rate constants of the desorption of radicals and their reactions on the surface, as well as their temperature dependencies, were determined experimentally. In other words, like desorption of methanol from the surface, if the time interval for heating the sample is shorter than that for the decomposition reaction of the adsorbed radicals on the surface, most of the radicals will desorb to the fluid phases.

We have seen that in an equilibrium state, at low temperatures, the desorption nearly always is an endothermic and activated process. Therefore, this stage may not be a spontaneous process at low temperatures. Apparently, the chemical system uses the thermal effect of other exothermic chemical acts on the surface for desorption of species. In other words, here the thermodynamic and kinetic conjugation of elementary chemical acts may take place. On the other hand, the desorption of radicals or other species strongly depends on coverage of the surface. For instance, the weakness of the chemical bonds of radicals on the surface in co-adsorption with other species was shown in the example of the co-adsorption of methyl radicals with oxygen on the surface on Rh(111).[28] Unfortunately, the enthalpies and rate constants of desorption of the radicals, as well as their reactions on the surfaces, are known only in rare cases.

In heating, the energy transfers to the different components of the system by different pathways. A certain quantity of the internal energy of the adsorbed radicals or other species on the solid surface, in nonequilibrium conditions and in certain circumstances, may be transformed into either kinetic energy of species, desorbing them, or into vibrational and rotational energy of the bonds, resulting in the dissociation and further reactions. The chemisorption of species on the surface in the majority of cases is an exothermic chemical reaction having about 70–200 kJ/mol, even up to a 400 kJ/mol thermal effect. The energy released in this process is the difference of the energies of the old and new chemical bonds, as always in the chemical reactions. Part of this energy may be transferred to the solid, and other parts may be distributed between the different degrees of freedom of the formed species or transformed into their kinetic energy. The energy accumulated in new species may also be transferred to the electronic or vibrational-rotational degrees of freedom. Moreover, the energy of the electronically excited states of species may be transferred into vibrational states. The heat transfer often leads to different physical phenomena and effects, such as emission of different lattice atoms to the gas phase or diffusion to the bulk of crystal of the adsorbed species, luminescence from the crystal, electronic emission from the surface, nonequilibrium conductivity in crystals, etc.

In the case of metals and metal oxides, the desorption of hydrocarbon radicals (R) has a high activation barrier; for example, for methyl radicals from (0001)-oriented α-Fe_2O_3, it is 84 or 133 kJ/mol in different UHV regimes.[29] Lunsford represented the experimental values obtained by him and by other researchers for desorption of CH_3 radicals from Li^+ promoted MgO, which are 22–24 kcal/mol.[129 in Ch. 1] The activation barrier of desorption in the case of methyl radical from metal, such as $Ga(Ga\text{-}CH_3)$, is 39 kcal/mol, according to the experimental data.[30] These and other data indicate that in the mentioned cases, in the heterogeneous–homogeneous reaction the methyl radical on the surfaces is mainly in the chemisorbed state.

Nonequilibrium (irreversible) states may appear due to the chemisorbed state and further reactions of the adsorbed species. *A priori*, it may be accompanied by the relaxation: reconstruction and reorganization of the surface structure, which also are preconditions of the appearance of nonstationary concentration of active centers and the autocatalysis phenomenon. An example of nonequilibrium desorption of the radical was given by Grankin et al.[31] By the method of modulated molecular beams, it was shown that the interaction of different molecules with excited atoms or radicals on the surface led to the oscillatory mode of the reaction. Obviously, this kind of reaction is a nonequilibrium process. The oscillatory mode of reaction was observed experimentally in the reaction

$$H + H_{(ads)} \rightarrow H_{2\,(ads)}$$

in molecular beam investigations on a Teflon-covered surface. High frequency vibrations of H_2^{γ} (about 4eV) appear as a result of this reaction.[31]

The essential property of heterogeneous catalytic reactions, and also that of heterogeneous–homogeneous reactions, is their cyclic nature. Obviously, the regeneration of the surface active site(s) in catalysis is an indispensable stage among the molecular events on the surface. The regeneration of the active site may take place when the adsorbed or chemisorbed chemical species leaves the active site. This creates the preconditions for "self-repair and reorganization" of the surface—in other words, the regeneration of the active site. The surface reorganization regenerating the active site is one of the exhibitions of self-organization in catalytic systems. Then, the repaired surface active site, in turn, regenerates the intermediate species. Without a doubt, all these phenomena are interconnected by different interactions in a complex system. The integrity of all interactions in the system is governed by a common phenomenon, self-organization, which may be exhibited as a combination of elementary steps of the stepwise reactions, composing the catalytic cycle or autocatalytic reaction, reconstruction or reorganization of the solid phase, and, finally, periodical (cyclic) regeneration of the active sites. Krilov[51 in Ch.1] described this state of the surface of the solid substance in catalysis as "excited," emphasizing the divergence from the equilibrium state. Note that the active site on the surface may exist before the reaction or may be generated during the reaction.[15]

Different theoretical models have been elaborated to describe the chemical process in equilibrium and nonequilibrium states in catalysis, which include a number of energy transfer coefficients, constants, and other parameters.[27, p. 52] Unfortunately, the determination of the majority of them is not available by the experimental pathway (for example, the so-called adiabaticity parameter describing the efficacy of energy transfer to the bonds in the adsorbed state,[27] which until now had not been determined experimentally).

Are the catalytic events synchronous in time and space (length) scale? The characteristic time of the majority of catalytic reactions is a second (length 0.1 micron); adsorption, desorption, and diffusion, 10^{-3} s; but reconstruction 10^{-9} s (nanoseconds). That is why reconstruction of the surfaces and regeneration of the active site(s) may often not be easily revealed in experimental studies. According to Krilov and Shub, the overall duration of the effect of the "excited" state may be a few orders higher than the "relaxation time" of the catalyst.[51 in Ch. 1] As regards the desorption process, there are a number of experimental methods for investigation of the desorption of radicals,[32, 33] such as temperature programmed desorption (TPD), isothermal desorption (ID), scanning tunneling microscopy (STM), laser-induced thermal desorption (LITD), field-ion microscopy (FIM), and molecular beam reactive scattering (MBRS). In the surface science literature, the theoretical methods of Monte Carlo, transition state theory, molecular dynamics, electronic configuration calculations,[32, 33] and others are well known.

6.5   HETEROGENEOUS–HOMOGENEOUS OXIDATION AND SELF-ORGANIZATION IN THE CHEMICAL SYSTEM

Self-organization is one of the most common phenomena in nature and society. That is why it is a subject of investigation not only in physics, chemistry, and biology, but also in mathematics, cybernetics, sociology, economy, etc. In chemistry, one of the simple (although nonexact and incomplete) definitions of the term is that self-organization is a phenomenon of the spontaneous formation of structures ("creation of order") from the interacting components (few or many) of the chemical system, in conditions far from the equilibrium state.[34]

From the viewpoint of the consideration of the entropy of the system, it is a move of the dynamic system from a "less ordered" to a "more ordered" state. In mathematical terminology, the emergence of self-organization is feasible by typically nonlinear interactions of components in systems. In the case of linear interactions, "the effects are proportional to their causes."[35] While, in the case of nonlinear interactions, the effects are not additive. Another important precondition for the emergence of this phenomenon is the complexity of the system. It would be specially noted that there is no controversy between the second law of thermodynamics and this phenomenon. The second law is applicable for steady-state or reversible processes in isolated system that does not exchange energy or matter with its surroundings. In consideration of these peculiarities, according to Heylighen,[35] self-organization is the spontaneous emergence of global coherence out of local interactions. In the thermodynamics of open systems, one of the important phenomena is multiplicity of the steady state for the same set of parameters.[36] Multiplicity of steady states arises in the nonlinear interactions of the components of a chemical system. Here, we should again mention the difference between the steady and equilibrium states. In steady state, although the system does not change its parameters with time, we cannot characterize this state as a thermodynamically equilibrium state. In this case, the processes are continuing, while in a system at the thermodynamic equilibrium state, the processes are stopped.

There exist some relations between the phenomenon of self-organization and Le Chatelier's principle, the well-known principle in "classical" chemistry[37,38] formulated in 1884 by French chemist and engineer Henry Louis Le Chatelier (1850–1936).[37] This principle states that, if the dynamic equilibrium in a chemical system is perturbed by changing the conditions of its existence, the system moves in the direction counteracting this change. Therefore, the system itself creates conditions to move to another equilibrium state that may be less stressed by changes in its environment. In this regard, Le Chatelier's principle is a simple and particular case of exhibiting the common phenomenon of self-organization in chemical systems. In fact, this is an example of "negative feedback" in a chemical system, which is known also in the theory of the self-organization phenomenon.[36]

Thus, under appropriate conditions, far from the equilibrium state, a system with chemical reactions may be spontaneously organized. Self-organization in chemical systems may be exhibited in molecular self-assembly, autocatalytic networks, reaction–diffusion systems, formation of spatiotemporal patterns, or oscillations of the concentrations and rates of reactions components. The existence of metastable states, multiple steady states, fixed points, hysteresis, and a number of critical phenomena in complex systems are indications of self-organizing systems.

As mentioned earlier, developing the thermodynamics of the irreversible processes, Prigogine created the theory of these phenomena in the 1950–1960s (the reaction–diffusion model named the Brusselator).[39,40] In a nonlinear system, far from equilibrium conditions, Prigogine hypothesized the formation of structures termed dissipative. They concern the structures arising as a result of the export, flux, or dissipation of entropy from the system, in fact, creating order via internal organization.

In a closed but not isolated system, as a driving force for the self-organization phenomenon may serve the minimization of free energy. For the appearance of dissipative structure, the decrease of free energy ($\Delta G < 0$) in chemical reactions is a necessary condition. If the reaction is endergonic, this condition may be achieved by coupling with another reaction with a larger exergonic effect. In this case of self-organization, the energy is produced within the system and is not taken from the surrounding environment. In an open system, self-organization takes place in the conditions of the continuous outcome and income of the energy and mass by continuous production and export of entropy to the environment. Some thermodynamic preconditions, which lead to self-organization in chemical systems, were given in Section 1.7.2.

The phenomenon of self-organization may be revealed in many real, multicomponent, and complex chemical systems. The classical example of the exhibition of self-organization in homogeneous catalytic reactions is the Belousov–Zhabotinsky reaction (see Section 6.7.1). Conditions leading to the emergence of oscillations, chemical waves, and self-organized spatiotemporal patterns, pulses, and spirals in homogeneous catalysis were given in Section 6.6. Analogous phenomena, according to Ertl,[35 in Ch. 3] were first observed by Wicke and co-authors in 1972, in investigations of the heterogeneous catalytic system of oxidation of CO on Pt.[41] However, the first well-proven demonstration of this phenomenon in heterogeneous catalytic reactions was presented by Ertl with co-workers on the example of CO oxidation on the crystalline metal samples and was represented in a very interesting form in his Nobel lecture in 2007. In this lecture, Ertl demonstrated a "molecular movie" for the first time for the wide scientific audience, "obviously confirming the emergence of the self-organization phenomena."[35 in Ch. 3]

The existence of reaction cycles in the complex chemical transformation is evidence of self-organization in the chemical system. The chemical system, completing one catalytic cycle, itself creates the conditions necessary for a new catalytic cycle. This is a natural peculiarity of both the chain-radical and catalytic reaction systems. However, this is considered a low-level emergence of this phenomenon.

In heterogeneous catalysis, involving also bimodal reaction sequences, the investigation of self-organization phenomena is linked with the notion of complexity of systems. Complexity arises[42] in multicomponent systems, components of which are interacting one with the other in conditions far from equilibrium. In chemical systems complexity is related to the chemical reactions (kinetic complexity), on the one hand, and to the molecular structures (structure complexity), on the other hand. In this regard, the majority of the heterogeneous catalytic reactions are complex systems in both aspects. The exhibition of self-organization in heterogeneous catalysis often connects with surface relaxation, reorganization, and reconstruction of the catalyst during the catalytic cycle. In their review article, Piumetti and Lygeros[43] examined four groups of catalysts: 1. single-site catalysts; 2. nanocatalysts, 3. zeolites, and 4. metal oxides (which operate as redox catalysts) in relation to self-organization phenomena. For the first group of catalysts, named single-site catalysts, nearly all mechanism suggestions involved the self-repairing of the single active site during the reaction. Different types of catalyst were mentioned within this group, such as ions, atoms, molecular complexes or clusters, immobilized organometallic species "ship-in-bottle structures," and microporous materials. In this case, the self-repairing of the active site was considered the lowest level of self-organization, as the interactions with neighboring sites are minimal and mass transfer phenomena are not essential. Nanoparticles, operating as nanocatalysts, have a higher number of coordinative unsaturated sites, and interactions of sites are more probable in this case. Under the changes via adsorption from the gas phase, the nanoparticle has tended to minimize surface energy. According to the authors' description, the metal nano-sized particles typically form convex polyhedral structures, exhibiting different surface atom densities, electronic structures, chemical reactivity and thermodynamic properties. These conditions lead to reconstruction and self-organization phenomena in nanocatalysis. For the group of zeolite catalysts, the reorganization and reconstruction of the Bronsted acidic sites take place on the surface. The specific feature of zeolites is exhibited in non-convex structures, which are able to discriminate both reactants and products molecules by size and shape. Evidently, the interactions of sites will be well exhibited in the catalytic reactions of this group. In the case of metal oxides catalysts, mainly operating by the Mars–van Krevelen mechanism, it was considered that active sites were not isolated and formed ensembles, a catalytically active network exhibiting high level self-organization of the surface layer in catalytic action. For these four groups of catalysts, the main characteristic is complexity of the structures leading to the formation of ensembles of active sites (in exception in the first group). Nonlinear combinations of complexity structures were considered a cause of diversity of the reaction pathways and products, and in some cases, even a cause of nonreproducibility of results in multicomponent systems. Note that all this description is qualitative rather than quantitative. However, Piumetti and Lygeros concluded that the level of self-organization increased at the transition from the first to the fourth group of catalysts. In addition, the mass transfer phenomena increased the complexity in all four groups of catalysts. The schematic representation of this dependence is given in Figure 6.2. Another generalized elucidation was done in the same work[43] in discussion of the reactions catalyzed by transition metal oxides. It was concluded that "both structural and chemical complexity are closely related, one to other, in transition metal oxide catalysis."[43]

Returning to the heterogeneous–homogeneous reactions, note that they completely involve the above-described peculiarities of surface processes in the exhibition of the phenomena of self-organization. In appropriate conditions, the heterogeneous–homogeneous mechanism of reaction is possible for all four above-mentioned groups of catalysts, the examples of which were given in Section 3.5. Moreover, the chemical interactions in heterogeneous–homogeneous systems are more complex than those in the case of the only heterogeneous reaction. It was shown by the theoretical analysis of the model reaction systems of increasing complexity, involving the cases of the bimodal reaction sequences (see Section 6.8). Indeed, heterogeneous–homogeneous reactions are cyclic processes connected with the periodic rebirth and disappearance of active sites on the surface and intermediates in the fluid phases, through the interactions of components between two phases and reconstruction or reorganization of solid-state substances. In any heterogeneous–homogeneous reaction, occurring far from the equilibrium state, the general decrease of free energy is accompanied

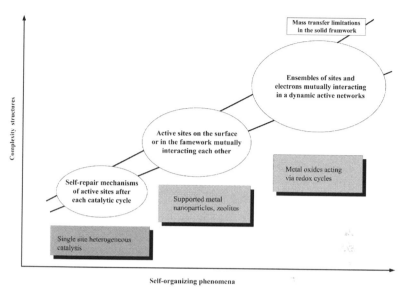

FIGURE 6.2 Scheme presenting the self-organization phenomena in dependence on the complexity of structure in different catalytic processes (Adapted and simplified from the work of Piumetti M., Lygeros N., Ref.[43])

by the flux of entropy out of the system. Self-organization in the bimodal reaction system emerges just in this manner.

6.6 WHY THE REACTION "CHOOSES" A HETEROGENEOUS–HOMOGENEOUS PATHWAY

As a rule, multistage and stepwise heterogeneous–homogeneous reactions are non-equilibrium processes in multicomponent and complex systems. How does a heterogeneous–homogeneous reaction progress? How does the reacting system overcome the restrictions imposed by thermodynamics?

From the point of view of reaction kinetic analysis, in general, thermodynamic restrictions may be overcome in two ways. One way is through creation of the superequilibrium concentrations of active intermediates (see Section 1.6). It may happen due to the utilization of part of the energy released in the primary oxidation reaction. It is in just this way that branched-chain reactions progress in the gas phase. The analogous phenomenon also exists in the heterogeneous catalysis. Like branched-chain reactions, the endothermic elementary acts in heterogeneous reactions become possible due to the accumulation of the superequilibrium concentrations of surface species, which may be stationary but not equilibrium concentrations. All this also applies to heterogeneous–homogeneous reactions. An example of the accumulation of superequilibrium concentrations of surface intermediates in the heterogeneous–homogeneous reaction was shown in the ethanol oxidation with air over the Pt-catalyst, discussed in Section 3.5.3; Ref.[123 in Ch. 1; 240 in Ch. 3]

The second way is the permanent removal of the intermediates or products from the local reaction zone by their diffusion or transfer to another phase. As a result, the overall reaction may be shifted to the right (see Section 1.7.2). In Section 6.4, it was shown that the important stage of the heteogeneous–homogeneous reaction, desorption of intermediates, usually is not a spontaneous process. It may occur either by the creation of superequilibrium concentrations of surface species, or by the permanent "removal" of active intermediates from the reaction zone.

In certain conditions, heterogeneous–homogeneous reactions are exhibitions of the spontaneous process of the spatial separation of stages in the complex process. The reaction chooses this pathway itself, by self-organization in the system. It arises by the spontaneous removal of the intermediate species from one phase to the other, shifting the reactions to the right. Via kinetic coupling, the thermodynamic restrictions may be overcome.

Thus, these conditions are essential requirements for the occurrence of the reaction in systems fare from equilibrium. In this regard, Krilov and Shub, in their book *Nonequilibrium Processes in Catalysis*,[51 in Ch. 1] wrote: *The concept of the reaction affinity introduced in thermodynamics of irreversible processes permits to state of necessary conditions for the reaction to become irreversible. In heterogeneous catalysis, the continuous shift of equilibrium of the intermediate stages may occur as a consequence of their spatial separation, so that different steps of the process may take place on different spatially separated active sites, or even in different phases. In addition to concentration gradients, temperature gradient can also occur in heterogeneous catalysis, resulting in the shift of intermediate equilibrium.*

Although this conceptual statement concerns heterogeneous catalytic reactions, and, here, there is no direct mention about heterogeneous–homogeneous reactions, it fully corresponds to the cases in which the heterogeneous reaction generates the active intermediates on the surface, the further reactions of which may occur on the other active site, in other reaction zones or in other phases, *spatially* separated in reaction conditions. The irreversibility and spontaneous shifting of the reaction to the right, therefore, will be expected in heterogeneous–homogeneous reactions.

6.7   CHEMICAL OSCILLATIONS IN HETEROGENEOUS AND HETEROGENEOUS–HOMOGENEOUS REACTIONS

6.7.1   OSCILLATIONS IN SOME HETEROGENEOUS CATALYTIC REACTIONS

The spontaneous appearance of oscillations in chemical reactions is one of the exhibitions of self-organization phenomena in chemical systems.[44–48; 53, 68 in Ch. 1] The existence of the oscillatory regime of oxidation of hydrocarbons and other organic compounds in chemistry were known beginning from the investigations conducted in the nineteenth century.[44] However, only in the second half of the twentieth century did chemical oscillations receive clarification in light of the perceptions of the self-organization phenomena.[48] The oscillations in chemical systems may be expressed as periodic changes in the concentration of components and their rates of transformation or other parameters (for instance, temperature or pressure) of systems across time, from one extreme limit to the other (maximum ↔ minimum).

In 1910, Lotka mathematically described the autocatalytic chemical reaction by differential equations corresponding to the simple reaction scheme $A \rightarrow B$, occurring via the formation of intermediates and having oscillatory behaviors.[45] Then, Volterra attempted to explain oscillatory behaviors in the "prey–predator system," expressed as changes in the population of prey and predators with feedback effect, by analysis of the corresponding differential equations. The scheme of the $A \rightarrow B$ reaction was presented as

$$A + X \rightleftarrows 2X$$

$$X + Y \rightleftarrows 2Y$$

$$Y \rightleftarrows B$$

where X was prey, Y predator. This scheme involves two interacting intermediates. The solution of the differential equations of the rates (known as the Lotka–Volterra equation) for certain parameters shows oscillating behaviors in the system. Although the existence of oscillations in this simple case was evident, its application in the kinetics of the complex chemical reactions was not completely realistic.

In Chapter 1 (Section 1.6), we mentioned the first reaction–diffusion model produced by Alan Turing, which permitted the description of oscillatory behaviors in chemical systems. Prigogine and co-workers[48] proposed another model of the chemical system, named the Brusselator that exhibited similar behaviors. It was a mathematical model of a hypothetical autocatalytic reaction occurring as a sequence of the following steps:[48]

$$
\begin{array}{ll}
A \rightarrow X & k_1 \\
B + X \rightarrow Y + D & k_2 \\
2X + Y \rightarrow 3X & k_3 \\
X \rightarrow E & k_4
\end{array}
$$

where X and Y are intermediates k_i (k_1, k_2, k_3, k_4) constants of the rate of reactions.

In the 1950s, Nicolis and Prigogine proposed the main concept of the appearance of chemical waves based on the idea of the so-called dissipative structures, which, in fact, were stable, regular structures appearing as a result of the self-organization of the reacting species. For his works in thermodynamics of irreversible processes Prigogine received the Nobel Prize in Chemistry in 1977. Here, we would not like to describe this theory with his voluminous mathematical analysis, as all details of the theory were well represented and commented in a great number of physical, chemical, and biological science textbooks. Only some examples related to the oscillations in heterogeneous and, especially, heterogeneous–homogeneous reactions will be represented below.

Chemical oscillations (also named chemical waves or chemical clocks) are found in homogeneous, heterogeneous, and heterogeneous–homogeneous reactions of oxidation of organic and inorganic compounds with molecular oxygen, in certain conditions and parameters of the chemical systems. The classical examples of chemical oscillations in homogeneous systems are Belousov–Zhabotinsky, Briggs–Rauscher, and Bray–Liebhafsky reactions.[68 and 53 in Ch. 1; 45; 48]

In Section 6.5, we mentioned that the first famous example of the periodic reaction and chemical oscillations was the Belousov–Zhabotinsky reaction. In 1958, Belousov observed periodic oscillations of the concentrations of cerium ions (Ce^{4+} and Ce^{3+}) in the bromation reaction of malonic acid $CH_2(COOH)_2$ to obtain $CHBr(COOH)_2$, in stirring conditions. However, the first attempt to publish these results was not successful: The paper was refused for publication. All that was published on the subject was a short report that appeared in 1967, in the Proceedings of a conference on radiation chemistry. Later, Zhabotinsky, from Moscow State University, confirmed the existence of the observed phenomenon and in 1980 (after Belousov's death) was awarded the Soviet Union's highest prize of that time: Hero of Socialist Labor.[44]

In heterogeneous catalysis, the most investigated reaction that may occur in the oscillatory regime is oxidation of CO with dioxygen on the surfaces of noble metals, although a great number of other examples are now known.[45] There are also a large number of examples of heterogeneous–homogeneous reactions in which chemical oscillations were observed (oxidation of methane and ammonia on the metals; acetaldehyde, ethylene glycol on the metal oxides and salts, alcohols on the transition metals, etc.[49–51; 115 in Ch. 2]).

All oscillating reactions share at least some common features. All are multicomponent systems, where the chemical reactions occur as stepwise processes, producing more than one active intermediate. On the other hand, all these reactions, without exception, occur in thermodynamically nonequilibrium conditions, which finally play a determining role in different exhibitions of self-organization phenomena. All oscillatory observations in chemical reactions, mathematically (kinetically) may be described as nonlinear processes. Nonlinearity is key to describing self-oscillations, including reaction rate oscillations in both heterogeneous and heterogeneous–homogeneous systems. Among the common features of oscillating chemical systems, Parmon[102 in Ch. 1] mentions the irreversibility of at least one stage (step) of the process.

In some cases, the appearance of oscillations in heterogeneous catalytic reactions may be caused by physical phenomena, such as heat and mass transfer processes and phase transition in the surface layer.[53 in Ch. 1] However, in the majority of cases, the oscillations in heterogeneous catalytic systems have chemical or kinetic origins related to the complex mechanism of chemical processes.

In 1978, Bykov, Yablonski, and Kim[52] gave the first simple kinetic model of the heterogeneous reaction, showing the possibility of chemical oscillations and multiplicity of the steady state:

$$A_2 + 2Z \rightleftarrows 2A\text{-}Z \qquad O_2 + 2Z \rightleftarrows 2O\text{-}Z$$
$$B + Z \rightleftarrows B\text{-}Z \qquad CO + Z \rightleftarrows CO\text{-}Z$$
$$A\text{-}Z + B\text{-}Z \rightarrow AB + 2Z \qquad O\text{-}Z + CO\text{-}Z \rightarrow CO_2 + 2Z$$
$$B + Z \rightleftarrows (B\text{-}Z)$$

where Z is surface active site and additional (B–Z), formed by the so-called buffer step, is a "unreactive" intermediate on the surface. Three differential equations in the form of a mass action law may be written from this scheme. Twenty-three "phase portraits" were found for seven parametral variables, involving oscillations, multiplicity of steady states, chaos, and other bifurcations.

The emergence of oscillations is linked with the phenomena of the phase transition observed experimentally in heterogeneous reaction systems. By application of surface science methods, self-oscillations in heterogeneous-catalytic reactions were experimentally investigated in the systems, such as $H_2 + O_2$; $CO + O_2$; $NO + H_2$; $NO + NH_3$; $CO + NO$ on Pt, Rh, Ir, and others.[53–55] In particular, in the $H_2 + O_2$ reaction on Pt(111) and Pt(100) on monocrystalline surfaces, periodical phase transitions by the formation of "subsurface" oxygen layer were observed. This means the transfer of the active sites from one state to another, more reactive state. Apparently, the transfer is a reversible process and plays an essential role in initiation of self-oscillations and chemical waves. The catalytic oxidation of hydrogen with oxygen on nanosized noble-metal monocrystals Pt(111) and Pt(100), as well as Rh and Ir, was investigated by Gorodetskii[56] applying the field emission microscopy (FEM), field ion microscopy (FIM), temperature programmed desorption (TPD), and high-resolution energy electron loss spectroscopy (HREELS). The main result of the work was *in situ* and "nearly visual" observation of the propagating waves of O_{ads} and H_{ads} layers. Under certain critical conditions (p, T), the oxidation was accompanied by propagating wavefronts possessing a sharp boundary between O_{ads} and H_{ads}. The formation of H_2O and spatial distribution of Pt-sites (atomic resolution about 5Å) also were "visualized" by the applied technique. The observed self-oscillations were investigated in more detail. It was concluded that the regular chemical waves were initiated on a Pt nanoplane surface by reversible-phase transition Pt(100)-hex $\leftrightarrow$ (1×1)Pt(100).

A well-known example of the observation of kinetic oscillations, accompanied by surface reconstruction, is the reaction of CO with oxygen in UHV conditions on a Pt monocrystal or crystallite surface.[54] The following surface reconstruction takes place in the oscillatory regime: Pt(100) nonreconstructed (1×1) phase transforms to the reconstructed hexagonal phase, and Pt(001) nonreconstructed (1×1) phase transforms to the reconstructed (1×2) missing row phase. These conclusions were based on measurements of the photoelectric work function.[54] It is noteworthy that in absorption, both the adsorbent and adsorbate layers undergo reconstruction simultaneously. Another example of the oscillation of the reaction rates is oxidation of CO on Pd(110) at $T = 450$ K at UHV conditions, on fixed concentrations of reactants, and other experimental conditions.[56] Here, as well as in the above-mentioned cases, surface reconstruction is apparently related to the self-organization of randomly offered molecules on the surface.[52, 55, 56]

6.7.2 Oscillations in Heterogeneous–Homogeneous Reactions

As mentioned earlier, oscillations were observed in a large number of heterogeneous–homogeneous reactions of oxidation with oxygen, of both organic and inorganic compounds: CH_4, NH_3, CO, H_2 on metals, alcohols and aldehydes on transition metals and its oxides, etc.[49–51; 115 in Ch. 2] Chemical oscillations in the oxidation or combustion of some hydrocarbons and oxygenated organic compounds were observed, even beginning with the very early investigations of cool flames after their discovery by Sir Humphry Davy (1815) in experiments with diethyl ether.[57] Cool flames have unusually low temperature (80–500°C) and release very small amounts of heat, and often the reaction is accompanied by light emission. In certain cases, in acetaldehyde/air (or oxygen) mixture, the cool flames appear in the regime of the chemical oscillations of the heterogeneous–homogeneous reaction.[115 in Ch. 2] Although the cool flame usually appears in the homogeneous-fluid phase, it is linked with the heterogeneous reactions on the walls of the reaction vessel or solid-phase catalyst. For example, in oxidation of the hydrocarbons with air of a certain composition of reactants, it was found that the appearance and position of the cool flame in a reactor depended on the nature of the reactor material.[49; 115 in Ch. 2] For example, quartz and Pyrex (borosilicate) reactors were tested in the oscillatory regime of cool flames of isooctane and n-heptane.[58, p.20] In the quartz reactor, the emission of light was not observed, while the Pyrex reactor stimulated cool-flame oscillations and emission of light. The assumption was that the HO_2 radicals formed in reaction had different rates of termination on the walls: the quartz vessel, resembling the base- or salt-coated surfaces, destroyed HO_2 more effectively than the Pyrex vessel, resembling the boric acid-coated surfaces.[58]

Barcicki and Wojnowski[49] observed chemical oscillations in NH_3 oxidation with oxygen on a Pt-Rh catalyst.[46] They detected periodic changes of temperature on the catalyst and periodic initiation of the gas-phase reaction of oxidation in flow conditions. The results were interpreted as a coupling of heterogeneous dissociation and homogeneous oxidation of NH_3.[49]

"Critical phenomena," particularly oscillations in the oxidation of CO with oxygen to CO_2, were observed over Pd or Pd-containing catalysts in heterogeneous–homogeneous reaction systems.[59, 60] In heterogeneous–homogeneous reactions of the oxidation of ethylene glycol with oxygen over the Ag-catalyst (650–700°C in air flow), Vogyankina et al.[51] observed temperature oscillations in the catalytic layer. According to these authors, glyoxal forms on the surface of the catalyst, while CO_2 from CO in the volume phase, between grains of the catalyst, related with the oscillatory regime of reaction.[51] Werner et al.[50] described oscillations in the ethanol/air mixture system over Cu chips at 660 K. They observed the oscillations with cycle periods of about 3.5 min. Oscillations were detected by maxima and minima of ethanol consumptions, producing corresponding amounts of formaldehyde and CO_2. In oscillatory regimes, depending on reaction conditions, the partial oxidation of methanol on the Cu-catalyst gives formaldehyde and the total oxidation CO and CO_2. Applying a combination of some experimental methods of surface science, the oscillations were observed on single-crystal Cu(110) at 573 K and Cu(111) at 623 K,[61] in oxidation of methanol. The main conclusion was that the dynamic phenomena related to the oscillations arose due to the chemical mechanisms, which were a result of the various exhibitions of self-organization phenomena. Oscillations were more often observed in the gas phase, near the surface of the catalyst. The Mars–van Krevelen reaction mechanism was suggested for surface reaction by two independent pathways related to the oscillatory regime. On the other hand, these authors highlighted the role of gas-phase thermal conductivity in this reaction. For instance, the replacement of He by Ar results in changes of the characteristics of oscillations. In mechanism suggestions, linked with the appearance of oscillations, it underlined the role of the oxidation state of cupper depending on the oxygen/methanol ratio.

As the above examples of heterogeneous–homogeneous reactions show, although the mass and energy transfer, as well as other physical phenomena, play an important role, the oscillatory regime is mainly chemical in origin. Here, as well as in "purely" heterogeneous or "purely" homogeneous oxidation, the mechanism suggestions are based on the existence of two competing and interacting channels of reaction, accompanied by the diffusion of intermediates and heat transfer phenomena. One of them generates intermediates, which are consumed by another channel. When the concentration of this intermediate decreases to certain values, the second channel becomes predominant. Then, repeatedly, the predominant channel of reaction changes. The peculiarity of the oscillations in heterogeneous–homogeneous reactions is that the intermediate formed by one of the reaction channel may be transferred from one phase to another. An example of the reaction scheme of the oscillating heterogeneous–homogeneous reaction, corresponding to the above-described scenario of chemical events, is given in the next section. It is acetaldehyde oxidation with air in certain conditions.[62] In the suggested scheme, the leading active center (radical RCO_3) produces vibrational excited formaldehyde $CH_2O(v)$, which plays a role of inhibitor for this reaction channel.[62]

As a main conclusion from the above-mentioned examples, the exhibition of self-organization phenomena is obvious in the systems far from equilibrium, where the irreversible processes become a source of creating order via dissipative structures. One of the interesting features of oscillations in the case of the heterogeneous–homogeneous reaction sequence is that they appear nearly synchronously in both the solid and gas phases. For example, in heterogeneous–homogeneous oxidation of propane in the presence of a nickel wire catalyst, the oscillation of the temperature and concentrations of reactants simultaneously appeared on the surface of the catalyst and in the gas phase at 650–750°C.[63]

Often, as in the case of oscillating reactions, we are able to phenomenologically or kinetically describe a chemical system from the thermodynamic viewpoint, even without understanding the details of the complete mechanism. However, only knowledge of the detailed mechanism, more or less proven experimentally, will permit control of the desired reaction pathway and use of it in practical aims. Presently, the oscillating heterogeneous–homogeneous reactions of oxidation have no important practical application, excluding some possible applications in analytical chemistry for determining trace components.[64] Nevertheless, it is possible that at any time they may be used for certain aims. In the history of the sciences, there are many examples of applications of phenomena to occur much later. For instance, lasers were not widely applied in practice until a very long time after their discovery.

6.8 MODELING HETEROGENEOUS–HOMOGENEOUS REACTIONS

Mathematical models of heterogeneous–homogeneous and homogeneous–heterogeneous reactions are based on the sets of reactions occurring as parallel and/or consecutive sequences on the solid surface and in the fluid phases, involving the reactions of intermediates, as well as energy and mass transfer processes.[65–72; 50 and 87 in Ch. 1; 189 in Ch. 3] Modeling of the heterogeneous–homogeneous reactions is complicated by certain factors.[189 in Ch. 3; 65, 70] A model of the heterogeneous–homogeneous reaction must combine two interacting constituents of the dynamic system: the model of the heterogeneous reaction and the model of the homogeneous reaction, including all mass and energy transfer processes between two phases. Unfortunately, for numerical modeling, the experimental kinetic and thermodynamic data are usually insufficient for both homogeneous and heterogeneous constituents in the overall reaction.[70] It is conditioned by the absence not only of the majority of the uptake coefficients, heterogeneous rate constants, thermal effects in interactions of radicals, and other active species on the surfaces, but also of other important parameters in heterogeneous–homogeneous reactions, as the diffusion coefficients of species on the surfaces and in different reaction media. Theoretical calculations of the set of these data, for example, by DFT or molecular dynamics simulations, also are connected with a number of difficulties. Some of them were elucidated by Deutschmann in his review article.[69]

On the other hand, the suggested mechanism of any reaction may never be considered as exactly and definitively determined. The development of chemical kinetics gives much evidence of this statement, which may be found in textbooks. For example, the history of the development of mechanism suggestions for the $H_2 + I_2$ reaction shows how the notions about the mechanism of reaction were periodically changed during more than one century, in light of new experimental data.[72] In general, the agreement between experimental data and a chosen scheme of the mechanism is not yet proof that it is the only right suggestion.

These limitations and complexities of reaction systems[70] are some of the obstacles to creating generalized mathematical models of heterogeneous–homogeneous or homogeneous–heterogeneous reactions that sufficiently describe the behaviors of this kind of system. In particular cases, the majority of the investigated mathematical models simulate the catalytic combustion or oxidative conversion reactions, for instance, of hydrogen[70, 71] or hydrocarbons.[50 in Ch. 1; 189 in Ch. 3; 73] General problems and desirable strategy of the modeling of the bimodal reaction sequences in oxidation were examined in the review articles of Sinev at al.[87 in Ch. 1; 189 in Ch. 3] devoted to the light alkanes C_1-C_4, as well as those of Deutschmann[69] involving other classes of compounds.

The complexity of the description[70 (p. 44)] and variety of interactions in the coupling of heterogeneous and homogeneous reactions may be hypothesized by the scheme presented in Figure 6.3. Let us consider that the same reactant(s) may react in the gas phase and on the surface of the solid phase(s) by parallel and consecutive reactions forming different intermediates and products. Let us consider also that there exist predominant reaction channels for the reactant(s) in primary reactions in both phases. All reactants and intermediates, *a priori*, may interact or react with others in or on different phases, forming a number of new (secondary) intermediate species. Moreover, all species formed, in their turn, may be transferred from one phase to the other and may react each with the other in the gas and on the solid phases. In this case, the composition of the final products of overall heterogeneous–homogeneous interactions may be quite complex, containing the products of the primary homogeneous and heterogeneous reactions, as well as further reactions in/on both phases by the participation of the species transferred either from the gas phase to the surface or

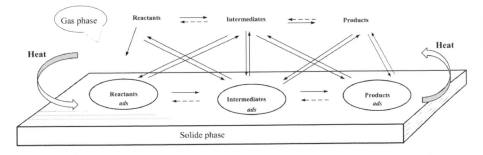

FIGURE 6.3   Scheme of a number of interactions of heterogeneous and homogeneous constituents in bimodal reaction sequences.

in reverse direction. Supplementary considerations arise in the system, taking into account the reversibility of certain steps.[65] On the other hand, in bimodal reactions, the catalytic or inhibiting surface undergoes reversible or irreversible changes during the reaction. In addition, the heat (energy) transfer and diffusion processes, in general, accompany nearly every elementary step. Fluids dynamics phenomena also are factors, playing an important role in bimodal reaction sequences.[69] Even in the case of one initial reactant, there may be a very large number of elementary reactions in models. For example, the kinetic model of the thermal decomposition of methyl decanoate ($C_{11}H_{22}O_2$) at 773–1123 K,[74] investigated also experimentally, only in the gas phase contains 3231 reactions of 324 species. In this case, the effect of a number of heterogeneous reactions may be "masked" by a high-temperature process, while, at relatively low temperatures, they also must be included in the reaction schemes.[74, 75] In the case of two or three initial reactants (one of which may be dioxygen or air) on multiphase catalysts, the number of elementary reactions may be increased from ten to several ten thousands.[76] For example, the kinetic scheme of the oxidation of n-alkanes (undecane and dodecane) consisted of 5864 reversible reactions of 1377 species.[75] However, there exist certain methods of minimizing the number of steps, and the simplified models also are useful for mathematical analysis of bimodal reaction sequences.[77]

Already the early attempts of mathematically modeling the heterogeneous–homogeneous reactions, even using simplified reaction schemes, showed the complexity of the kinetic description of these reactions in comparison with only homogeneous or heterogeneous reactions. Such a conclusion was made by Holton et al.,[65] modeling the bimodal reactions in decreasing complexity: as two dimensions—two phases; two dimensions—one phase; and one dimension—one phase models on the examples of both the consecutive and concurrent reactions in two phases. As a model of consecutive reaction, the authors[65] examined the heterogeneous catalytic oxidation of methanol to formaldehyde, coupled to the gas-phase oxidation of formaldehyde to carbon dioxide and, as a model of concurrent reaction, the decomposition of nitrous oxide in the gas phase or over gold catalysts. Mathematical analyses, even of the simplified systems of bimodal reactions, reveal the emergence of synergistic effects in coupled processes, some exhibitions of which are given in the below examples.[115 in Ch. 2; 54, 78]

Song et al.[78] attempted to describe the dynamic behaviors of a simple homogeneous–heterogeneous reaction, suggesting that the heterogeneous reaction occurs at a planar surface and the homogeneous reaction in a stagnant boundary layer above it. The simplified model proposed by Song et al. was based on two nonlinear parabolic differential equations related to the rates of homogeneous and heterogeneous reactions, as well as coupled homogeneous–heterogeneous reactions involving different parameters. It was shown that for a homogeneous–heterogeneous reaction, five steady states might be found for two coupled first-order reactions, but only three steady states for uncoupled reactions. The oscillatory behaviors were found in all given cases. It was found that the coupled reaction could involve larger parameter regions. Song et al.'s work also described the crucial role of the so-called Lewis number, which was the ratio of two constants: mass transport time constant and thermal transport time constant, in the appearance of oscillations or, in mathematical terminology, of the so-called Hopf bifurcations. Investigating the appearance of the bifurcations in dynamic systems, Wilhelm and Heinrich[79] found that the "smallest chemical reaction system" could have a Hopf bifurcation (in a three-dimensional system) containing at least three intermediates for a bimolecular chemical reaction. Mathematical analysis has established another interesting fact: a two-component system with only mono- and bimolecular reactions cannot "show limit-cycles of oscillations."[79] Therefore, the system must contain more than two "agents" or at least a trimolecular reaction. These examples of mathematical analysis, even on the simplified models, reveal the feasibility of multiple steady states and oscillatory behavior of the systems with heterogeneous–homogeneous or homogeneous–heterogeneous reactions.

The reaction kinetic model involving both homogeneous and heterogeneous stages at low-temperature oxidation of acetaldehyde with molecular oxygen or air was investigated by Sargsyan et al.[115 in Ch. 2] The mechanism suggestions of the oxidation of acetaldehyde they presented by the following scheme:

1. $RCO + O_2 \rightarrow RCO_3$
2. $RCO_3 + RCHO \rightarrow RCO_3H + RCO$
3. $RCO_3H + wall \rightarrow (OH)_{ads} + RCO_2$
4. $RCO_3 + S \rightarrow RCO_3\text{-}S$
5. $(OH)_{ads} + RCO_3\text{-}S \rightarrow CH_2O(v) + CO_2 + H_2O + hv + S$
6. $CH_2O(v) + RCO_3 \longrightarrow CH_2O + RO + CO_2$

$$RCO_3H + HCO$$

   a. $RO + O_2 \rightarrow CH_2O + HO_2$

   b. $RO + RCHO \rightarrow RCO + ROH$

7. $RCO_2 + RCHO \rightarrow RCO_2H + RCO$
8. $CH_2O(v) + M \rightarrow CH_2O + M$
9. $RCO + wall \rightarrow$ stabilization

where R is CH_3; M the gas molecule, and S the active site on the wall. This is a scheme that corresponds to the mechanism suggestions of a chain-radical reaction. The initiation stage of radicals was not included in the scheme. It might occur by any known way. It was considered that the leading active centers of the reaction in the gas phase, RCO_3 radicals, produced vibrational-excited molecules of formaldehyde $(CH_2O(v))_S$ on the surface, which played the role of inhibitor by feedback.[115 in Ch. 2] There was considerable experimental evidence that RCO_3 radicals were leading active centers of the reaction in the gas phase. They were obtained in a number of kinetic investigations using different methods, including the EPR matrix isolation method (the relevant references were given in Section 3.5.4). The elementary reactions 3, 4, 5, and 9 were heterogeneous reactions. Among them reaction 3 was the stage of the degenerate branching of chains on the surface of the reaction vessel, at low-temperature oxidation. The appearance of vibrational excited formaldehyde $CH_2O(v)$ was assigned to the heterogeneous reaction 5. It was assumed that electronically excited formaldehyde $CH_2O(v)$, formed on the surface, desorbed from the surface and diffused to the gas phase. The quenching, according to the given scheme, took place in their interaction with the leading active centers in the gas phase. Among the experimental evidence was luminescence emission in acetaldehyde oxidation or in heterogeneous decomposition of peroxyacids in certain conditions.[112 in Ch. 2] The fast quenching transfer of the internal energy to RCO_3 radicals, as well as surface reactions, led to the auto-oscillations at certain reaction parameters.[115 in Ch. 2] For instance, the phase portraits (geometric representation of the trajectories of the dynamic system in phase plane), obtained by mathematical analysis, showed the appearance of the oscillatory pattern at T = 450–500 K, P ≤ 10 torr. It was also considered that the solid surface might be modified during the reaction and that the appearance of different bifurcations was related to them. Sargsyan et al.[115 in Ch. 2] mentioned a good agreement between the results obtained by mathematical analysis of the above model and certain experimental data. For instance, the values of the frequency of oscillations ($\omega = 1$–3 s^{-1}), obtained in nonisothermal oxidation of acetaldehyde[80] may be predicted by these calculations.

Analyzing the more general cases of bimodal reactions, occurring by the transfer of active intermediates in excited states from one phase to another, Sargsyan attempted to reveal a number of generalized behaviors for this kind of system, involving also the possibility of predicting the appearance of the oscillations, spatiotemporal pattern, etc.[62] The following generalized scheme of the heterogeneous–homogeneous (and/or homogeneous–heterogeneous) reaction was taken into consideration

$$Z + A \rightarrow X + C \qquad k_1$$
$$X + B \rightarrow Z + L \qquad k_2$$
$$L + S \rightarrow X'S + Y \qquad k_3$$
$$Y + B \rightarrow Z + D \qquad k_4$$
$$X + S \rightarrow XS \qquad k_5$$
$$XS + X'S \rightarrow In^* + 2S \qquad k_6$$
$$In^* + X \rightarrow Prod. \qquad k_7$$
$$In^* \rightarrow Stab. \qquad k_8,$$

where A, B, C, D were initial reactants and the final products, X, Y, Z, were active intermediate species, S surface active centers, XS, X'S active species adsorbed on the surface, L intermediate product (in nonsteady-state concentrations), and In^* excited inhibitor. Figure 6.4 showed one of the main results obtained on the basis of this hypothetical scheme, showing the possibility of the concentration oscillations of radicals in the boundary layer above the solid surface. Mathematical analysis of this system helped determine the conditions of the appearance and propagation of oscillations, as well as the changes of the mode of reaction in dependence on the chosen parametric set.

There are a number of examples of bimodal reaction sequences in oxidation or dehydrogenation of some hydrocarbons, where the models were based primarily on the mechanical combination of existing mechanism suggestions for gas-phase reactions with the experimental data obtained in heterogeneous catalysis.[189 in Ch. 3; 73, 74] Sinev et al.[189 in Ch. 3] combined a heterogeneous catalytic reaction scheme with the existing gas-phase scheme for oxidative coupling reaction of light alkanes, involving about 50 elementary reactions. For the kinetic models of the heterogeneous constituent in oxidative coupling of methane, the redox nature of catalytic active sites, producing radicals, was taken into consideration. The base of the scheme for the

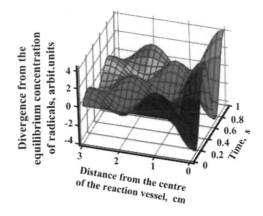

FIGURE 6.4 Spatial-temporal changes of the concentration of radicals (oscillations) in the heterogeneous–homogeneous reaction corresponding to the reaction scheme and calculations obtained in the work.[62] (Adapted by permission of the author Sargsyan G. N.)

heterogeneous reaction was the mechanism of Mars–van Krevelen, the main stages of which may be presented as in the following simplified form:

$$ZO + X \rightarrow Z + XO$$

$$Z + 1/2O_2 \rightarrow ZO$$

Z is the active site of the surface, whereas X and XO are reduced and oxidized organic molecules. According to the results, it was found that the general scheme was in good agreement with experimental data of the process. Unfortunately, these and many other modeling investigations did not lead to new crucial predictions, although, often, they successfully describe the process, being in generally good agreement with the experimental data. Examples are a great number of modeling investigations of the oxidative coupling of methane by application of different metal oxide catalysts. The recommendations followed from the modeling investigations, at best case, lead to some optimization of the process but do not predict the pathways of the solution of the main problems, which arise in attempts to create an industrial-scale process.[73; 192 in Ch. 3]

Different simplified schemes of the combinations of heterogeneous and homogeneous reactions may be derived from the above generalized, though of course incomplete, scheme, by coupling some parallel and consecutive stages in/on both phases. For instance, three different schemes of bimodal reactions (Figure 6.5),

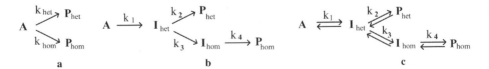

FIGURE 6.5 Three simplified schemes of the bimodal reaction sequences: (a) irreversible and parallel reaction; (b) irreversible consecutive-parallel reaction; and (c) reversible consecutive-parallel reaction, where I is an intermediate. (Adapted from Ref.[5 in Ch. 1])

involving two parallel and irreversible reactions and, in a more complex case, parallel-consecutive and reversible reactions, were analyzed in order to reveal the optimal amount of the catalyst and the desired yield of the reaction products.[5 in Ch. 1] According to Poissonnier et al., the adequate coupling of heterogeneous and homogeneous reactions, "potentially, allows achieving synergistic effects" for fixed bed reactor models."[5 in Ch. 1]

In the case of the consecutive-parallel bimodal reaction sequence containing 1–2 reversible reactions (Figure 6.5c), the concurrent adsorption of intermediates plays a predominant role, affecting the yields of products in both phases. As Poissonnier et al.[5 in Ch. 1] showed, maximizing or minimizing the amount of catalyst, the yields of either homogeneously or heterogeneously produced products might be changed. Note that, even theoretically, it is hard to imagine a fully reversible, complex chemical reaction, having a bimodal sequence, as it is indispensably related to heat transfer and diffusion processes, usually creating nonequilibrium concentrations of intermediates. In this regard, the scheme (Figure 6.5c) may be considered an idealized formal kinetic model.

The modeling investigations of the bimodal reactions, such as the reforming of hydrocarbons, catalytic combustion, and reduction of pollutants by some oxidation processes, have great importance in chemical engineering.[69] They are directed mainly toward optimization of the chemical processes, which also involve the design of reactors and engineering of operations by control of certain parameters of the processes. In several cases, the models permit calculation of some parameters, overcoming their determination by the experimental pathway. For example, the ignition or extinction parameters, optimal concentrations of initial reactants, appropriate parameters, and types of reactors in catalytic combustion of propane/air over Pt may be effectively calculated on the basis of bimodal reaction schemes, as shown by Olsene et al.[68] A relatively large number of modeling works of chemical engineering interests may be found for catalytic combustion reactions of hydrogen and hydrocarbons, where the catalyst plays different roles, being the sink of radicals for the homogeneous reaction and also the supplier of them to the gas phase.[69-72] In his review article devoted to modeling the interactions in coupling between the heterogeneous and homogeneous reactions, Deutschmann[69] presents a dozen examples of the process simulations, taking into consideration the role of the fluid dynamics and coupling the chemical model with the heat and mass transfer processes.

All the above examples of analyses of mathematical models and numerical simulations of the oxidation of organic compounds with O_2, containing both the heterogeneous and homogeneous stages as unique complex processes, indicate the possibility of the emergence of certain critical phenomena, depending on the nature of the solid surfaces and the given set of parameters. As mathematical analysis of models shows, the appearance of bifurcations, hysteresis phenomena, autocatalysis, spatiotemporal patterns, oscillations of the concentrations and reaction rates may be predicted as results of the nonlinear interactions in heterogeneous–homogeneous, as well as homogeneous and heterogeneous reaction systems.

Returning to the problem of the theoretical explanation of bimodal reaction sequences, from the point of view of the self-organization phenomena, we note that they govern the overall process. The above described phenomena and effects in the heterogeneous–homogeneous reactions may be considered expressions of the self-organization phenomenon in chemical systems in the form of macrokinetic phenomena. The majority of the kinetic schemes, presenting the simplified mechanism of heterogeneous–homogeneous reactions, are in agreement with the experimental observations. Not only the combination of interconnected chemical reactions and physical (diffusion and energy transfer) processes, as shown by the above modeling investigations, but also the chemical nature and structure of the solid phase play a determining role. In mathematical analysis, it sometimes seems that the role of solid phases is "masked" in different constants of equations, though they are obvious in numerical analysis of bimodal reaction models.

6.9   CONCLUDING NOTES

The majority of the oxidation reactions of organic and inorganic compounds with dioxygen in the chemical industry and in the laboratory, as well as in Earth's conditions, in nature, occurs as irreversible processes, far from thermodynamic equilibrium conditions. Therefore, understanding their mechanism is indispensably linked with the kinetic peculiarities of the systems, in these conditions. As has been observed in this work the nonlinearity of chemical interactions within systems and the nonequilibrium state of overall systems were main preconditions for the emergence of self-organization phenomena. It may take place only in multicomponent and complex systems. According to Paradisi et al.,[81] "The multicomponent system can be considered 'complex,' when its dynamics trigger the emergence of self-organizing structures." The complexity of the chemical systems is one of the preconditions and, at the same time, a result of the emergence of self-organization phenomena. In this regard, the system becomes more complex and, therefore more "ordered" (but less "ordered" with its environment), due to the self-organization phenomena, particularly in the spontaneous coupling of heterogeneous and homogeneous reactions. The phase separation of stages (steps) in the heterogeneous–homogeneous reaction also is a spatial separation of species, and it leads to the creation of "spatial order." Therefore, the system in this state is more ordered than the initial system. At the same time, the overall system, together with its surrounding in this state, is closer to the maximum of entropy and to the minimum of Gibbs energy within the system than in initial conditions.

The emergence of the temperature and concentration gradients of active species essentially changes the notion about chemical dynamics in these kinds of systems. Active intermediates, radicals, excited species, "hot" atoms and molecules may be formed in superequilibrium concentrations in the system. Here, the statement of the classical theories in heterogeneous catalysis that the catalyst opens a new chemical pathway in comparison with the noncatalytic pathway, by the expression of Krilov,[15] "does not exhibit all aspects of the mechanism of catalytic action" in irreversible processes occurring far from the equilibrium state. Truly, this statement "does not take into account the probable role of the nonequilibrium in catalytic conditions";

therefore, we have the "overcoming of endothermic stages forming the active centers with high energy." The overcoming of thermodynamic restrictions was exemplified by a number of catalytic reactions, occurring as irreversible processes, far from equilibrium conditions.[15]

All these fundamental problems in the theory of heterogeneous catalysis also concern the complex heterogeneous–homogeneous reactions, where the new reaction pathway is realizable via a specific reaction sequence and finally governed by the self-organization of the chemical system. The important stages of the surface generation of excited species, radicals, or radical-like species with their further partial diffusion to other phase (gas or liquid) is an exhibition of this kind of evolution of the chemical system by feedback in kinetics of the system.

Here, again, it will be noted that the generation of radical or other active intermediate species on the surfaces, in bimodal sequences, is typically a heterogeneous reaction performed by kinetic coupling. Boudart and Djega-Mariadassou in their work *Kinetic Coupling in and between Catalytic Cycles*[97 in Ch.1; 82; 83] showed the basic analogies and causes of the origination of "self-assembly" (self-organization) in chemical reactions, the main idea of which is the feasibility of the kinetic coupling of catalytic cycles of the heterogeneous catalytic and homogeneous chain radical reactions. In this regard, the bimodal (heterogeneous–homogeneous and homogeneous–heterogeneous) reactions may be considered as processes, based on this kind of kinetic peculiarities.

According to the classical notions of the heterogeneous–homogeneous reaction, the primary reaction, at low temperatures, usually begins on the surface of the solid substance. The boundary surface layer produces radicals and other active intermediates, transferring them to the gas phase. Therefore, the boundary surface layer plays the role of source, and then also of the sink of radicals for the homogeneous reaction. This perception is incomplete without consideration of the energy (heat) transfer processes, changing the behaviors of reactions in both phases, due to interactions between them. They often play a pivotal role in this class of reactions. For example, if the released energy and accumulated concentration of radicals are sufficient, an ignition may be observed on the boundary surface layer. The latter strongly changes the rate and occurrence of the reactions in both phases. We have also seen that the system of the heterogeneous–homogeneous reactions is more complex than that of the heterogeneous and homogeneous, taken separately. There is no universal unity of the complexity of the chemical system. However, as a measure of the complexity, in particular cases may be considered the quantity of information in system, related with the produced entropy. Presently, we do not know how many bits of information the simplest heterogeneous–homogeneous system contains. (The thermodynamic science states that one bit information is equivalent to the entropy $k \ln 2 = 10^{-23}$ J/K, where k is the Boltzmann constant).[103 in Ch. 1] However, we do know that, according to the universal theory, every evolution, including chemical reactions, is the creation of the new information. The heterogeneous–homogeneous reactions are not an exception to this common rule. In this regard, the bimodal reactions are an important part of the chemical evolution.

Relations between the outcome of entropy and accumulation of information in chemical reactions, in very easy and popular form, were presented in a short but very interesting book by Volkenstein, Entropy and Information,[103 in Ch. 1] published about 30 years ago. It was a wonderful publication then, and it remains useful even, today.

Modern theories in chemistry, based on entropy, information, complexity, evolution, and irreversibility of time, are interrelated.[84] In this context, the occurrence of homogeneous–heterogeneous or heterogeneous–homogeneous reactions of oxidation may be understood as nonequilibrium processes.

REFERENCES

1. Prigogine I. Introduction. p. 23. In: Nicolis G., Baras F. (eds). Chemical Instabilities, Proceedings of the NATO Advanced Research Workshop on Chemical Instabilities. Application in Chemistry, Engineering, Geology and Material Science. Austin, Texas, USA, March 14–18, 1983. D. Reidel Publishing Company, Dordrecht, 420 pages.
2. Imbihl R. Non-linear dynamics in catalytic reactions. Chapter 9. In: Holloway S., Richardson N. V. (eds). Dynamics (Handbook of Surface Science). 2008, Elsevier, Amsterdam, 998 pages. p. 345.
3. Dewar R. C., Lineweaver C. H., Niven R. K. (eds). Beyond the Second Law: Entropy Production and Non-equilibrium Systems. 2013, Springer, 434 pages.

6.1

4. Robertson A. J. B. The development of ideas on heterogeneous catalysis, progress from Davy to Langmuir. Platinum Metals Rev., 1983, v. 27, 1, p. 31–39.
5. Medford A. J., Vojvodic, A., Hummelshøj J. S., Voss J., Abild-Pedersen F., Studt F., Bligaard T., Nilsson A., Nørskov J. K. From the Sabatier principle to a predictive theory of transition-metal heterogeneous catalysis. Journal of Catalysis. Special Issue: The Impact of Haldor Topsøe on Catalysis, 2015, v. 328, p. 36–42.
6. Sabatier P. *La catalyse en chimie organique, Librairie Polytechnique, Paris et Liège*, 1920.
7. Anton A. B., Cadogan D. C. The mechanism and kinetics of water formation on Pt(111). Surface Science Letters, 1990, v. 239, 3, pages L548–L560.
8. Anton, A. B., Cadogan D. Kinetics of water formation on Pt(111). Journal of Vacuum Science and Technology, a-Vacuum Surfaces and Films, v. 9, 3, 1991, p. 1890–1897.
9. Taylor H. S. A Theory of the Catalytic Surface. Proc. R. Soc. London, Ser. A. 1925, v. 108, p. 105–111.
10. Taylor H. S. Advances in Catalysis, Vol. 1, Academic Press Inc., New York, 1948.
11. Li C., Liu Y. Bridging Heterogeneous and Homogeneous Catalysis: Concepts, Strategies, 2014, Wiley, 656 pages.

6.2

12. Christiansen A. J. The Elucidation of reaction mechanisms by the method of intermediates in quasi-stationary concentrations. Advances in Catalysis, 1953, v. 5, p. 311–353.
13. Kolasinski K. W. Physical Chemistry: How Chemistry Works. John Wiley and Sons VCH 2017, Chichester, 728 pages.
14. Barteau M. A., Madix R. J. Acetylenic complex formation and displacement via acid-base reactions on Ag(110). Surface Science, 1982, v. 115, p. 355–381.
15. Krilov O. V. About some analogies between catalysis and chain reactions, *Kinetika i Kataliz* (Kinetics and Catalysis, in Russian), 1985, v. 26, 2, p. 263–273.

16. Boudart M. Kinetics of Chemical Processes: Butterworth-Heinemann Series in Chemical Engineering.1991, Stoneham, 245 pages.

6.3

17. Duprez D., Cavani F. (eds). Handbook of Advanced Methods and Processes in Oxidation Catalysis From Laboratory to Industry. 2014, World Scientific (USA), 1036 pages, p. 364.
18. Patience G. S., López Nieto J. M. (eds). International VPO Workshop: Butane oxidation technology and catalyst characterization Applied Catalysis A: General, 2010, v. 376, 1–2, p. 1–104.
19. Ballarini N., Cavani F., Cortelli C., Ligi S., Pierelli F., Trifirò F., Fumagalli C., Mazzoni G., Monti T. VPO catalyst for n-butane oxidation to maleic anhydride: A goal achieved, or a still open challenge? Topics in Catalysis, 2006, v. 38, 1, p. 147–156.
20. Huang X-F., Li C-Y., Chen B-H., Silveston P. L. Transient kinetics of n-butane oxidation to maleic anhydride over a VPO catalyst. AIChE Journal, 2002, v. 48, 4, p. 846–855.
21. Schlogl R. Introduction, p. 2. In: N. Mizuno (ed). Modern Heterogeneous Oxidation Catalysis: Design, Reactions and Characterization, 2009 Wiley, Weinheim, 340 pages.
22. Bakhtchadjyan R., Manucharova L. A., Tavadian L. A. Photocatalytic selective oxidation of DDT (dichlorodiphenyltrichloroethane) to dicofol. Chapter 3, p. 85–106. In: DDT: Properties, Uses and Toxicity. Series: Environmental Remediation Technologies, Regulations and Safety. Editor: Kathleen Sanders. Nova Science Publishers (USA), 2016. ISBN: 978-1-53610-009-9.
23. Lawless E. W., Ferguson T. L., Meiners A. F. Guidelines for the Disposal of Small Quantities of Unused Pesticides, volume 1, section 7. Review of the chemistry of pesticide disposal. DDT. p. 104–105. Environmental Protection Agency, National Environmental Research Centre, Office of Research and Development, Cincinnati, Ohio, United States, 1975, 335 pages.

6.4

24. Elnashaie S. S. E. H. Modelling, Simulation and Optimization of Industrial Fixed Bed Catalytic Reactors. Topics in Chemical Engineering. CRC Press., 1994, 478 pages, p. 26.
25. Okkerse M., De Croon M. H. J. M., Kleijn C. R., Marin G. B., Akker H. E. A. A surface and gas phase mechanism for the desorption of growth on the diamond surface in an oxy-acetylene torch reactor. Journal of Applied Physics, 1998, v. 84, 11, p. 6387–6398.
26. Gudmundsdóttir S., Skúlason E., Weststrate K-J., Juurlink L., Jónsson H. Hydrogen adsorption and desorption at the Pt(110)-(1 × 2) surface: Experimental and theoretical study. Phys. Chem. Chem. Phys., 2013, v. 15, p. 6323–6332.
27. Zenobi R., Zare R. N. Two step lazer spectroscopy (Desorption mechanism, Chapter 2. In: Lin S. H. (ed.). Advances in Multi-photon Processes and Spectroscopy, Volume 7, World Scientific Publishing, 1991, 311 pages, p. 50.
28. Walter J. E., Rappe A. M. Coadsorption of methyl radical and oxygen on Rh(111). Surface Science, 2004, v. 549, p. 265–272.
29. Liu L., Quezada B. R., Stair P. C. Adsorption, desorption and reaction of methyl radicals on surface terminations of α-Fe$_2$O$_3$. J. Phys. Chem. C, 2010, v. 114, 40, p. 17105–17111. DOI: 10.1021/jp1039018
30. Stringfellow G. B. Organometallic Vapor-Phase Epitaxy: Theory and Practice. Academic Press, Boston, 1989, 397 pages, p. 329.

31. Grankin V. P., Styrov V. V., Tyurin Yu. I. Self-oscillatory heterogeneous recombination of hydrogen atoms and non-equilibrium desorption of molecules from the surface (Teflon), particles, and their interaction. Journal of Experimental and Theoretical Physics. 2002, v. 94, 2, p. 228–238.

32. Suryanarayana C. (ed.). Non-equilibrium Processing of Materials, Volume 2, 1st Edition, Pergamon Materials Series, 1999, Non-equilibrium Processing of Materials, Volume 2, Pergamon Materials Series, Pergamon, Amsterdam, 437 pages.

33. Chiesa M., Giamello E., Che M. EPR characterization and reactivity of surface-localized inorganic radicals and radical ions. Chem. Rev., 2010, v. 110, 3, p. 1320–1347. DOI: 10.1021/cr800366v

6.5

34. Epstein I. R., Pojman J. A., Steinbock O. Introduction: Self-organization in non-equilibrium chemical systems. Chaos: An Interdisciplinary Journal of Nonlinear Science, 2006, v. 16, 037101. http://dx.doi.org/10.1063/1.2354477

35. Heylighen F. The science of self-organization and adaptivity. The Encyclopedia of Life Support Systems, EOLSS, 2001, v. 5, 3, p. 253–280.

36. Zilbergleyt B. Discrete thermodynamics of chemical equilibria and classical limit in thermodynamic simulation. Israel Journal of Chemistry. Special Issue: Mathematical Modeling of Physicochemical Processes, 2008, v. 47, 3–4, p. 265–272.

37. Le Chatelier H. L. General statement of the laws of chemical equilibrium. *Comptes rendus*, 1884, v. 99, p. 786–789.

38. Eschenmoser A. Chemistry of potentially prebiological natural products. Origins of life and evolution of the biosphere. 1994, v. 24, 5, p. 389–423.

39. Prigogine I., Stengers I. Order Out of Chaos. Man's Dialogue with Nature. Bantam Books, London, 1984, 349 pages.

40. Prigogine I., Lefever R. Symmetry-breaking instabilities in dissipative systems. J. Chem. Phys., 1968, v. 48, 4, p. 1695–1700.

41. Beusch H., Fieguth P., Wicke E. *Thermisch und kinetisch verursachte Instabilitäten im Reaktionsverhalten einzelner Katalysatorkörner.* Chem. Inog. Techn., 1972, v. 44, 7, p. 445–451.

42. Charbonneau P. Natural Complexity: A Modeling Handbook, Princeton University Press., 2017, Princeton, 355 pages.

43. Piumetti M., Lygeros N. Complexity structures of active sites in catalysis. Chimica Oggi - Chemistry Today, 2013, v. 31, 6, p. 48–52.

6.7

6.7.1

44. Mortimer M., Taylor P., Lesley G., Smart E., Clark G. (eds). Chemical Kinetics and Mechanism, Volume 1. Royal Society of Chemistry, 1st edition 2002, 250 pages, p. 102.

45. Imbihl R., Ertl G. Oscillatory kinetics in heterogeneous catalysis. Chem. Rev., 1995, v. 95, 3, p. 697–733. DOI: 10.1021/cr00035a012.

46. Bykov V. I., Yablonsky G. S., Kim V. F. On the simple model of kinetic self-oscillations in catalytic reaction of CO oxidation. *Dokl. Akad. Nauk SSSR* (Reports of the Academy of Sciences of the USSR, in Russian), 1978, v. 242, 3, p. 637–639.

47. Zhabotinsky A. M. A history of chemical oscillations and waves. Chaos, 1991, v. 1, p. 379–386.

48. Nicolis N. G., Prigogine I. Self-Organization in Non-equilibrium Systems. 1977, Wiley, New York, 491 pages.
49. Barcicki J., Wojnowski A. Oscilliatory phenomena during ammonia oxidation in the presence of Pt-Rh catalyst. React. Kinet. Catal. Letters, 1978, v. 9, 1, p. 59–64.
50. Werner H., Herein D., Schulz G., Wild U., Schlögl R. Reaction pathways in methanol oxidation: kinetic oscillations in the copper/oxygen system. Catalysis Letters, 1997, v. 49, 1, p. 109–119.
51. Vodyankina O. V., Kurina L. N., Izatulina G. A. Volume stages in the process of the ethylene glycol catalytic oxidation to glyoxal on silver. Reaction Kinetics and Catalysis Letters, 1998, v. 65, 2, p. 337–342.
52. Bykov V. I., Yablonski G. S., Kim V. F. On the simple model of kinetic self-oscillations in catalytic reaction of CO oxidation. Dokl. Akad. Nauk SSSR (Reports of the Academy of Sciences of the USSR, in Russian), 1978, v. 242, 3, p. 637–639.
53. Gorodetskii V. V. Experimental evidence of propagating waves in hydrogen oxidation on platinum-group metals (Pt, Rh, Ir). Kinetika i Kataliz (Kinetics and Catalysis, in Russian), 2009, v. 50, 2, p. 304–313.
54. Ertl G. Oscillatory catalytic reactions at single-crystal surfaces. Advances in Catalysis. Edited by D.D. Eley, Herman Pines, and Paul B. Weisz, 1990, v. 37, p. 213–277.
55. Werner H., Herein D., Schulz G., Wild U., Schlögl R. Reaction pathways in methanol oxidation: kinetic oscillations in the copper/oxygen system. Catalysis Letters, 1997, v. 49, 1–2, p. 109–119.
56. Gorodetskii V. V., Matveev A. V., Kalinin A. V., Neuwenhuys B. E. Mechanism of CO oxidation and oscillatory reactions of Pd tip and Pd(110) surface: FEM, TPR, XPS studies. Chemistry for Sustainable Development, 2003, v. 11, p. 67–74.

6.7.2

57. Griffiths J. F., Inomata T. Oscillatory cool flames in the combustion of diethyl ether. J. Chem. Soc. Faraday Trans., 1992, v. 88, p. 3153–3158. DOI: 10.1039/FT9928803153
58. Wong G-H. Experimental investigation of initiation of cool flame in flow reactor, 2010, Thesis, Iowa University, 40 pages. Iowa Research Online: http://ir.uiowa.edu/etd/1279
59. Ostapyuk V. A., Boldyreva N. A., Korneichuk G. P. Oxidation of carbon monoxide on palladium under heterogeneous and heterogeneous–homogeneous conditions. Theoretical and Experimental Chemistry, 1982, v. 17, 5, p. 562–563.
60. Tsetskhladze D. T., Wolfson V. Ya., Vlasenko V. M. Mechanism of the oxidation of carbon monoxide on palladium—containing catalysts. Theoretical and Experimental Chemistry, 1983, v. 19, 1, p. 86–90.
61. Bottger I., Pettinger B., Schedel-Nierdrig Th., Knop-Gericke A., Schlogl R. Self-sustained oscillations over copper in the catalytic oxidation of methanol. p. 57. In: Studies in Surface Science and Catalysis, v. 133. Froment G. F., Waugh K. C. (eds). Reaction Kinetics and the Development and Operation of Catalytic Processes, 2001, Elsevier Science.
62. Sargsyan G. N. Excited species in chemical chain gas phase processes. Chemical Journal of Armenia, 2010, v. 63, 3, p. 297–308.
63. Gladky A. Yu., Ermolaev V. K., Parmon V. N. Oscillations during catalytic oxidation of propane over a nickel wire. Catalysis Letters, 2001, v. 77, 1–3, p. 103–106.
64. Jie R., Zhang X., Gao J., Yang W. The application of oscillating chemical reactions to analytical determinations. Central European Journal of Chemistry. 2013, v. 11, 7, p. 1023–1031.

6.8

65. Holton R. D., Trimm D. L. Mathematical modelling of heterogeneous–homogeneous reactions. Chemical Engineering Science. 1976, v. 31, 7, p. 549–561.
66. Merkin J. H. A model for isothermal homogeneous–heterogeneous reactions in boundary-layer flow. Mathematical and Computer Modelling, 1996, v. 24, 8, p. 125–136.
67. Chaudhary M. A., Merkin J. H. A simple isothermal model for homogeneous–heterogeneous reactions in boundary-layer flow. I. Equal diffusivities. Fluid Dynamics Research, 1995, v. 16, p. 311–333.
68. Olsen R. J., Williams W. R., Song X., Schmidt L. D., Aris R. Dynamics of homogeneous–heterogeneous reactors. Chem. Engineering Science, 1992. v. 47, 9–11. p. 2505–2510.
69. Deutschmann O. Modeling of the interactions between catalytic surfaces and gas-phase. Catal. Lett, 2015, v. 145, p. 272–289. DOI 10.1007/s10562-014-1431-1
70. Mantsaras G. Chemical combustion of hydrogen, challenges and opportunities, p. 97–158. In: Dixon A. G. (ed). Advances in Chemical Engineering Modeling and Simulation of Heterogeneous Catalytic Processes, v. 47, Elsevier, Amsterdam, 2014 (first edition) 300 pages.
71. Appel C., Mantzaras J., Schaeren R., Bombach R., Inauen A., Kaeppeli B., Hemmerling B., Stampanoni A. An experimental and numerical investigation of homogeneous ignition in catalytically stabilized combustion of hydrogen/air mixtures over platinum. Combust. Flame, 2002, v. 128, p. 340–368.
72. Anderson J. B. The hydrogen-iodine reaction. 100 years latter, p. 167. In: Wolfrum J., Volpp. H-R., Rannacher R., Warnatz J. (eds). Gas phase Reaction Chemical Systems, Experiments and Models, 1996, Springer-Verlag, Berlin, 343 pages.
73. Reyes S. C., Iglesia E., Kelkar C. P. Kinetic-transport models of bimodal reaction sequences-I. Homogeneous and heterogeneous pathways in oxidative coupling of methane. Chemical Engineering Science, 1993, v. 48, 14, p. 2643–2661.
74. Herbinet O., Glaude P-A., Warth V., Battin-Leclerc F. Experimental and modeling study of the thermal decomposition of methyl decanoate. Combust Flame, 2011, v. 158, 7, p. 1288–1300. DOI: 10.1016/j.combustflame.2010.11.009
75. Mzé-Ahmed A., Hadj-Ali K., Dagaut P., Dayma G. Experimental and modeling study of the oxidation kinetics of n-undecane and n-dodecane in a jet-stirred reactor. Energy and Fuels, 2012, v. 26, 7, p. 4253–4268. DOI: 10.1021/ef300588j
76. Turányi T., Tomlin A. S. Analysis of Kinetic Reaction Mechanisms. Springer-Verlag Berlin, Heidelberg, 2014, 363 pages, p. 8.
77. Conejeros R., Vassiliadis V. S. Analysis and optimization of biochemical process reaction pathways. 2. Optimal selection of reaction steps for modification. Ind. Eng. Chem. Res., 1998, v. 37, 12, p. 4709–4714. DOI: 10.1021/ie980411c
78. Song X., Schmidt L. D., Aris R. Steady states and oscillations in homogeneous–heterogeneous reaction systems. Chemical Engineering Science, 1991, v. 46, 5–6, p. 1203–1215.
79. Wilhelm T., Heinrich R. Smallest chemical reaction system with Hopf bifurcation. Journal of Mathematical Chemistry, 1995, v. 17, 1, p. 1–14.
80. Gray P., Griffiths J. F., Hasko S. M., Lignola P. G. Oscillatory ignitions and cool flames accompanying the non-isothermal oxidation of acetaldehyde in a well stirred, flow reactor. Proc. Roy. Soc. Lond. A, 1981, v. 374, p. 313–339.

6.9

81. Paradisi P., Kaniadakis G., Scarfone A. M. The emergence of self-organization in complex systems. Preface. p. 407–411. In: Paradisi P., Kaniadakis G., Scarfone A. M. (eds). Chaos, Solitons and Fractals, v 81. Part B, 2015, Elsevier, p. 407–588.

82. Boudart M., Djéga-Mariadassou G. Kinetic coupling in and between catalytic cycles. Catalysis Letters, 1994, v. 29, 1, p. 7–13.
83. Butt J. B. Catalyst deactivation and regeneration. In: Anderson J. R., Boudart M. Catalysis: Science and Technology. Springer, 1984, p. 41.
84. Mikhailovsky G. E., Levich A. P. Entropy, Information and Complexity or Which Aims the Arrow of Time? Entropy (journal), 2015, v. 17, p. 4863–4890; Open Access. ISSN1099-4300, Website: www.mdpi.com/journal/entropy. DOI: 10.3390/e17074863

Background

Bimodal reaction sequences manifest as diverse "surface effects" in thermal oxidation and decomposition reactions, combustion processes, catalysis, photocatalysis, enzymatic reactions, electrochemical processes, nanocatalysis, etc. They are unified processes consisting of homogeneous and heterogeneous constituents. We have outlined only some of the more important behaviors of the heterogeneous–homogeneous or homogeneous–heterogeneous reactions of oxidation on a limited number of examples. In this work we have not discussed all of the diversity and prevalence of this complex phenomenon. The heterogeneous–homogeneous reactions seem to be more widespread than our present knowledge about them, for instance, their existence was hypothesized, even, in some astrochemical processes.[1]

Before the 1990s and the 2000s, it was thought that the number of reactions via the bimodal sequence (when a catalyst or solid substance initiates a chain reaction in the fluid phases supplying radicals to the volume) was "insignificant" in comparison with the "purely" heterogeneous catalytic or "purely" homogeneous reactions. However, by the appearance of the heterogeneous photocatalytic reactions of oxidation, this opinion gradually changed. From today's point of view, the former generalized statement of many authors about a very limited spread of these reactions is, to say the least, disputable. During the last few decades, a significant number of reactions having a bimodal sequence of occurrence in complex chemical transformations have been documented. Moreover, as it was shown in Chapter 5, the complex heterogeneous–homogeneous and homogeneous-heterogeneous reactions, observed in a number of atmospheric processes, are evidence of the existence of the bimodal reaction sequences in natural processes as well. On the other hand, biological oxidation, which is not discussed in this work is a domain containing a very great number of examples of mutually interacting heterogeneous and homogeneous reactions, occurring at ambient temperatures and in specific conditions. It should be especially mentioned that a living cellule is a natural heterogeneous–homogeneous system.

One of the main challenges of the modern chemistry is to reveal the real bridges between the heterogeneous, homogeneous, and enzymatic types of catalysis. The modern strategy of the syntheses and design of new effective processes is more often based on the combination of the advantages of the heterogeneous catalysis with the achievements of the homogeneous catalysis, using the efficiency and selectivity of the homogeneous catalysts and the relatively better stability of the heterogeneous catalysts. One of the best examples of the controlled combination of homogeneous and heterogeneous oxidation catalysis is the creation of the heterogenized homogeneous catalysts in nanoscale dimensions, permitting the discovery of new catalytic processes of high efficiency in oxidation processes, some examples of which were given in the different sections. On the other hand, this approach reduces the existing gap between the enzymatic oxidation and "traditional" catalytic processes (e.g., heterogeneous or homogeneous).

The bimodal reaction sequences are not easily controllable. To achieve a high selectivity of the desired product and the necessary conversion of the initial reactant(s) for the majority of bimodal oxidation processes is one of the main challenges for researchers of this area. Examples of the oxidative coupling of methane, discussed in Chapter 3, show how difficult the controllable combination of the heterogeneous and homogeneous constituents in the bimodal reaction sequence can be. Even after more than three and a half decades of intensive investigation in this area, the research is not yet sufficient for to recommend the large-scale industrial production of ethylene by the oxidative coupling of methane.

Interesting cases of the overall heterogeneous–homogeneous reaction are processes of the coupling of the homogeneous reaction with heterogeneous electrochemical, as well as enzymatic reactions (the so-called enzymatic built-in heterogeneous reactions). Several strategies were proposed for the control of this kind of heterogeneous–homogeneous reaction. They include careful choice of not only of the catalysts, but also of the appropriate reaction conditions, permitting the combination of homogeneous and heterogeneous constituents in enzymatic oxidation reactions. An example of this is the combined heterogeneous and homogeneous reactions in the electrochemical oxidation of ethanol to acetaldehyde, when a graphite or platinum electrode continuously produces nicotinamide adenine dinucleotide (NAD^+) from NADH, performing selective and controllable oxidation.[2]

The reader may find a number of examples of the coupling of heterogeneous reaction with the homogeneous enzymatic reactions, including reactions involving nanosized catalysts, in certain works presented in the International Symposiums on the Relations between Heterogeneous and Homogeneous Catalysis (held every two or three years since 1973). These international meetings of researchers working in the different areas of catalysis aim to unify efforts for the creation of highly effective catalytic processes. Nevertheless, these efforts, mainly, are directed toward the creation of either homogeneous or heterogeneous processes separately, rather than toward unified heterogeneous–homogeneous processes. Discussing the relationships between heterogeneous and homogeneous catalysis, recently, R. A. van Santen in his book, *Modern Heterogeneous Catalysis: An Introduction*, wrote "The sciences of homogeneous catalysis and surface-related heterogeneous catalysis have become close partners not only in practice, but also in the scientific sense," and the "increasing knowledge about each of two systems has led to cross-fertilization, which was significantly aided the advances of the past 20 years on the molecular level understanding of heterogeneous catalytic reactions."[3]

The investigations of the heterogeneous–homogeneous reactions may largely contribute to resolving some main challenges facing humanity in domains such as chemical industry, energy production, environmental problems, water purification, biotechnology, pharmaceutical industry, medicine, etc. In the chemical industry, the heterogeneous catalytic processes, which are tools of the manufacturing of the thousands products, may not be developed without control of the feasibility of bimodal reaction sequences. Today, the practical application of new and sustainable chemical technologies, especially photocatalytic processes, strongly relates to the old problems of the heterogeneous–homogeneous reactions. One of the main challenges facing humanity in climate change may be understood through involving also the knowledge just in this area. The living cellule that is a central subject of the investigation in modern biological science and medicine is a natural heterogeneous–homogeneous medium, involving an enormous number of chemical reactions via the participation of intermediates transferring and reacting in or on different phases.

All this diversity of domains related to the bimodal sequences of reactions are evidence that the heterogeneous–homogeneous or homogeneous–heterogeneous processes are an individual class of reactions. Nevertheless, the wider scientific society does not consider them to constitute a specific class of chemical reactions, despite research showing that the heterogeneous–homogeneous reaction is not simply the sum of the heterogeneous and homogeneous constituents. Combining the energy transfer and diffusion of the intermediates between the phases, the bimodal reactions gain properties, which have neither homogeneous nor heterogeneous constituents when taken separately. The emergence of synergistic effects is characteristic in the coupling of heterogeneous and homogeneous reactions. It has been shown that only in multiphase and multicomponent systems involving chemical reactions, as a result of non-linear interactions of intermediates, a number of phenomena might arise, such as autocatalysis, critical phenomena, a multiplicity of steady states, bifurcations, chemical oscillations, and waves. In some aspects, being more complex than only homogeneous or only heterogeneous systems, the heterogeneous–homogeneous reactions exhibit qualitatively different properties. In this regard, due to their macro- and microkinetic peculiarities, the heterogeneous–homogeneous reactions may be considered a specific and separate class of chemical reactions. Unfortunately, you will not find any definition of the term "heterogeneous–homogeneous reaction" or "bimodal reaction sequence" in the IUPAC *Golden Book*[8 in Ch. 1] or in the main encyclopedias of natural sciences, including chemistry. However, as it was shown in this work, this lack is not proof of the insignificance of the bimodal reaction sequences in chemistry. Rather, it is an omission that may be corrected in the near future.

Chemical reactions, particularly the oxidation reactions, involving a bimodal sequence of occurrence are a part of the evolution on our planet. The theory of evolution, presently, is related to the self-organization concept and synergistic science, which have become a central idea in all natural sciences. Chemical evolution manifests as the "self-complexation" and "self-evolution" of catalysts, creating biological enzymes and corresponding processes. The chemical reactions, especially their energy, are the fundamental aspect of this evolution. Simply put, the energy of a chemical reaction may be used by the self-organizing system for useful work. The level of the organization of catalysts gradually increases during the chemical evolution. Mother Nature creates simple crystals, then more complex, ideal crystals involving metal organic complexes, which in turn produce complex homogeneous and heterogeneous catalysts. What is the role of the heterogeneous–homogeneous reactions in this development? We have mentioned that the living cell is a micro-heterogeneous–homogeneous system. Continuing this thought, the micro-heterogeneous–homogeneous catalytic systems at the nanoscale level become the base of the creation of biochemical structures and processes, which are the basis of life.[4] The specific micro-heterogeneous–homogeneous character of the inorganic (non-biological) reactions in nanoscale and colloidal dimensions has already been established. The nanoscale catalysis may be considered a specific exhibition of the micro-heterogeneous–homogenous reaction, bridging the two main classes of reactions. One of the best examples of the successful catalysis of oxidation reaction may be the application of gold nanoparticles as a catalyst of alcohol oxidation.[5] In more complex systems, the more complex structures, "ordered complex catalysts," enzymes, and reactions may be created in biological conditions.

The wonderful idea of creating a "factory on the scale of a chip"[6] may be one of the future ways to perform bimodal reactions in completely controllable conditions. Researchers of this domain predict that, in a micro-reactor of this kind, "it will be possible to separately regulate the reaction rate and the transport rate" or "to apply the necessary heat in a precise way," preventing the formation of side products.[4]

As a central science interconnecting all natural sciences, chemistry is a tool applied to change nature for human needs. It is not only an art in terms of the creation of new materials and new processes, but it is also a philosophy of new perceptions, new ideas, and new worldviews intended to change the old world. The more profound knowledge of the heterogeneous–homogeneous reactions of oxidation undoubtedly will have an important role in how the chemical sciences will serve the building of the future world.

Reactions occurring by the transfer of active intermediate species from one phase to another still conceal many secrets. Unraveling these secrets will add many new colors to the canvas of our knowledge.

REFERENCES

1. Fioroni M. Astrochemistry of transition metals? The selected cases of [FeN]⁺, [FeNH]⁺ and [(CO)₂FeN]⁺: pathways toward CH_3NH_2 and HNCO. Phys. Chem. Chem. Phys., 2014, v. 16, 44, p. 24312–24322.
2. Bauer R., Friday D. K., Kirwan D. J. Mass transfer and kinetic effects in an electrode-driven homogeneous reaction. Ind. Eng. Chem. Fundamen., 1981, v. 20, 2, p. 141–147.
3. Van Santen R. A. Modern Heterogeneous Catalysis. An Introduction. 2017, Wiley-VCH, Weinheim, Germany, 592 pages, p. 50.
4. Granatov G. G. Concepts of the Modern Natural Sciences. System of Fundamental Perceptions (Handbook), 2014, Flinta, Moscow, 577 pages (in Russian).
5. Skrzynska E., Ftouni J., Girardon J.-S., Capron M., Jalowiecki-Duhamel L., Paul J.-F., and Dumeignil F. Oxidation of glycerol to glycolate by using supported gold and palladium nanoparticles. Chem Sus Chem., 2012, 5, p. 2065.
6. About the "factory on the scale of a chip," see: www.vermeer.net/pub/.../future-perspectives-in-cata.pdf. (Bram Vermeer Journalistiek). In: van Santen R. (ed). Future Perspectives in Catalysis (Catalysis Shapes Our World), Dutch National Research School Combination Catalysis Controlled by Chemical Design (NRSC-Catalysis), Eindhoven (Netherlands), 2009, 72 pages.

Subject Index

Bold page numbers represent pages of tables or figures in the text.